ADVANCES IN ENZYMOLOGY

AND RELATED AREAS OF MOLECULAR BIOLOGY

Volume 40

CONTRIBUTORS TO VOLUME 40

CLINTON E. BALLOU, *Department of Biochemistry, University of California, Berkeley, California*

ESTHER BRESLOW, *Department of Biochemistry, Cornell University Medical College, New York, New York*

ANTHONY CERAMI, *The Rockefeller University, New York, New York*

FRANK G. DEFURIA, *Cornell University Medical College, New York, New York*

ALAN D. ELBEIN, *Department of Biochemistry, The University of Texas Health Sciences Center, San Antonio, Texas*

OL'GA O. FAVOROVA, *Institute of Molecular Biology of the Academy of Sciences, Moscow, U.S.S.R.*

PETER N. GILLETTE, *The Rockefeller University, New York, New York*

LEV L. KISSELEV, *Institute of Molecular Biology of the Academy of Sciences, Moscow, U.S.S.R.*

JAMES M. MANNING, *The Rockefeller University, New York, New York*

DENIS R. MILLER, *Cornell University Medical College, New York, New York*

RICHARD L. SOFFER, *Albert Einstein College of Medicine, New York, New York*

HENRY J. VOGEL, *Department of Pathology, Columbia University College of Physicians and Surgeons, New York, New York*

RUTH H. VOGEL, *Columbia University College of Physicians and Surgeons, New York, New York*

ADVANCES IN ENZYMOLOGY

AND RELATED AREAS OF MOLECULAR BIOLOGY

Founded by F. F. NORD

Edited by ALTON MEISTER

CORNELL UNIVERSITY MEDICAL COLLEGE, NEW YORK

VOLUME 40

1974

AN INTERSCIENCE ® PUBLICATION

JOHN WILEY & SONS
New York · London · Sydney · Toronto

An Interscience® Publication

Copyright © 1974, by John Wiley & Sons, Inc.

Library of Congress Catalog Card Number: 41–9213

ISBN 0–471–59175–0

Printed in the United States of America.

10 9 8 7 6 5 4 3 2 1

F. F. NORD
1889–1973

This volume of Advances in Enzymology is dedicated to the memory of Friedrich F. Nord, who founded the series in 1941 and edited the first 34 volumes. Nord was Professor of Organic Chemistry and Enzymology at Fordham University. He was an international authority on the biosynthesis and the degradation of lignin. His work included important contributions to the metabolism and enzymology of molds, and key discoveries on the biochemistry of wood. Nord published more than 400 scientific papers and wrote three books. Fordham University honored him by establishing the F. F. Nord Lectures in Biochemistry. His contributions to science are continued, not only through his numerous doctoral students, many of whom are now noted scientists, but also through the publications that he founded—the Archives of Biochemistry and Biophysics and the Advances in Enzymology.

Alton Meister

CONTENTS

ADVANCES IN ENZYMOLOGY
AND RELATED AREAS OF MOLECULAR BIOLOGY

Volume 40

BIOCHEMICAL AND PHYSIOLOGICAL PROPERTIES OF CARBAMYLATED HEMOGLOBIN S

By JAMES M. MANNING, ANTHONY CERAMI, PETER N. GILLETTE, FRANK G. DE FURIA,* and DENIS R. MILLER, *New York*

CONTENTS

I. Introduction

Sickle-cell anemia, first described by Herrick in 1910 (1), is a genetic disease that results in the synthesis of an abnormal hemoglobin molecule (2,3). The substitution of a valine residue for a glutamic acid residue at the sixth position of the β-chains of hemoglobin S (4) leads to profound changes in the solubility of the deoxygenated protein (5). Erythrocytes from individuals with sickle-cell disease do not retain their biconcave discoid shape on partial deoxygenation (6); such cells appear in the peripheral circulation in a variety of distorted forms, including the sickle-shaped cell. The clinical manifestation of sickle-cell

* Deceased.

1

disease occurs when these sickled cells occlude the capillaries, thus depriving the tissues of their necessary supply of oxygen.

The reasons for the striking difference between the properties of deoxyhemoglobins A and S, where only 2 of the 574 amino acid residues of the tetrameric molecule have been altered, are not understood. Several proposals suggesting that complementary stacking of tetramers of deoxyhemoglobin S within the red cell results in cell sickling have been made (7,8), but definitive experimental data supporting these proposals are not available. Insight into the three-dimensional structure of deoxyhemoglobin S by X-ray diffraction techniques has been delayed by difficulty in obtaining material suitable for study, but recent studies (9,10) offer some hope that the structure of deoxyhemoglobin S may soon be solved.

Cyanate prevents the sickling of cells from patients with sickle-cell disease by carbamylating hemoglobin S at its NH_2-terminal valine residues (11). This chapter describes the properties of carbamylated hemoglobin S: the specificity of cyanate for the NH_2-terminal residues of hemoglobin, the biological functions of the red cell after treatment with cyanate *in vitro*, and the physiological properties of the carbamylated red cell *in vivo*.

II. Effects of Cyanate on the Solubility of Deoxyhemoglobin S and the Sickling of Erythrocytes

Deoxyhemoglobin S is about 100 times less soluble than deoxyhemoglobin A in concentrated phosphate buffers (5). Allison (12) has shown that the viscosity of isolated deoxyhemoglobin S, at a concentration approaching that in the intact erythrocyte, is so much greater than that of deoxyhemoglobin A that deoxyhemoglobin S forms a gel under these conditions (Fig. 1). This and other similar studies have formed the basis for the hypothesis that the abnormal form of the deoxygenated S/S cell (Fig. 2) can be traced directly to the insolubility of the abnormal protein that forms gellike aggregates within the deoxygenated red cell.

The proposal (8) that the deoxygenated hemoglobin S tetramer contains an additional hydrophobic bond between the valine residue at the NH_2-terminus of the β-chain and the valine residue at the sixth position of the β-chain of hemoglobin S led to the clinical trials of large amounts of urea as a treatment for sickle-cell disease (13). The rationale for the clinical use of urea (prevention of the formation of

Fig. 1.   Hemoglobin S (200 mg/ml) was deoxygenated by equilibration with a mixture of 90% N_2 and 10% CO_2 for 5 min at 0°C. The tube was incubated at 37°C, and after 3 hr the presence or absence of a gel was determined. From Cerami and Manning (11).

the hydrophobic bond peculiar to deoxyhemoglobin S) has not been borne out by the trials reported to date (14).

The formation of cyanate in urea solutions and the potential reactivity of cyanate with functional groups of proteins (15), which prompted us to investigate the possible role of cyanate as an inhibitor of red-cell sickling, has been the main stimulus for our studies (11). We found that incubation of oxyhemoglobin S *in vitro* with low concentrations of cyanate inhibited the subsequent gelling of the protein on deoxygenation (Fig. 3) and prevented the sickling of 60–80% of the deoxygenated red cells *in vitro* (Fig. 4). Thus carbamylation of sickle-cell erythrocytes results in preservation on deoxygenation of normal cell form in most of the cells, approaching that found in a population of oxygenated sickle cells (Fig. 5).

From a comparison of the relative effects of cyanate and urea *in vitro* (11) we concluded that about 10–100 times more urea than cyanate is necessary to prevent both sickling of red cells and gelling of isolated deoxyhemoglobin S (Tables I and II). In addition, carbamylation of hemoglobin with cyanate is a time-dependent, irreversible reaction, whereas the effect of urea of these cells is immediate and completely

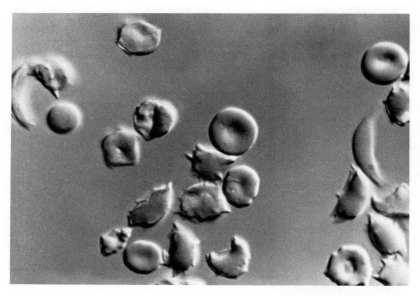

Fig. 2. A suspension of sickle-cell erythrocytes in phosphate-buffered saline solution was deoxygenated by evacuation at 30 mm Hg with a water aspirator for 7 min at 37°C. After an additional 5 min at 37°C the cells were fixed rapidly by dilution with buffered formalin (11). The micrographs were taken by Dr. James Jamieson of Rockefeller University with a Zeiss microscope with Nomarski differential interference contrast optics (×800). From Cerami and Manning (11).

reversible. Thus the mechanism of the inhibition of red-cell sickling *in vitro* is different for both compounds, and we chose to investigate in detail the mechanism by which the carbamylation of hemoglobin S by cyanate prevents the gelling of the deoxygenated protein and the subsequent sickling of the red cell.

III. Carbamylation of Hemoglobin by Cyanate

A. SPECIFICITY OF CYANATE FOR THE NH_2-TERMINAL RESIDUES OF HEMOGLOBIN

The reactive tautomer of cyanate is isocyanic acid, $HN{=}C{=}O$ (16), and the electrophilic carbon atom of this compound can undergo nucleophilic attack by several functional groups of proteins. Stark and

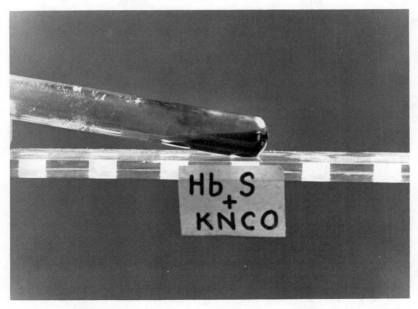

Fig. 3.   Oxyhemoglobin S was incubated with 1 mM KNCO for 1 hr at 37°C and then deoxygenated as described in Fig. 1. From Cerami and Manning (11).

Smyth (17–20) made a comprehensive study of the carbamylation of the functional groups of amino acid residues in proteins. The sulfhydryl groups of cysteine residues react with cyanate, as does the phenolic oxygen of tyrosine residues, the imidazole nitrogen of histidine residues, and the carboxyl groups of aspartic and glutamic acid residues. However, at pH 7.4 with low concentrations of reactants, the equilibria of these reactions are not in favor of the product, but rather toward the free functional group of the amino acid residue. In general, only the amino groups of proteins are irreversibly carbamylated. [The hydroxyl group of a serine residue at the active site of chymotrypsin is irreversibly carbamylated, leading to inactivation of the enzyme, but this is the only reported example of the irreversible carbamylation of hydroxyl groups (21).]

Stark (16) and Smyth (19) have carried out extensive studies on the mechanism of the carbamylation of NH_2-groups, and their results indicate that it is the unprotonated form of the NH_2-group that is carbamy-

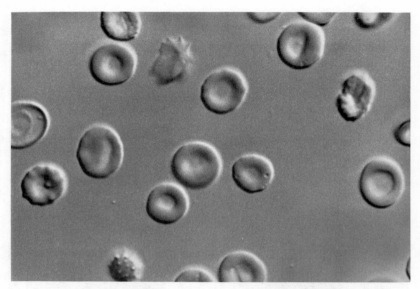

Fig. 4.   Oxygenated sickle-cell erythrocytes were treated with 30 mM KNCO for 1 hr at 37°C and then deoxygenated as described in Fig. 2. From Cerami and Manning (11).

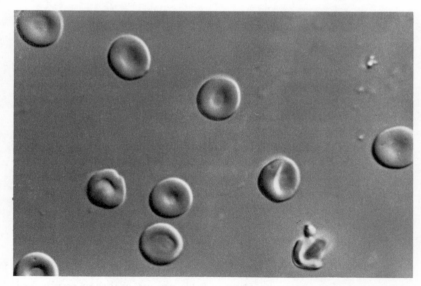

Fig. 5.   Oxygenated sickle-cell erythrocytes. From Cerami and Manning (11).

TABLE I

Effect of Urea and KNCO on the Sickling of Deoxygenated Erythrocytes[a,b]

Experiment	Compound	Concentration during incubation (M)	Concentration during deoxygenation (M)	Normal deoxygenated cells (%)
1	—	—	—	17
2	Urea	1.0	1.0	69
3	Urea	—[c]	1.0	70
4	Urea	1.0	0.1	21
5	Urea	0.1	0.1	22
6	KNCO	0.1	0.01	72
7	KNCO	0.01	0.001	34

[a] From Cerami and Manning (11).
[b] Oxygenated erythrocytes (2 μmoles hemoglobin S per milliliter) were incubated for 1 hr at 37°C. The cells were then diluted into phosphate-buffered saline solution and deoxygenated by evacuation at 30 mm Hg at 37°C. An oxygenated sample had 93% normal cells.
[c] The cells in experiment 3 were deoxygenated immediately.

TABLE II

Effect of Urea and KNCO on the Gelling of Deoxyhemoglobin S[a,b]

Compound	Concentration (mM)	Gelling
—	—	+
KNCO	1	−
Urea	10	+
Urea	50	+
Urea	100	−

[a] From Cerami and Manning (11).
[b] Oxygenated hemoglobin S (200 mg/ml; 0.2 ml) was incubated with KNCO or with urea for 1 hr at 37°C. The contents of the tubes were gassed with a mixture of 90% N_2 and 10% CO_2 for 5 min at 0°C. The tubes were incubated at 37°C; after 3 hr the presence or absence of a gel was determined.

lated by isocyanic acid (Diagram I). It follows then that the lower the
pK_a of the NH$_2$-group, the greater will be its rate of carbamylation.

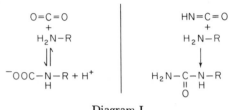

Diagram I

The pK_a of the NH$_2$-terminal valine residues of the α-chain of
hemoglobin A has been reported to be 6.7 (22). This is an unusually
low value for the pK_a of the NH$_2$-terminal residue of a protein; the
reported values have been in the range 7.5–8.0. Thus, at a physiological
pH of 7.4, about 10 times as many of the NH$_2$-terminal groups of
hemoglobin would be in the unprotonated form compared with the
NH$_2$-terminal residues of a protein, where such groups had a pK_a of 7.7.
It is the anomalously low pK_a of the NH$_2$-terminal amino groups of
hemoglobin that confers on this protein its special affinity for carbon
dioxide and, as will be discussed below, for cyanate as well. The ϵ-NH$_2$-
groups of lysine residues have pK_a values in the range 9–10, so that at
physiological pH practically all of these NH$_2$-groups would be in the
protonated form.

The incorporation of [^{14}C]cyanate into hemoglobin S parallels the
number of cells that maintain their normal discoid form on deoxygenation. This relationship was found to be a function of the concentration
of cyanate (Fig. 6) as well as the extent of time that cyanate was in
contact with the cells (Fig. 7), but the amount of carbamylation necessary to prevent 50% of the cells from sickling varied with the cells
from different patients (11); this phenomenon is still under study.
Once cyanate had been incorporated into hemoglobin in the red cell,
it was not removable by extensive washing or dialysis (11). These
results led us to consider the irreversible carbamylation of amino groups
as being responsible for the antisickling property of cyanate. Indeed,
we were able to show (Table III) that at relatively low levels of carbamylation, 80–90% of the [^{14}C]cyanate incorporated into the cells
could be accounted for by carbamylation of the NH$_2$-terminal valine

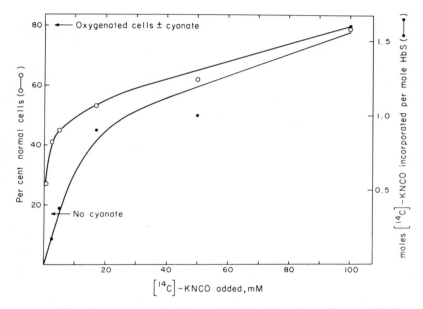

Fig. 6. The effect of KNCO on deoxygenated sickle-cell erythrocytes. Suspensions of oxygenated erythrocytes (0.5 ml) were incubated at 37°C with the designated amount of [^{14}C]KNCO (3.5×10^4 dpm/μmole). At the end of 1 hr, aliquots were removed for deoxygenation and determination of radioactivity. The percentage of normal oxygenated cells was the same (80%) in the presence or absence of KNCO; the remaining 20% of these cells are irreversibly sickled; that is, they are of abnormal form after oxygenation. On deoxygenation, 17% of the cells are normal in form. From Cerami and Manning (11).

TABLE III

The Site of Carbamylation of Hemoglobin S by Cyanate[a]

Experiment	[^{14}C]KNCO incorporated (mole/mole HbS)	Carbamylation (mole/mole HbS) at	
		NH$_2$-terminal valine	ϵ-NH$_2$ of lysine
1	1.6	1.4	0
2	1.5	1.2	0

[a] From Cerami and Manning (11).

9

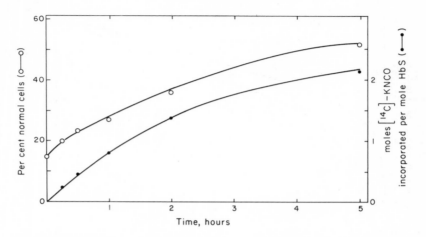

Fig. 7.   The kinetics of carbamylation and the increase in normal deoxygenated sickle-cell erythrocytes. Suspensions of oxygenated erythrocytes (2.0 ml) were incubated at 37°C with 0.01 M [^{14}C]KNCO (3.1 $\times$ 10^4 dpm/μmole). At the indicated times aliquots were removed for deoxygenation and determination of radioactivity. From Cerami and Manning (11).

residues of hemoglobin S and that there was no detectable carbamylation of the ϵ-NH$_2$-groups of lysine residues.

B. KINETIC STUDIES ON THE CARBAMYLATION OF HEMOGLOBIN S

The difference in the rates of carbamylation of the amino groups of the NH$_2$-terminal valine residues of hemoglobin S and ϵ-NH$_2$-groups of lysine residues is clearly shown by the results of the experiment described in Figure 8. *In vitro* carbamylation of isolated oxyhemoglobin at pH 7.4 and 37°C with fairly high concentrations of cyanate reveals a distinct triphasic rate profile when examined over an extensive incubation period (23). The initial phase of the reaction can be ascribed almost predominantly to carbamylation of the NH$_2$-terminal valine residues of the α- and β-chains of hemoglobin up to a level of about 1.5 carbamyl groups per hemoglobin tetramer. The third phase of the reaction represents carbamylation of the ϵ-NH$_2$-groups of lysine residues up to a point where nearly half the total protein NH$_2$-groups have been carbamylated. The intermediate stage of the reaction repre-

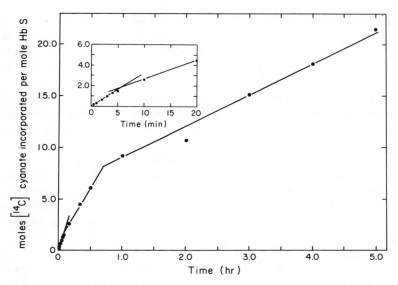

Fig. 8.   Carbamylation of the α- and ϵ-NH$_2$-groups of hemoglobin S. Oxyhemo-globin S (final concentration 0.48 mM) was mixed with [^{14}C]NaNCO (final con-centration 0.19 M) at pH 7.4 and 37°C. Portions of the reaction mixture were removed at the indicated intervals and placed into cold 5% trichloroacetic acid. The amount of radioactivity was determined after oxidation of the samples. From Lee and Manning (23).

sents carbamylation of two to three NH$_2$-terminal valine residues and nonspecific carbamylation of a few of the lysine residues. Autoradiography of a tryptic digest of oxyhemoglobin containing five [^{14}C]carbamyl groups per hemoglobin tetramer revealed only two radioactive peptides, whose amino acid compositions indicated that they were derived from the NH$_2$-terminal segments of the α- and β-chains of the protein (23). Hence the lysine residues that are carbamylated are distributed randomly throughout the hemoglobin molecule. If one takes into account the fact that there are 44 lysine residues and 4 NH$_2$-terminal valine residues in hemoglobin, then it can be calculated from the first phase of the reaction described in Figure 8 that the NH$_2$-terminal valine residues of hemoglobin S are carbamylated 50–100 times more rapidly than the ϵ-NH$_2$-groups of lysine residues. This result is in close agreement with what one would have predicted about the relative rates of carbamylation from knowledge of the pK_a values of the two types of NH_2-group. The fact that one can achieve this specificity *in vitro* with whole cells (11) at 1–10 mM concentrations of cyanate is probably an indication that the intracellular concentration of cyanate never reaches a level at which lysine residues can be significantly carbamylated.

The specificity of the carbamylation of the NH_2-terminal valine residues of hemoglobin with cyanate can also be demonstrated *in vivo* with experimental animals (Table IV). For a single dose of [^{14}C]cyanate injected intraperitoneally into a mouse, the most extensive carbamylation takes place with the hemoglobin in the red cell; the total serum proteins are carbamylated about one-fifteenth the extent of total hemoglobin *in vivo*. The preferential carbamylation of the NH_2-terminal residues of hemoglobin is undoubtedly due to the anomalously low pK_a of these residues in the hemoglobin tetramer; the serum proteins have pK_a values that do not favor extensive carbamylation *in vivo*.

Njikam et al. (24), in studies with whole blood (S/S) *in vitro*, have found that with oxygenated erythrocytes both the α- and β-chains of hemoglobin are carbamylated to nearly the same extent. With deoxyhemoglobin, however, the α-chain was carbamylated 1.7 times more than the β-chain *in vitro*. When the blood samples from patients on oral cyanate therapy were examined for the distribution of carbamyl

TABLE IV

Extent of Carbamylation in Mice *in Vivo*[a,b]

Organ	Percentage of Injected Dose	Organ	Percentage of Injected Dose
Heart	0.08	Bones	3.5
Lungs	0.05	Muscle	2.1
Stomach	0.06	Red blood cells	7.5
Spleen	0.01	Serum proteins	0.5
Kidneys	0.05	Subtotal	13.6
Intestine	0.4		
Liver	0.7		
Brain	0.17	Urine	7.0
Skin	0.8	Expired as $^{14}CO_2$	72.2
Subtotal	2.3	Total recovery	95.1

[a] From Cerami et al. (39).
[b] Female mouse (B_6D_2) injected with 10 μmoles of [^{14}C]NaNCO.

TABLE V

Distribution of Carbamyl Groups on the NH_2-Terminal Residues of Hemoglobin
After Carbamylation *in Vitro* and *in Vivo*[a]

Percentage of O_2 saturation	Ratio of α/β chain carbamylation	
	In vitro	*In vivo*
95	0.8	
50	1.7	1.6

[a] From Njikam et al. (24).

groups between the hemoglobin chains, Njikam et al. concluded that most, if not all, of the carbamylation that occurred *in vivo* was with deoxyhemoglobin (Table V).

C. ANALOGY OF ISOCYANATE AND CARBON DIOXIDE

The reactive species of cyanate, isocyanic acid (HN=C=O), bears a close chemical similarity to the structure of carbon dioxide (O=C=O). As shown in Diagram I, both compounds can undergo nucleophilic attack by an unprotonated amino group. The work of Rossi-Bernardi and Roughton (25) has suggested that from 30 to 50% of the blood carbon dioxide is carried in the form of carbamino compounds on the NH_2-terminal residues of hemoglobin. From our studies it has been clear that cyanate has a specificity for these same NH_2-terminal residues of hemoglobin in that these residues are carbamylated preferentially both *in vitro* and *in vivo* compared with the other NH_2-groups of hemoglobin. These experimental findings led us to the consideration of cyanate as a structural analog of carbon dioxide (11). However, there are three notable differences in these reactions. First, whereas carbamino formation is a reversible chemical reaction because of the relative ease of carbon dioxide regeneration, the carbamylation of amino groups with cyanate is an irreversible reaction. Indeed the stability of the carbamyl bond is the basis of a method developed by Stark and Smyth (26) for the quantitative determination of the NH_2-terminal residues of a protein with cyanate. The second essential difference in the product of these reactions is that carbamino formation results in the formation of a negatively charged product, whereas carbamylation yields an uncharged derivative. The third difference, as

14   JAMES M. MANNING ET AL.

discussed below, is that the carbamino derivative of hemoglobin has a
lowered affinity for oxygen, whereas carbamylated hemoglobin has an
increased oxygen affinity (27,28).

Cyanate preferentially carbamylates deoxyhemoglobin compared
with liganded hemoglobin (Fig. 9). Since carbon dioxide has been
shown to bind preferentially to the deoxy form of hemoglobin (29),
these results provide direct experimental support for the suggestion
that cyanate and carbon dioxide are indeed structural analogs. In the

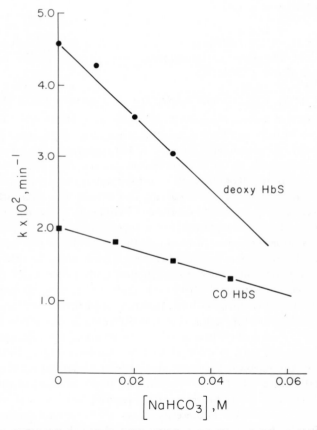

Fig. 9.   Carbamylation rates of liganded hemoglobin and deoxyhemoglobin S.
Inhibition of carbamylation by CO_2. From Lee and Manning (23).

presence of increasing amounts of carbon dioxide as bicarbonate, carbamylation of deoxyhemoglobin is decreased much more than that of liganded hemoglobin (Fig. 9). Thus carbon dioxide competes more effectively with the carbamylation of deoxyhemoglobin than it does with liganded hemoglobin—further evidence of the chemical analogy of isocyanate and carbon dioxide.

IV. Biological Functions of Erythrocytes Partially Carbamylated *in Vitro*

The relative specificity of cyanate for the NH_2-terminal amino groups of hemoglobin S described in Section III.A has raised two main questions concerning the possible deleterious physiological effects of cyanate on erythrocyte function and metabolism:

1. How are the three main functions of the NH_2-terminal amino groups of hemoglobin affected by partial carbamylation (i.e., capacity to carry CO_2, binding of 2,3-diphosphoglycerate, and uptake of Bohr protons on deoxygenation)?

2. What is the effect of cyanate on the levels of essential metabolites and enzymes within the erythrocyte?

We have carried out studies designed to answer these questions. The *in vitro* studies were done under conditions (1-h incubation with 10 mM cyanate at pH 7.4 and 37°C) that yielded 0.7–1.0 carbamyl group per hemoglobin tetramer. These conditions had been shown to be effective in the inhibition of sickling for 60–80% of the erythrocytes *in vitro* at 6 mm of O_2 tension (11). Thus an average of three NH_2-terminal valine residues per hemoglobin tetramer would remain free for maintaining the essential functions of hemoglobin.

The carbon dioxide–cyanate analogy described in the preceding section led to the question of whether the partially carbamylated erythrocyte would retain its capacity to transport carbon dioxide. As shown in Table VI, there is a slightly decreased capacity of the partially carbamylated red cell to carry carbon dioxide, but the plasma can compensate for this decrease by carrying more than its usual load of bicarbonate. Thus the ability of whole blood to carry carbon dioxide *in vitro* (either as CO_2 or bicarbonate) is not reduced by partial carbamylation.

TABLE VI

Carbon Dioxide Capacity of Blood After Carbamylation[a,b]

Parameter	Control	After carbamylation
p_{CO_2}	45.4	44.8
pH	7.23	7.26
Red cell CO_2, mM	7.59	6.70
Plasma CO_2, mM	17.65	19.20
Total CO_2, mM	18.18	18.23

[a] From de Furia et al. (28).
[b] Values were determined at 50% oxygen saturation.

Since the studies of Bunn and Briehl (30), of Perutz (31) , and of Arnone (32), have shown that the positively charged NH_2-terminal residues of the β-chain of deoxyhemoglobin are the primary binding site for 2,3-diphosphoglycerate, we thought that it was important to investigate whether carbamylation of about 25% of the NH_2-terminal residues would lead to a significant decrease in the binding of 2,3-diphosphoglycerate. As shown in Table VII, the reduction in the oxygen affinity of hemoglobin mediated by 2,3–diphosphoglycerate is not significantly reduced by partial carbamylation. When inosine,

TABLE VII

Synthesis of 2,3-Diphosphoglycerate and Its Binding to Hemoglobin After Carbamylation[a]

Parameter	Control	Cyanate-treated	After incubation of cells with inosine, pyruvate, and phosphate Control	After incubation of cells with inosine, pyruvate, and phosphate Cyanate-treated
P_{50}, mm Hg	18.2	14.8	28.8	24.7
DPG, μmoles/ml cells	1.23	1.13	3.56	3.37
ATP, μmoles/ml cells	1.37	1.40	1.43	1.44

[a] From de Furia et al. (28).

pyruvate, and phosphate are added to an *in vitro* suspension of partially carbamylated erythrocytes, there is a net synthesis of 2,3-diphosphoglycerate and a concomitant increase in P_{50}, the pressure of O_2 (in mm Hg) necessary to maintain the hemoglobin at half-saturation with oxygen. We conclude then that the synthesis and function of this important red-cell metabolite are not affected by low levels of carbamylation.

Kilmartin and Rossi-Bernardi (27) had shown that horse hemoglobin with its four NH_2-terminal residues carbamylated was able to take up about 75% of the normal complement of Bohr protons on deoxygenation. We would expect that the carbamylation of only one of the NH_2-terminal residues of hemoglobin would result in a decrease of about 6% of the Bohr protons at most. Such a decrease in the Bohr effect is within the limits of precision of the procedures that we used, and we found, as shown in Figure 10, that the Bohr effect of sickle-cell erythrocytes is not demonstrably reduced by partial carbamylation. It would appear that the level of carbamylation necessary to inhibit erythrocyte sickling *in vitro* does not result in a decreased ability of hemoglobin to carry out its other vital functions.

The effect of cyanate on red-cell metabolism and function was also evaluated. As shown in Figure 11, there was no decrease in the levels of several key erythrocyte metabolites after incubation *in vitro* with levels of cyanate sufficient to achieve *in vitro* 60–80% inhibition of red-cell sickling. Most of the enzymes of the Embden–Meyerhof pathways were not affected after exposure of the erythrocyte to cyanate (Fig. 12). There is, however, a slight decrease in the levels of pyruvate kinase. The level of the important red-cell enzyme glucose-6-phosphate dehydrogenase is not significantly affected at this level of carbamylation.

The most significant change in erythrocyte function after treatment with cyanate is the increase in the oxygen affinity of the cells. Fresh, untreated erythrocytes from sickle-cell-anemia patients have a decreased oxygen affinity, with P_{50} values reported in the range 30–35 (33). As shown in Figure 13, with increasing concentrations of cyanate *in vitro*, the P_{50} value decreases (i.e., the oxygen affinity increases) concomitant with the increased levels of hemoglobin carbamylation. The conditions necessary to prevent 60–80% of the cells from sickling *in vitro* are precisely those that result in a decrease in the P_{50} value from 31 to a normal range of 26–27 (28), corresponding to the incorporation of 0.7 carbamyl group per hemoglobin tetramer.

18   JAMES M. MANNING ET AL.

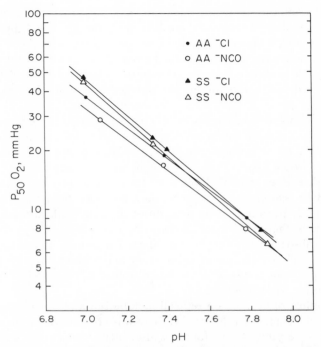

Fig. 10. The effect of pH on oxygen affinity after carbamylation. Normal and sickle erythrocytes were incubated with either 10 mM KNCO or 10 mM KCl at 37°C. After 1 hr the cells were washed and suspended in a solution containing 0.15 M NaCl and 0.015 M tris–HCl buffer. The P_{O_2} at 50% saturation and the pH were measured. The ($\Delta\log P_{50}$)/ΔpH (Bohr effect) of normal (-0.48) and sickle (-0.53) cells remained in the normal range (-0.50) after carbamylation. From de Furia et al. (28).

V. Mechanism of the Antisickling Effect of Cyanate

The salient features of the carbamylation of hemoglobin S are (1) the inability of the uncharged carbamyl group of the NH$_2$-terminus of the polypeptide chains to form salt linkages either positively charged (—NH$_3^+$) or negatively charged after carbamino formation (—HNCOO$^-$), and (2) the increased oxygen affinity of the carbamylated protein.

A schematic representation of the oxyhemoglobin and deoxyhemoglobin tetramers is shown in Diagram II. The presence of aggregates of

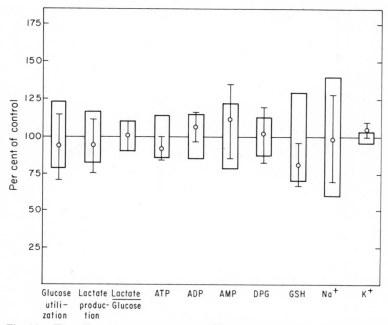

Fig. 11. The effect of cyanate on glycolysis and intracellular metabolites. Mean (○) and standard deviation (I) of values for sickle-cell blood from six patients after incubation with 10 mM KNCO for 1 hr at 37°C are expressed as percentages of control values (treated with 10 mM KCl for 1 hr). Standard deviation of the control value is denoted by the open bars. From de Furia et al. (28).

deoxyhemoglobin S tetramers would serve to shift this thermodynamic equilibrium to the right in sickle cells, so that the net oxygen affinity of the cells would be decreased. The low oxygen affinity of cells from sickle-cell-anemia patients (33) is consistent with this model.

From the work of Perutz (31) we know that the tetramer of deoxy-hemoglobin is in a more constrained conformation due to presence of

$$[\text{oxy}] \rightleftharpoons [\text{deoxy}] \rightleftharpoons [\text{deoxy}]_n$$

Diagram II

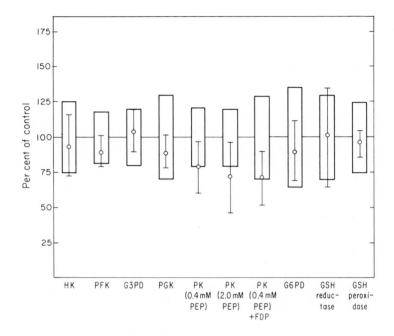

Fig. 12. The effect of cyanate on glycolytic enzymes. The incubation conditions *in vitro* were the same as those described in Fig. 11. From de Furia et al. (28).

salt bridges between charged side chains of amino acid residues with each other and with 2,3-diphosphoglycerate. X-Ray diffraction studies indicate that several of the six to eight salt bridges that are essential for maintaining the deoxyhemoglobin conformation involve the charged NH₂-terminal residues of the polypeptide chain. Our own studies (28), those of Kilmartin and Rossi-Bernardi (27), and those of others (34–36) have shown that blocking these NH₂-terminal residues of hemoglobin results in an increase in the oxygen affinity of the protein. We propose that derivatization of some of these NH₂-terminal residues by carbamylation (less than one carbamyl group per hemoglobin tetramer) would serve to prevent the formation of some of these salt bridges, resulting in a shift of the thermodynamic equilibrium shown in Diagram II to the left. Thus the increased number of morphologically normal deoxygenated cells at a given physiological oxygen tension may be partially the result of an increased oxygen affinity of

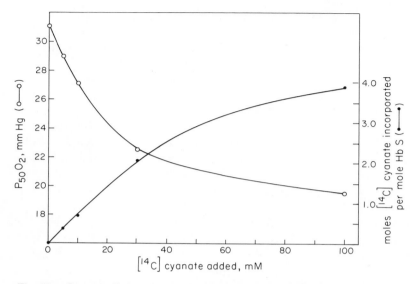

Fig. 13. Oxygen affinity as a function of carbamylation. The P_{50} of the cyanate-treated cells and the amount of [^{14}C]cyanate incorporated into acid-precipitable protein are plotted as a function of the concentration of cyanate during the 1-hr incubation at 37°C. From de Furia et al. (28).

the carbamylated hemoglobin S molecule. Carbamylation of hemoglobin S could also have a direct effect on the solubility of the deoxygenated protein as studies on the gelling of deoxyhemoglobin S suggest and the results of others have indicated (35,37). Further studies will be needed to assess the relative contribution of each mechanism to the antisickling effect of cyanate under physiological conditions. In addition, it should be noted that the role of intracellular factors such as 2,3-diphosphoglycerate, carbon dioxide, protons, and ionic strength in modulating the solubility and extent of oxygen saturation of hemoglobin S under physiological conditions remains to be elucidated.

VI. Physiological Studies with Cyanate

A. STUDIES ON ERYTHROCYTE SURVIVAL AFTER CARBAMYLATION *IN VITRO*

Before the clinical evaluation of cyanate as a possible treatment for sickle-cell disease was initiated, we carried out a series of studies to determine whether carbamylated cells would have an increased survival time in the patient. Sickle-cell erythrocytes have a shortened life span, presumably due to the changes in cell morphology and rigidity that accompany deoxygenation. It seemed reasonable to test whether carbamylated cells, many of which do not sickle *in vitro*, would last longer in the circulation of the patient.

The initial clinical studies were carried out by carbamylating *in vitro* 20 ml of cells from patients with sickle-cell disease with sodium cyanate; [^{51}Cr] Na_2CrO_4 was used for following cell survival *in vivo* (38). The level of carbamylation achieved in these experiments was two to three NH_2-terminal carbamyl groups per hemoglobin tetramer. After being washed for removal of the unreacted cyanate, the cells were reinfused into the patient. As shown in Figure 14, the cyanate-treated cells from two patients showed an increased survival time *in vivo*. For comparison, a survival study of normal adult red cells as measured by the chromate method is also shown. The nonspecific elution of ^{51}Cr from the cells results in values for apparent half-survival times for normal cells that are less than the true half-life of 60 days. In recent studies with experimental animals it has been shown that the presence of carbamyl groups on the hemoglobin molecule does not result in a different rate of elution of [^{51}Cr]chromate from the red cell that would lead to an apparent increase in cell survival. The results from similar studies on the cyanate-treated cells of seven patients are shown in Table VIII; carbamylated erythrocytes from patients with sickle-cell disease have a survival time that is increased by a factor of 2 to 4 *in vivo*.

B. *IN VIVO* STUDIES WITH CYANATE

Experiments with mice, rats, dogs, and monkeys have established that there are no apparent adverse effects on animals that have been administered cyanate five times weekly for 6–12 months (39). The LD_{50} for a single injection of sodium cyanate in mice and rats is 275 mg/kg, and the oral LD_{50} is approximately 1500 mg/kg in rats. The single effect observed in the animals is an increase in oxygen affinity of the blood as discussed above. It is important to point out that the concentration of cyanate attained and maintained during *in vivo* administration is probably considerably lower than that in a typical *in vitro* study. A sensitive assay for the presence of cyanate in whole blood, recently devised by Nigen et al (40), should prove useful in the evaluation of conditions whereby optimal carbamylation can be attained *in vivo*.

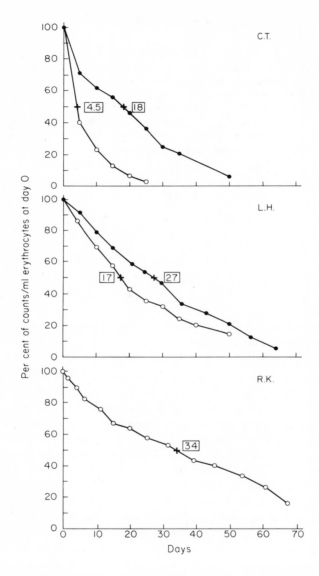

Fig. 14. The ^{51}Cr-labeled-erythrocyte survival curve for untreated (open circles) and cyanate-treated (solid circles) cells in two patients with sickle-cell disease (upper and middle panels) and in a normal subject (lower panel). Fifty percent of the radioactivity on day zero was reached at the day noted on each graph. From Gillette, Manning, and Cerami (38).

23

TABLE VIII

Apparent 50% Survival Times of Chromate-Labeled Erythrocytes
Untreated and Treated with Cyanate *in Vitro*[a]

Subject	Untreated cells (days)	Cyanate-treated cells (days)
Patients with sickle-cell disease		
RP	4.5	18
CT	4.5	18
MW	7	21
TJ	10	19
CW	13	20
MD	13	22
LH	17	27
Mean	9.9	20.7
Normal subjects		
MS	27	—
RK	34	—

[a] From Gillette, Manning, and Cerami (38).

Cyanate has been administered to patients with sickle-cell disease at The Rockefeller University either orally, intravenously, or by carbamylation of erythrocytes extracorporeally. Each route of administration yielded significant carbamylation of red cells, and no deleterious effects were observed. A dosage of about 20 mg/kg day orally has been found to yield a carbamylation level of 0.3 carbamyl group per hemoglobin tetramer. The hematologic data on two patients that have been maintained at this dose of cyanate orally for 6 months are shown in Table IX. The survival time of the erythrocytes carbamylated *in vivo* was increased; these results confirm the earlier findings on the increased cell survival after *in vitro* carbamylation of erythrocytes with cyanate (38). The increased levels of hemoglobin, the increase in the hematocrit, and the decreased levels of serum bilirubin and serum lactic dehydrogenase are indications that there has been a decrease in the hemolytic anemia in these patients after oral administration of cyanate. These data have been confirmed and extended for 10 patients over a period of 4–8 months, and there have been no short-term toxic effects due to cyanate in this period. Information of the clinical utility of

TABLE IX

Hematologic Data for Two Patients with Sickle-Cell Disease
Before and After Treatment with Cyanate

	Patient A		Patient B	
Parameter	Control	Treated	Control	Treated
Red-cell survival $T_{1/2}$, days (chromate method)	4.5	14	10	15
Hemoglobin, mean g/100 ml	8.0	10.0	9.3	12.0
Hematocrit, mean, %	24	30	29	36
Serum bilirubin, total, mg/100 ml	1.9	0.8	1.8	1.4
Serum lactic dehydrogenase, international units	175	60	225	70
Carbamyl group per molecule of hemoglobin	—	0.30	—	0.35

cyanate as a treatment for sickle-cell disease must await the results of controlled studies.

VII. Summary

The effects of carbamylation on the inhibition of erythrocyte sickling *in vitro*, the increased cell survival of partially carbamylated erythrocytes, the absence of any deleterious effect of cyanate on red-cell metabolism and function, and the positive hematologic effects of cyanate observed so far *in vivo* offer some hope that cyanate may become a useful therapeutic agent in the treatment of sickle-cell disease. However, though short-term toxic effects with the doses used to far have not been apparent, it remains to be seen whether long-term toxic effects will develop. This is an especially important consideration since each new erythrocyte that is synthesized would have to be carbamylated if the prophylactic treatment were to be successful. Nevertheless, we feel that the studies with cyanate have helped in the understanding of some of the underlying problems associated with the deoxyhemo-

26 JAMES M. MANNING ET AL.

globin S molecule. Removal of the charge on some of the NH_2-terminal valine residues of hemoglobin S by carbamylation results in such a modification of the abnormal gene product that the deoxygenated protein now has chemical, biological, and physiological properties similar to those of deoxyhemoglobin A.

Acknowledgments

The experiments described in this review were supported in part by the National Institutes of Health, National Science Foundation, the Childrens' Blood Foundation, and the National Foundation—March of Dimes. We wish to thank Drs. A. Kappas, M. McCarty, S. Moore, E. Reich, and W. H. Stein for their encouragement and support.

References

1. Herrick, J. B., *Arch. Intern. Med.*, *6*, 517 (1910).
2. Neel, J. V., *Science*, *110*, 64 (1949).
3. Pauling, L., Itano, H. A., Singer, S. J., and Wells, I. C., *Science*, *110*, 543 (1949).
4. Ingram, V. M., *Nature*, *180*, 326 (1957).
5. Perutz, M. F., and Mitchison, J. M., *Nature*, *166*, 677 (1950).
6. Hahn, E. V., and Gillespie, E. B., *Arch. Intern. Med.*, *39*, 233 (1927).
7. Perutz, M. F., and Lehmann, H., *Nature*, *219*, 902 (1968).
8. Murayama, M., *Science*, *153*, 145 (1966).
9. Finch, J. T., Perutz, M. F., Bertles, J. F., and Döbler, J., *Proc. Natl. Acad. Sci., U.S.*, *70*, 718 (1973).
10. Wishner, B. C., and Love, W. E., *Fed. Proc.*, *32*, 457 (1973).
11. Cerami, A., and Manning, J. M., *Proc. Natl. Acad. Sci. U.S.*, *68*, 1180 (1971).
12. Allison, A. C., *Biochem. J.*, *65*, 212 (1957).
13. Nalbandian, R. M., Henry, R. L., Barnhart, M. I., Nichols, B. M., and Camp, F. R., Jr., U.S. Army Medical Research Laboratory Report No. 896, 1970.
14. Opio, E., and Barnes, P. M., *Lancet*, *160* (1972).
15. Stark, G. R., Stein, W. H., and Moore, S., *J. Biol. Chem.*, *235*, 3177 (1960).
16. Stark, G. R., *Biochemistry*, *4*, 1030 (1965).
17. Stark, G. R., *J. Biol. Chem.*, *239*, 1411 (1965).
18. Stark, G. R., *Biochemistry*, *4*, 588 (1965).
19. Smyth, D. G., *J. Biol. Chem.*, *242*, 1579 (1967).
20. Stark, G. R., *Biochemistry*, *4*, 2767 (1965).
21. Shaw, D. C., Stein, W. H., and Moore, S., *J. Biol. Chem.*, *239*, PC671 (1964).
22. Hill, R. J., and Davis, R. W., *J. Biol. Chem.*, *242*, 2005 (1967).
23. Lee, C. K., and Manning, J. M., *J. Biol. Chem.*, *248*, 5861 (1973).
24. Njikam, N., Jones, W. M., Nigen, A. M., Gillette, P. N., Williams, R. C., Jr., and Manning, J. M., *J. Biol. Chem.*, *248*, 8052 (1973).

25. Rossi-Bernardi, L., and Roughton, F. J. W., *J. Physiol.*, *189*, 1 (1967).
26. Stark, G. R., and Smyth, D. G., *J. Biol. Chem.*, *238*, 214 (1963).
27. Kilmartin, J. V., and Rossi-Bernardi, L., *Nature*, *222*, 1253 (1969).
28. De Furia, F. G., Miller, D. R., Cerami, A., and Manning, J. M., *J. Clin. Invest.*, *51*, 566 (1972).
29. Kernohan, J. C., and Roughton, F. J. W., *J. Physiol.*, *197*, 345 (1968).
30. Bunn, H. F., and Briehl, R. W., *J. Clin. Invest.*, *49*, 1088 (1970).
31. Perutz, M. F., *Nature*, *228*, 726 (1970).
32. Arnone, A., *Nature*, *237*, 146 (1972).
33. Wintrobe, M. M., *Clinical Hematology*, Lea and Febiger, Philadelphia, 1967.
34. May, A., Bellingham, A. J., Huehns, E. R., and Beaven, G. H., *Lancet*, 658 (1972).
35. Jensen, M. C., Bunn, H. F., Halikas, G. C., and Nathan, D. G., *Adv. Exp. Med. Biol.*, *28*, 297 (1972).
36. Diederich, D., *Biochem. Biophys. Res. Commun.*, *46*, 1255 (1972).
37. Williams, R. C., Jr., *Proc. Natl. Acad. Sci. U.S.*, *70*, 1506 (1973).
38. Gillette, P. N., Manning, J. M., and Cerami, A., *Proc. Natl. Acad. Sci. U.S.*, *68*, 2791 (1971).
39. Cerami, A., Allen, T. A., Graziano, J. H., de Furia, F. G., Manning, J. M., and Gillette, P. N., *J. Pharmacol. Exp. Ther.*, *185*, 653 (1973).
40. Nigen, A., Peterson, C. M., Gillette, P. N., and Manning, J. M., *J. Lab. Clin. Med.*, in press.

INTERACTIONS OF POLYNUCLEOTIDES AND OTHER POLYELECTROLYTES WITH ENZYMES AND OTHER PROTEINS

By ALAN D. ELBEIN, *San Antonio, Texas*

CONTENTS

I. Introduction

In recent years a significant number of proteins have been shown to interact with, and in many cases to form complexes with, various types of polyelectrolyte. The interaction may be quite specific with regard to the polyelectrolyte, as in the case of various enzymes that contain polynucleotides or form complexes with them. These interactions of polynucleotides and proteins appear to be more specific than can be accounted for solely on the basis of their polyelectrolytic nature. The interaction may also be more general, as in the case of a number of

29

enzymes that are activated or inhibited by a number of different polyanions or polycations.

For purposes of discussion, a rather arbitrary classification has been set up in which activations of a general polyelectrolytic nature will be discussed first, followed by more specific types of interactions involving polynucleotides. Finally, inhibitions of enzymes by polyelectrolytes will be covered as well as the effects of these charged compounds on various other proteins and tissues. It should be pointed out that this arbitrary classification is for purposes of convenience and does not necessarily imply any similarity in mechanism of action. Although the interactions are probably all of a general ionic nature, the outcome of the interaction (i.e., activation or inhibition) may be due to a variety of causes. For example, the activation of the enzyme may be real or apparent; that is, it may be due to an actual change in the conformation of the protein or to the association or dissociation of subunits, or it may be due to a stabilization (protection) of the enzyme, or to the removal of inhibitors, or some such phenomenon.

Several excellent reviews have been published on the activation (1) or inhibition (2) of enzymes by polyelectrolytes. In addition, protein–nucleic acid interactions have been well covered in a number of timely reviews that will be cited later. Some protein–protein interactions bear similarities to polyelectrolyte phenomena and therefore may also involve ionic interactions, but these will be only briefly mentioned here since they have been well reviewed in the last few years (3–5).

II. Polyelectrolyte Activations (General)

A number of enzymes show an increase in activity on the addition of a polyelectrolyte. Depending on the enzyme in question, the polyelectrolyte "activator" may be a polyanion or a polycation. Thus such polyanions as polynucleotides, sulfated glycosaminoglycans, or glycoproteins may be activators of some enzymes, whereas such polycations as polylysine, polyornithine, or protamine are activators of others. Two general requirements that appear to be necessary for any of these compounds to be enzyme effectors are (1) the macromolecular nature of the polyelectrolyte and (2) the charged nature of the polymer. Although it is not known whether these interactions occur *in vivo*, it seems likely that such phenomena do have physiological significance because of the fact that polyelectrolytes, especially sulfated glycosaminoglycans, are widespread in nature. However, this section will deal only with *in vitro* studies on enzyme–polyelectrolyte activations. Section V will discuss studies on the effects of polyelectrolytes on tissues and cells.

Table I presents a list of enzymes that are activated by various polyanions and polycations. These enzymes are discussed individually in the following subsections.

A. POLYANION ACTIVATIONS

1. Lipoprotein Lipase

Probably the best studied example of an enzyme–polyanion interaction is that of the activation of lipoprotein lipase by heparin. Early observations of this system were made by Hahn (6) and Weld (7), who found that the plasma of animals with gross alimentary lipemia was rapidly cleared when these animals were transfused with blood containing the sulfated glycosaminoglycan heparin. Apparently the clearing factor is bound to tissue and is released into the blood by heparin (8). Korn (9) was able to isolate this clearing factor, which he called lipoprotein lipase, from acetone powders of rat heart. The isolated enzyme catalyzed the hydrolysis of the neutral fat of chylomicrons. This lipoprotein lipase was activated by small amounts of heparin (0.1–5 μg) and was inhibited by polycations like protamine; protamine inhibition could be reversed by the addition of heparin. Korn postulated that the enzyme contained bound heparin that could be competitively displaced by polyvalent cations.

The lipoprotein lipase from adipose tissue was later found to be inactivated by treatment with a bacterial heparinase as well as by pyrophosphate and other substances with properties suggesting that they might displace heparin, which is normally bound to the enzyme (10). The enzyme in postheparin plasma appears to depend on heparin for stability. Thus Robinson (11) showed that when plasma was passed through an ion-exchange resin to remove heparin, lipoprotein lipase activity in the effluent was lost on incubation at 37°C for 10 min. Activity was maintained, however, if heparin was added before incubation. Robinson and French (12) suggested that the inhibitory effect of protamine sulfate and other similar substances on the activity of lipoprotein lipase in postheparin plasma may be due to their ability to combine with heparin and therefore reduce the stability of the enzyme.

TABLE I

Activation of Enzymes by Polyelectrolytes

Enzyme (or reaction catalyzed)	Polyelectrolyte	Mode of activation	References
	Polyanion activations		
Lipoprotein lipase	Heparin, sulfated glycosaminoglycans (to some extent)	Stabilizes enzyme; heparin may be part of enzyme	6–21
Trehalose phosphate synthe-tase	RNA, sulfated glucosaminoglycans	May affect enzyme conformation; also stabilized enzyme.	22, 23
Rabbit-muscle aldolase	Glycorproteins containing sialic acid	May stabilize conformation of protein	24
Pepsinogen	heparin, RNA	Probably binds and removes inhibitory protein	25–27
Oxidative phosphorylation	RNA, chondroitin sulfate, heparin	Involved in binding soluble factors to particles	28–33
ATPase	DNA	Not known	34
	Polycation activations		
β-Glucuronidase	Chitosan, protamine	Prevents dissociation of subunits	35
Hyaluronidase	Polylysine, protamine sulfate, chitin	Prevents dissociation of subunits	36
Aldolase	Protamine sulfate, polylysine	Prevents dissociation of subunits	37
Lactic dehydrogenase	Protamine sulfate, polylysine	Prevents dissociation of subunits	37
α-Amylase and β-amylase	Protamine sulfate, polylysine	Prevents dissociation of subunits	37
Acid phosphatase	Proteins, long-chain polyamines	Stabilizes enzyme	38
Polynucleotide phosphorylase	Polylysine, polyornithine	Generalized ionic effect; possibly stabi-lizes conformation	39–44
Phosphorylase b	Protamine	Increases K_m for adenylic acid	45
Sialyl transferase	Histones, poly-l-lysine	Unknown	46

Reber and Studer (13) found that after the injection of ^{35}S-labeled heparin intravenously, a purified preparation of lipoprotein lipase from plasma contained 40 times more radioactivity than did other plasma proteins. *In vivo* a number of other charged compounds, called heparinoids, are able to increase the activity of lipoprotein lipase when injected into the blood (8,14). These include such materials as dextran sulfate, laminarin sulfate, chondroitin sulfate, and sulfopolyglucin. Not all of the active compounds are sulfated, but all have a strong negative charge.

The mechanism of heparin activation of lipoprotein lipase *in vitro* has been studied by a number of workers. Bernfeld and Kelley (15) found that heparin, at 1 $\mu g/ml$ of digest, activated lipoprotein lipase, but a number of other sulfated polysaccharides were potent inhibitors of the enzyme. All those polysaccharides that do not contain sulfamino groups but have at least 0.6 sulfate ester group per repeating unit proved to be potent inhibitors. These investigators concluded that sulfonic or *O*-sulfatyl groups occurring in a polymeric structure were responsible for the inhibitory effects and that this inhibition was independent of the following factors: (*1*) the chemical nature of the polysaccharide, (*2*) the presence or absence of branching, (*3*) the type and nature of the glycosidic bond, (*4*) the presence or absence of carboxyl or *N*-acetyl groups, and (*5*) the molecular weight. Thus it appears that the presence of *O*-sulfate groups in a polymeric structure is responsible for inhibition. Such a phenomenon has been termed "macroanionic inhibition" by Spensley and Rogers (16) because of its apparent lack of requirement for a specific chemical structure of the inhibitor. Bernfeld and Kelley found that only a few substances besides heparin were activators of lipoprotein lipase, and all of these, including such compounds as sulfated deacylated chitin, contain sulfamino groups. However, low-molecular-weight substances with sulfamino groups showed no activation, which suggests that the polymeric nature as well as the presence of *N*-sulfatyl groups are necessary for activation.

The experiments of Patten and Hollenberg (17) using NH_4OH–NH_4Cl extracts prepared from acetone powders of fat cells suggested that heparin stimulation of lipoprotein lipase is due to increased binding of the enzyme to chylomicrons. Heparin had no effect on the stability of the enzyme or on its activity once the complex was formed. It was not clear whether the complex also contained heparin. Protamine sulfate and NaCl had little effect on the binding of lipoprotein lipase to chylomicrons in the absence of heparin, but they did inhibit enzymatic activity after binding to substrate had occurred. These compounds also prevented the stimulatory effect of heparin on the enzyme–substrate binding. Presumably these inhibitors act on the enzyme at the catalytic site rather than at the binding site. Whayne and Felts (18) found that high-density lipoproteins added to lipoprotein lipase assay systems in increasing amounts appeared to increase the effective substrate concentration. In the absence of heparin, increasing concentrations of high-density lipoprotein increased the reaction rate, which approached V_{max} and produced a hyperbolic curve that conformed to Michaelis–Menten kinetics. In the presence of heparin, increasing concentrations of high-density lipoprotein produced an S-shaped curve and increased the V_{max} value. Their results suggested that heparin may function as a specific ligand that acts as an allosteric modifier of lipoprotein lipase and alters the kinetics of interaction of enzyme and substrate.

Olivecrona and Lindahl (19) found that a heparin-like glycosaminoglycan and lipoprotein lipase were concentrated in the same fraction of bovine milk that had been purified on DEAE-cellulose, which suggests the presence of this glycosaminoglycan as part of that enzyme. Olivecrona et al. (20) prepared agarose gels containing covalently bound heparin. Lipoprotein lipase from bovine milk bond to this column and was eluted with a salt gradient. The eluted enzyme was activated by heparin and inhibited by protamine and NaCl. Iverius et al. (21) further investigated the properties of lipoprotein lipase purified on heparin–agarose columns and compared them with crude enzyme. Serum, high-density lipoprotein (HDL), or very-low-density lipoprotein (VLDL) were used as activators. The activity of the enzyme varied with the ionic strength of the incubation medium, with maximal activity at physiological salt concentration (0.16 M). At higher ionic strengths the activity of the enzyme was depressed. This depression of activity, which was due to inhibition along with inactivation of the enzyme, was impeded in the presence of heparin. The purified enzyme was activated to about the same extent by HDL, VLDL, or serum, but the addition of heparin to these mixtures had no effect. On the other hand, the crude enzyme was extensively inhibited by serum, and this inhibition was abolished by the addition of small amounts of heparin. Heparan sulfate and dermatan sulfate were also able to stimulate the crude enzyme, but the required concentration exceeded that of heparin by a factor of 1000; chondroitin sulfate was ineffective. The data indicated that heparin may increase the activity of lipoprotein lipase, but only under otherwise suboptimal conditions. These findings suggested that the polysaccharide does not actually stimulate the enzyme, but rather prevents inhibition or inactivation. A further indication of this was the fact that heparin decreased the rate of heat inactivation of the enzyme. However, Iverius et al. found that both the purified lipoprotein lipase and the crude preparation of the enzyme contained small amounts of endogenous heparin-like material, further indicating that heparin may be part of, or bound to, the protein.

Although the precise role of heparin in lipoprotein lipase metabolism is not known, the above experiments indicate the following effects of this polyanion. *In vivo* heparin apparently causes the release of lipoprotein lipase from tissue. It has been assumed that endogenous heparin, which is synthesized by, stored in, and released from mast cells, acts in the same way in the circulation as does exogenous heparin. *In vitro* heparin activates the enzyme, but the exact nature of the activation is not clear. Heparin does stabilize the enzyme and protect it from heat inactivation, but it also appears to play a role in linking the enzyme to its lipid substrate. However, purified preparations of lipoprotein lipase do appear to be closely associated with or to contain heparin or some closely related sulfated glycosaminoglycan.

2. Trehalose Phosphate Synthetase

Liu et al. (22) found that crude extracts of *Mycobacterium smegmatis* catalyzed the synthesis of trehalose phosphate from glucose 6-phosphate and either GDP–glucose or UDP–glucose. When this crude extract was fractionated on DEAE-cellulose, a protein peak was eluted from the column which catalyzed the synthesis of trehalose phosphate from GDP–gluose but was inactive with UDP–glucose. However, the addition of later column fractions to this protein peak resulted in the restoration of UDP–glucose:trehalose phosphate synthetic activity. The active component in these later column fractions was found to be RNA. The RNA was not used up in the reaction and therefore appeared to "activate" the enzyme Other RNAs and polynucleotides could also stimulate UDP–glucose activity to varying degrees.

Lapp, Patterson, and Elbein (23) further investigated this enzyme system and found that a number of different high-molecular-weight polyanions were able to activate the synthetase to use UDP–glucose as a glucosyl donor. The best activator was found to be heparin.

Heparin concentrations of 0.5–1 $\mu g/ml$ were optimal for activation, whereas higher concentrations (5 μg) were inhibitory. Various chondroitin sulfate preparations, mycobacterial RNA, polyuridylic acid (mol. wt. 7500), and dermatan sulfate were also good activators, although less effective than heparin. High-molecular-weight (50,000) polynucleotides of uracil and adenine were less effective, and polyanions whose charge was due to carboxyl groups (hyaluronic, polyaspartic, and polygalacturonic acids) were ineffective in stimulating activity. Thus the best activators of this system may be polyanions containing N-sulfamino groups, as described by Bernfeld and Kelley (15) for lipoprotein lipase. However, in this system other polyanions are also good activators, whereas with the lipoprotein lipase, other sulfated glucosaminoglycans were inhibitors.

With the use of ^{3}H-labeled RNA, the polyanion could be shown to bind to the enzyme (by filtration or ultracentrifugation), and the pH optimum for binding was about 7.0, which suggests the involvement of histidine residues in the protein. Activation of the enzyme by polyanion was inhibited by high salt concentrations (0.05 M) and was also inhibited by microgram amounts of polyornithine. Thus the activation (and presumably the binding) involves some kind of electrostatic interaction between the polyanion and the protein. Further evidence for binding was the fact that the RNA was no longer susceptible to RNase digestion once it was mixed with the synthetase. Although these polyanions did protect the enzyme to some extent from heat denaturation, the activation did not appear to be simply a protection of the enzyme. In an experiment designed to determine the nature of the activation, enzyme was incubated with UDP–glucose (and other reaction components) in the presence or absence of polyanion. In the presence of polyanion, formation of trehalose phosphate began immediately and proceeded in a linear fashion for about 20 min, whereas in the absence of polyanion no trehalose phosphate was formed. However, when polyanion was added to these incubation mixtures, trehalose phosphate formation began immediately and proceeded at the same rate as in those tubes that contained polyanion from the start. It thus appears that the polyanion binds to the enzyme and may alter its conformation. Thus could involve an association or dissociation of subunits such as has been suggested by Bernfeld for polycation activation of other enzymes (see below).

An interesting study on the UDP–glucose:trehalose phosphate synthetase of *M. tuberculosis* was done by Goldman and Lornitzo (47), who found that an inhibitor of the enzyme was present in crude extracts. These authors (48) purified this inhibitor, called mycoribnin, from extracts of this organism and found that it is a polynucleotide with a molecular weight of 4000 and is composed of adenine and guanine. The inhibitor was found in the avirulent strain of this organism, but not in the virulent strain. Its inhibition of enzymatic activity was noncompetitive. Lapp et al. (23) found that mycoribnin also inhibited the trehalose phosphate synthetase of *M. smegmatis*, but this inhibition could be overcome by the addition of increasing amounts of *M. smegmatis* RNA activator.

3. Rabbit-Muscle Aldolase

Tandon, Saxena, and Krishna Murti (24) found that the glycoprotein orosomucoid stimulated and stabilized the activity of dilute solutions of five times recrystallized rabbit-muscle adolase. The stimulatory effect seemed to be related to the sialic acid content of the mucous preparation. Sialic acid alone stimulated 40% at 10 μg per 6 μg of enzyme, whereas 10 μg of orosomucoid that contained 1 μg of sialic acid stimulated 50–150%. The effect of orosomucoid was appreciably greater than that of bovine–serum albumin. Although these experiments suggested an effect of bound sialic acid, apian orosomucoid, which contained the lowest amount of sialic acid, produced the maximum stimulation. The authors suggested that exogenously added glycoproteins exerted some protective effect on the conformation of the protein. The activation, however, was not thought to be due to the prevention of loss of activity during preincubation. On the other hand, Bernfeld et al. (37) found that various polycations activated rabbit-muscle aldolase presumably by preventing the dissociation of the enzyme (see Section I.B.3). These authors reported that a variety of polyanions (sulfated polymers) inhibited this enzyme.

4. Activation of Pepsinogen by Polyanions

Polyanions are known to exert an inhibitory effect on the hydrolysis of substrates by pepsin (see Section IV), but have been shown to produce an activating effect on the conversion of pepsinogen to pepsin. Preliminary observations by Pamer (25) indicated that chondroitin 4-sulfate promoted the activation of pepsinogen. Anderson (26) also found that heparin and chondroitin 4-sulfate activated pepsinogen. Horowitz, Pamer, and Glass (27) found that the polyanions RNA, chondroitin 4-sulfate, heparin, and a sulfated glycoprotein fraction from human gastric juice promoted the activation of pepsinogen. Heparin was the most active, as expected, since it has sulfate groups of lowest pK_a and the highest charge density. Surprisingly, RNA, which is of lower charge density than chondroitin sulfate, was more effective in promoting pepsinogen activation. It therefore appears that charge density is not the only factor involved in the interaction, but that the conformation and molecular weight of the polyanion are also important. These polyanions probably exert their effect by binding and removing the basic polypeptide that is released during the activation process. Perlman (49) found that the configuration of pepsinogen was stabilized by side-chain interactions between basic amino acids on the peptide segment released during activation and carboxyl groups on the protein. Katchalski, Berger, and Neumann (50) showed that polylysine forms a complex with pepsin, rendering the enzyme inactive.

5. Oxidative Phosphorylation System

Pinchot (28,29) fractionated the enzyme system from *Alcaligenes faecalis*, which catalyzes the formation of ATP linked to the oxidation of NADH, into three components. These components were identified as a particulate NADH oxidase, a soluble protein-coupling factor, and a specific RNA oligonucleotide containing adenine, guanine, and uracil. In order to link phosphorylation to oxidation, the two soluble components of this system had to be added to the particles several minutes before the substrate was added. Synthetic polyadenylic acid was effective in replacing the RNA fraction from the organism, as were polymers synthesized from UDP, CDP, or GDP by polynucleotide phosphorylase. However, polycytidylic acid of molecular weight 100,000 was not effective in this system. The polynucleotides could be isolated intact at the end of the incubation, which indicates that they were not used up during the reaction. Further studies on this system by Shibko and Pinchot (30) indicated that the polynucleotide in combination with Mg^{2+} bound the coupling factor to the particles to make an active phosphorylating particle.

Studies on an analogous system were done by Allfrey and Mirsky (31), who found that cell nuclei that has been pretreated with DNase lost their ability to synthesize ATP as well as their capacity to incorporate amino acids into proteins and purines or pyrimidines into nucleic acid. The latter two effects seemed to be the result of impaired aerobic phosphorylation. The ability to synthesize ATP as well as the other synthetic capacities was restored on the addition of DNA to these nuclei. These workers (32) also found that a variety of polyanions, including polyethylene sulfonates, heparin, chondroitin sulfate, and RNA, could replace DNA in restoring the ATP synthesis. Polycations like polylysine and protamine did not restore function to depleted nuclei; in fact addition of polylysine to untreated nuclei inhibited the incorporation of amino acids into protein. Sekiguchi and Sibatani (33) found that yeast RNA and chondroitin sulfate also restored much of the ^{32}P incorporation into RNA and DNA in these DNase-treated nuclei.

6. ATPase

Debreceni, Behine, and Ebisuzaki (34) purified an ATPase from *Escherichia coli* infected with bacteriophage T4. The enzyme had a molecular weight of about 15,000 and was not detected in unifected cells. The formation of ADP by this purified enzyme was completely dependent on the presence of DNA and Mg^{2+}. Different DNAs varied in their ability to stimulate the reaction, but in all cases heat-denatured DNA was better than native DNA. Thus heat-denatured DNAs from bacteriophage T4 and T7, *E. coli*, and calf thymus stimulated the reaction more effectively than did the native DNA from *E. coli* or calf thymus, whereas the native DNAs from bacteriophage T4 and T7 were without effect, as was *E. coli* SRNA. Since the type of DNA present had a marked effect on the extent of the reaction, the ability of DNA to stimulate ADP formation did not seem to be due simply to its polyanionic nature. A number of other DNA-dependent ATPase activities have been reported, but in all of these other cases the enzyme also had endonuclease or exonuclease activity. Thus Anai, Hirahashi, and Takagi (51) found that *Micrococcus lysodeikticus* endonuclease degraded three molecules of ATP for every phosphodiester bond broken. Other workers, including Buttin and Wright (52), Barbour and Clark (53), and Nobrega et al. (54) described an ATP-dependent exonuclease that may play a role in genetic recombination. When ATPase activity was studied with these enzyme preparations, it was found to be dependent on the presence of native DNA. However, DNA was also degraded in these reactions. However, the phage-induced *E. coli*

ATPase contains no endonuclease or exonuclease activity. Although this enzyme appears to be quite specific for DNA (rather than for polyanions in general), it is included here because DNA seems to cause an activation of ATPase activity, but there is no evidence of the enzyme's being intimately associated with DNA.

B. POLYCATION ACTIVATIONS

1. β-Glucuronidase

Bernfeld et al. (35) studied the effects of dilution on highly purified preparations of β-glucuronidase from calf liver and calf spleen. On dilution of these enzymes, there was a considerable reduction in the specific activity with a decrease in the monomolecular rate constant. The authors suggested that this loss of specific activity was due to dissociation of the enzyme since the experimental data corresonded to theoretical curves derived from the law of mass action. The possible relevance of other factors, such as the presence of inhibitors in the substrate preparation or the instability of the enzyme, was ruled out by various experiments. However, the authors did not identify possible inactive products resulting from the dissociation of the enzyme.

The "dissociation" of β-glucuronidase could be reversed by a number of substances, which thus functioned as activators of the enzyme. Both the dissociation and the association of the enzyme followed the law of mass action. The following compounds were found to enhance the activity of dilute solutions of β-glucuronidase (in order of decreasing potency): chitosan, protamine, crystalline bovine-serum albumin, DNA from salmon milt, 1,10-diaminodecane, DNA from fish sperm, gelatin, crystalline chymotrypsin, thymus DNA, other α,ω-diaminopolymethylenes, spermine, spermidine, yeast, RNA, lysine, and ornithine. The latter compounds required milligram quantities for 50% activation, whereas 2 µg of chitosan gave the same effect. A variety of monoamino acids and mononucleotides failed to stimulate the enzyme. Activation required at least two basic groups per molecule, with the distance from each other markedly influencing potency. However, the size of the molecule did not appear to be an important factor.

2. Hyaluronidase

The influence of dilution on a commercial sample of testicular hyaluronidase and on a preparation of the enzyme that had been purified fivefold by zone electrophoresis was studied by Bernfeld, Tuttle, and Hubbard (36). On dilution, hyaluronidase was transformed into an inactive form, as demonstrated by the progressive loss in specific activity. As in the case of β-glucuronidase, the authors suggested that the loss in activity was due to a reversible dissociation of the enzyme. The loss in activity of the hyaluronidase could be reversed by the addition of a number of cations, with poly-L-lysine being the most effective, followed by protamine, chitin, and 1,10-diaminodecane. A number of proteins were not effective in preventing the dissociation. In all cases of activation the specific activity of the hyaluronidase could be restored to that of the undiluted enzyme, but never exceeded that activity. A number of macromolecular polyanions, such as polystyrene sulfate, heparin, sulfated pectic acid, amylopectin sulfate, and DNA, were found to be inhibitors of hyaluronidase. The inhibition was competitive since it could be overcome by increasing the concentration of activator (polycation). Thus these polyanions appeared to bind the activator and thereby promote the dissociation of the enzyme.

3. Aldolase, Lactic Dehydrogenase, α-Amylase, and β-Amylase

Bernfeld, Berkeley, and Bieber (37) also studied the effect of dilution on crystalline samples of rabbit-muscle aldolase, rabbit-muscle lactic dehydrogenase, pancreatic α-amylase, and sweet-potato β-amylase. All of these enzymes showed a similar decrease in specific activity on dilution as seen for β-glucuronidase and hyaluronidase. The concentrations at which the specific activity fell to one-half its maximal value were as follows: 2 $\mu g/ml$ for aldolase, 1.5 $\mu g/ml$ for lactic dehydrogenase, 0.005 $\mu g/ml$ for α-amylase and 1 $\mu g/ml$ for β-amylase. Various polycations, as well as some proteins acting as polycations, were able to restore the activity of these enzymes. However, each enzyme acted differently to different cations; that is, a polycation activator of one enzyme did not necessarily activate another enzyme. In all cases of activation, restoration of enzymatic activity reached that of the original undiluted sample, but did not exceed it. These various cations did not stimulate the undiluted enzyme. The effect of polycations was not simply due to stabilization since the addition of protamine to a sample of adolase that had been incubated without substrate produced a substantial increase in activity beyond that of the enzyme in the absence of polycation. A number of polyanions, particularly polysaccharide sulfate esters, inhibited these enzymes, presumably by promoting the dissociation of the protein. The effectiveness of the inhibitor depended largely on the chemical nature of the negative groups and to a lesser extent on molecular size.

4. Acid Phosphatase

Jeffree (38) found that acid phosphatase from the prostate gland was irreversibly inactivated when subjected to high dilution in the absence of protective substances. This inactivation appeared to be the result of surface denaturation. Enzyme activity, however, was maintained in the presence of certain proteins, polypeptides, and amino compounds, particularly long-chain polyamines. The protective effect of the polyamines increased with chain length between terminal amino groups, but disappeared when the amino groups were destroyed with nitrous acid or were converted to amides. Furthermore, complete hydrolysis of the proteins destroyed their protective action while enzymatic digestion reduced it significantly. These studies indicated that both charge and size are involved in stabilization of this enzyme.

5. Polynucleotide Phosphorylase

Polynucleotide phosphorylase represents an interesting case of polycation stimulation that has been well studied. Dolin (39,40) found that the rate of polyadenylate synthesis from ADP by both a crude and a partially purified enzyme preparation from *Clostridium perfringens* was stimulated twenty-fold by various basic polypeptides, but not by acidic polypeptides. The phosphorylase eluted from calcium phosphate gel (and purified about tenfold) showed a K_s for ADP of $1.6 \times 10^{-2}\ M$, whereas in the presence of saturating concentrations of poly-L-lysine, the K_s decreased thirtyfold to $5.8 \times 10^{-4}\ M$. This enzyme also catalyzed polymer formation from UDP and CDP, but both of these reactions were inhibited 90% by polylysine, as was the phosphorolysis of polyadenylate. Further purification of the enzyme by electrophoresis on starch agar gave a hundredfold purified enzyme that behaved like an acidic protein. This preparation had no polymerase activity in the absence of activator, but it catalyzed the incorporation of ADP in the presence of polylysine. However, this enzyme preparation did not catalyze polymer formation from UDP or CDP. While the tenfold purified enzyme was stimulated maximally by polylysine, polyornithine, and polyvinylamine but also reacted to protamine and histone, the purified enzyme was stimulated only by polylysine and polyornithine.

Dietz and Grunberg-Manago (41) found that polynucleotide phosphorylase existed in two different forms, which could be separated by sucrose-density-gradient centrifugation. The heavier form, which had a molecular weight of 192,000, had polymerase activity for ADP in the absence of polylysine. However, the activity of this enzyme was increased by a factor of 10–15 by the addition of $(NH_4)_2SO_4$ and 30–40 times by polylysine. This fraction also catalyzed the phosphorolysis of polyadenylate. The lighter band of molecular weight 62,000 showed an absolute requirement for polylysine and did not catalyze phosphorolysis. This lighter enzyme also had a requirement for β-mercaptoethanol, which was not shown by the heavier enzyme. Fitt, Dietz, and Grunberg-Manago (42) purified the enzyme further to obtain a preparation with a very low nucleotide content. With this preparation polylysine stimulated GDP and ADP polymerization, but inhibited the polymerization of UDP and CDP. The authors studied the effect of salt and pH on ADP polymerization in the presence and absence of polylysine. Various inorganic salts at high concentration were able to stimulate polymer synthesis markedly, but they prevented the stimulation by polylysine. The effect of polylysine was dependent on pH and on ionic strength, which indicates that the activation by the polycation was due to some sort of electrostatic interaction. Polylysine was found to protect the enzyme from thermal inactivation, but inorganic salts were not effective in this regard. The polybase was also found to interact with the product of the reaction (polyadenylic acid) and to remove it from solution, but this finding did not appear to be sufficient to explain the effect of polylysine. Fitt and Wille (43) treated polynucleotide phosphorylase with trypsin for short periods of time. This treatment resulted in the rapid and preferential loss of the activity stimulated by polylysine (and polyornithine) as well as by salt. Although polyornithine protected the enzyme from trypsin digestion, this could be due to trypsin inhibition by polybase since Katchalski et al. (103) found that polylysine inhibited pepsin. Fitt and Wille suggested that trypsin partially degraded the phosphorylase so that it no longer was capable of interacting with the polybase. They also examined the effect of various sized activators on the polymerization of ADP (44). The effectiveness of the activators (oligomers of lysine) increased with increasing chain length up to about 20 or 30 residues. Polymers of the

L- and DL-series were equally effective. Lysine, diethylamine, and triethylamine were also effective as long as their concentration was high enough, but their effect seemed to be similar to that of NaCl. The authors suggested that the stimulation by low-molecular-weight compounds probably involves a generalized ionic effect on the enzyme molecule, causing it to adopt a more active configuration. The increased efficiency of polymers could be due to a certain spacing of charged groups, which interact more efficiently with acidic groups on the enzyme surface.

6. *Phosphorylase b*

Krebs (45) found that the protamine salmine activated phosphorylase b in the presence of AMP by increasing the affinity of the enzyme for this nucleotide. In the absence of salmine the K_m for AMP was 6×10^{-5} M, whereas in the presence of this cation (40 μg/ml) it was $5 \times 10^{-6} M$. The V_{max} was also somewhat higher in the presence of salmine. Polylysine also activated the enzyme, but histone and globin had no effect. Phosphorylase b could also be produced from phosphorylase a by the action of trypsin, but the phosphorylase b formed in this way was insensitive to saline. Phosphorylase a, on the other hand, was inhibited by microgram quantities of salmine either in the presence or absence of AMP. This inhibition could be overcome by insulin.

7. *Sialyl transferase C*

A particulate enzyme preparation from embryonic chicken brain was shown by Kaufman, Basu, and Roseman (46) to catalyze the transfer of sialic acid from CMP-N-acetylneuraminic acid to monosialoganglioside to form disialoganglioside. The rate of this reaction was stimulated by a factor of 2.5 by optimum concentrations of histone (4.2 mg/ml) or by lower concentrations of poly-L-lysine. Bovine-serum albumin or globulin did not affect the reaction. The mechanism of activation by polycations is not known, but it may involve interactions of various components with the membranes.

III. Specific Interactions of Enzymes with Polynucleotides

The literature dealing with protein–nucleic acid interactions is much too voluminous to be covered in this chapter. However, some general examples and a few selected examples of enzymes that either contain or interact with a polynucleotide will be discussed. Although all of these interactions of proteins with nucleic acids resemble the polyelectrolyte–protein phenomena in that they undoubtedly involve electrostatic attractions and ionic binding, they show an additional factor of specificity that is not usually seen in the more general polyelectrolyte interactions discussed so far.

The readers are directed to a number of excellent reviews dealing with selected areas of protein–nucleic acid interactions (55–58). Only a few general examples from these reviews will be cited here. Not surprisingly, basic polyamino acids like polylysine and polyornithine form complexes with DNA, and these complexes are dependent on the ionic strength of the medium. In the presence of 1 M salt, the combination of DNA and polylysine is reversible (59). Even here some specificity is displayed since polylysine preferentially binds to AT-rich regions of DNA (55). Histones, of course, also interact with DNA, and these interactions are probably involved in the regulation of transcription and replication (56). Numerous studies have been done on the lac operator, which is a DNA segment that interacts with the lac-repressor protein. The binding of lac repressor to the lac operator depends on ionic strength (57). The RNA polymerase of *E. coli* is an interesting enzyme that binds to DNA, and this binding is dependent on ionic strength and template conformation (58). Both the complete enzyme (or holoenzyme) and the core enzyme (holoenzyme without σ-factor) bind to polyanionic materials. The polyanion heparin, which effectively competes with DNA for binding to the enzyme, also forms complexes with the β'- and σ-subunits (58).

Several protein factors that appear to play a role in binding mRNA to ribosomes have been described. A protein factor that shows a high affinity for polyuridylic acid was recently isolated from ribosomes by Smoralsky and Tal (60). Finally, it should be mentioned that it has been suggested by some workers that certain proteins may have dual roles and functions both as enzymes and as repressors (61). Several enzymes that seem to fit this category are phosphoribosyl–ATP synthetase and threonine deaminase, which interact with RNA and may function in regulation. These enzymes will be discussed here without regard to any possible involvement in regulation.

The enzymes discussed in this section are listed in Table II.

TABLE II

Specific Interactions of Polynucleotides with Enzymes

Enzyme (or reaction catalyzed)	Polynucleotide	Mode of action	References
Enzymes containing poly-nucleotides:			
Thiol–disulfide exchange protein	RNA (oligonucleotide)	Bound to enzyme; necessary for activity	62
Phosphofructokinase	RNA (oligonucleotide)	Bound to enzyme; may be modulator of activity	63–65
Enzymes activated by, or forming complexes with, polynucleotides:			
Glutamine-synthetase-deadenylating enzyme	rRNA	Stimulates deadenylation	66, 67
Phosphoribosyl transferase	His-tRNA	Forms complex with enzyme	68
Threonine deaminase	Leucyl-tRNA	Forms complex with enzyme	69, 70
Nucleotidyl transferase	tRNA	Forms complex with enzyme	71
Aminoacyl synthetases	tRNA	Forms complex with enzyme	(72–75

A. ENZYMES THAT CONTAIN BOUND POLYNUCLEOTIDES

1. Thiol–Disulfide Exchange Protein

Sakai (62) isolated an enzyme from sea-urchin eggs that catalyzes electron exchange between the contractile protein of the cell cortex and the sulfhydryl groups of a protein from the mitotic apparatus. This enzyme was purified about 40-times, at which point it was homogeneous by ultracentrifugation and gel electrophoresis. The protein contained about 20% RNA. About 10 molecules of nucleotide (3 molecules each of guanylic and cytidylic and 2 of adenylic and uridylic) are attached to a polypeptide of 10,500 molecular weight. The nucleotide was resistant to RNase but was cleaved by phosphodiesterase. The nucleotide could be separated from the protein by phenol extraction, and the protein was rid of its nucleotide by phosphodiesterase treatment. Removal of the nucleotide from the enzyme resulted in a complete loss of catalytic activity, but when the protein was mixed with the isolated nucleotide in the presence of glutathione, ascorbic acid, and tris buffer at 25°C, activity was largely restored. There is no information available in this system as to whether the nucleic acid is directly concerned with the enzymatic site or whether it merely provides support for the tertiary structure of the protein that is essential for enzymatic activity. Sakai postulated that the nucleic acid could control initiation and termination of cell division by controlling the activity of this enzyme.

2. Phosphofructokinase

Hofer and Pette (63,64) purified phosphofructokinase from rabbit skeletal muscle by several different procedures. One purification method gave a protein essentially free of nucleic acid, whereas the other gave a protein fraction that contained nucleic acid. The nucleic acid could be separated from the protein by heat denaturation and was found to be a polynucleotide. About 20 moles of nucleotide were present for 100,000 g of protein. The effect of AMP was studied in both the nucleic-acid-rich and the nucleic-acid-free enzyme preparations. Stimulation of the ATP-inhibited enzyme by AMP only occurred with the nucleic-acid-rich enzyme. There was a linear relationship between the effect of 5'-AMP and the nucleic acid content of the preparation, which suggests that different enzyme preparations represent different mixtures of the two forms of phosphofructokinase. Hofer and Pette (65) suggested that the action of different effectors is related to the presence of the nucleic acid and that the nucleic acid might function as a modulator of enzyme activity.

B. ENZYMES THAT ARE ACTIVATED BY, OR FORM COMPLEXES WITH, POLYNUCLEOTIDES

1. *Glutamine-Synthetase-Deadenylating Enzyme*

The glutamine synthetase of *E. coli* may have from 0 to 12 equivalents of adenyl groups bound per mole of enzyme. Shapiro (66) showed that the adenyl residues are hydrolyzed from the enzyme by the glutamine-synthetase-deadenylating enzyme. The deadenylation was stimulated considerably by specific nucleotides; thus a 45-fold stimulation was afforded by a combination of UTP (0.8 mM) and ATP (40 μm). A ribosomal RNA preparation from *E. coli* could replace ATP in the synergistic stimulation of the reaction. Polyuridylic acid and other polynucleotides stimulated the reaction with a less purified enzyme preparation, but were inactive with a more highly purified enzyme. However, 16S plus 23S rRNA also stimulated the purified enzyme, which suggests that RNA might have a more specific effect. Anderson and Stadtman (67) separated the deadenylating enzyme into two protein fractions, PI and PII. The PI component alone, when adsorbed on manganese prosphate, could catalyze a slow deadenylation of the synthetase. They found that tRNA markedly stimulated some preparations when deadenylation was catalyzed by the combined action of PI and PII, but it inhibited deadenylation by PI alone. The mechanism of this action is not known.

2. *Phosphoribosyl Transferase*

Kovach et al. (68) purified phosphoribosyl transferase from *Salmonella typhimurium* so that the enzyme was more than 90% pure. Using tritium-labeled histidyl-tRNA, these workers were able to demonstrate the binding of this RNA to the enzyme either on Sephadex G-100 gels or by filtration studies. The binding of histidyl-tRNA was favored over that of other aminoacylated tRNAs, but the enzyme did have a high affinity for tRNA. Binding of RNA was dependent on Mg^{2+} concentration. Kovach et al. suggested that the complex of enzyme and histidyl-tRNA might act as a repressor for enzymes involved in histidine synthesis.

3. Threonine Deaminase

Although L-theronine deaminase from *Salmonella typhimurium* is composed of four identical polypeptide chains, it possesses only two binding sites for each stereospecific ligand, Therefore, although the enzyme is structurally a tetramer, it functions as a dimer. Hatfield and Burns (69) studied the kinetics of reassociation of the dimers into the tetrameric form and found that the tetramer remained catalytically inactive (immature) unless it was exposed to a ligand for which the native or mature enzyme possessed a stereospecific site (L-threonine, L-valine, L-isoleucine). When L-isoleucine and L-valine were present together, the ligand-induced maturation of the enzyme was blocked, but if either one of these amino acids was removed, the immature enzyme became active. Hatfield and Burns (70) found that the immature form of the enzyme bound leucyl-tRNA, but this binding site was lost on maturation of the enzyme. It appeared from their studies that the immature enzyme bound two molecules of leucyl-tRNA per protein molecule. However, no binding of valyl-tRNA or isoleucyl-tRNA was observed. Furthermore, uncharged leucine-tRNA did not compete with charged leucyl-tRNA, which indicates that binding was highly specific. The authors suggested that threonine deaminase containing bound leucyl-tRNA might function as a repressor.

4. Yeast Nucleotidyl Transferase

Nucleotidyl transferase was purified more than a thousandfold from yeast by Morris and Herbert (71). When the purified enzyme was sedimented in sucrose density gradient, the enzymatic activity was found as a single symmetric peak (4.3S). However, in the presence of tRNA the enzymatic activity sedimented 50% faster ($\sim$ 6S). That the enzyme forms a stable complex with the tRNA was shown by sucrose-density-gradient centrifugation and by the binding of radioactive tRNA to nitrocellulose filters in the presence, but not in the absence, of enzyme. The enzyme bound very tightly to any tRNA that was missing part of the terminal pCpCpA triplet. However, if the terminal triplet was complete, the complex was much weaker. The transferase did not bind small fragments from T_1 RNase digestion nor did it bind polyuridylic or polycytidylic acid. These polynucleotides, however, did inhibit the binding of enzyme to tRNA.

5. Binding of Aminoacyl Synthetases to tRNA

The binding of tRNA to the aminoacyl synthetases has been demonstrated by a number of workers. Thus Langerkvist and Waldernström (72) showed binding to yeast valyl-tRNA synthetase by Sephadex columns; Yarus and Berg (73) demonstrated binding to yeast leucyl-tRNA synthetase by Millipore filtration, and Makman and Cantoni (74) showed binding to seryl-tRNA synthetase. Small modifications of the tRNA at its CCA end prevented the association. Letendre, Humphreys, and Grunberg-Manago (75) found that polyuridylic acid could associate with lysyl-tRNA synthetase and prevent the binding of tRNA. Polyuridylic acid was a competitive inhibitor of tRNA.

IV. Inhibition of Enzymes by Polyelectrolytes

A large number of enzymes have been shown to be inhibited by either polycations or polyanions. As might be expected, enzymes that are activated by polyanions are usually inhibited by polycations (or at least activation is prevented) and vice versa. Thus lipoprotein lipase was inhibited by protamine, trehalose phosphate synthetase by polyornithine, and so on, whereas hyaluronidase, β-glucuronidase, aldolase, and amylase were inhibited by polyanions.

The experiments of Gelman, Rippon, and Blackwell (76) showed that polylysine interacts with chondroitin sulfate at neutral pH, so that one would anticipate a competition between enzyme and polycation (or polyanion) for the opposite polyelectrolyte. In addition to these enzymes, many more that do not show polyelectrolyte activation have been studied and found to be inhibited by either polycations or polyanions. Inhibition by the polyelectrolyte was generally reversible by the addition of the oppositely charged polyvalent material. Table III lists a number of enzymes that show this inhibition. Because of limitations in time and space, and because not much is known about the mechanism of inhibition, these enzymes will not be discussed individually. For more comprehensive coverage the reader is referred to extensive reviews by Bernfeld (1,2) and by Sela and Katchalski (59).

V. *In Vivo* Effects of Polyelectrolytes

In addition to the *in vitro* activations and inhibitions of polyelectrolytes already discussed, many polyelectrolytes exert a profound effect on various cells and tissues. This is not unexpected since one would anticipate that these compounds would interact with various proteins (or other charged compounds) either within cells or at the surface of cells in a somewhat analogous manner to the *in vitro* situation. That such interactions do indeed occur is indicated by some of the studies outlined in this section.

A. EFFECTS OF POLYANIONS ON CELLS

Most of the polyanions (e.g., heparin, chondroitin sulfate, hyaluronic acid) that have been discussed so far are naturally occurring materials found in most mammalian cells. The presence of glycosaminoglycans in the intercellular material or ground substance has been well established *in vivo* (109), and many cell surfaces are known to contain a wide variety of complex carbohydrates, including glycoproteins, glycolipids, and glycosaminoglycans (110). Kraemer, for example, showed the presence of heparan sulfate in cell membranes (111), and Kojima and Yamagata (112) reported the occurrence of chondroitin sulfate at the cell surfaces of rat ascites hepatomas. Lippmann (113) proposed a role for mucopolysaccharides in the initiation and control of cell division since these materials appear to be universal components of cell surfaces.

Mascona (114) demonstrated the presence of specific factors that are associated with the cell surfaces and intracellular spaces, and function as cell ligands and mediate the histologic attachment and aggregation of cells. The reaggregation of liver and cartilage cells was studied by Khan and Overton (115), who found that a lanthanum staining material that appeared to be a mucopolysaccharide was present between the cells at all stages of reassociation, but was absent in freshly isolated cells. During the development of the chick axial region, Kvist and Finnegan (116) observed an increase in sulfated glycosaminoglycan in the extracellular matrix of all axial areas. The accumulation of various mucopolysaccharides during the development of embryos of the Pacific great skate was shown by McConnachie and Ford (117). Thus glycosaminoglycans have been implicated as playing a key role in development as well as in cell aggregation.

A direct effect of mucopolysaccharides on cells in culture has been demonstrated in several laboratories. Pessac and Defendi (118) isolated from chick embryo and mammalian cell cultures factors that induced cell aggregation. These factors, which were released from these cells

TABLE III
Polyelectrolyte Inhibitors of Enzymes

Enzyme	Polyelectrolyte	Reference
Polyanion inhibitors:		
Alkaline phosphatase	Heparin	77
Acid phosphatase	Heparin	78
	Chondroitin 4-sulfate	78
Hyaluronidase	Heparin	79
	Chondroitin 4-sulfate	80
	Chitin sulfate	81
	Amylopectin sulfate	36
	Cellulose sulfate	81
	Synthetic polyglucose sulfate	82
	Pectic acid sulfate	36
	DNA	36
	Polystyrene sulfonate	16
	Polymethacrylic acid	36
Ribonuclease	Heparin	83
	Chitin sulfate	84
	Amylase sulfate	84
	Dextran sulfate	85
	Cellulose sulfate	81
	Carboxymethylcellulose sulfate	84
	Synthetic polyglucose sulfate	82
	Pectic acid sulfate	84
	Pectin sulfate	86
	Galacturonan sulfate	87
	Polyaspartic acid	88
	Tetrametaphosphate	88
	Polyvinyl sulfate	84
	Silicotungstate	86
	RNA, polynucleotides	89
Amylase	Heparin	90
	Amylopectin sulfate	37
	Amylose sulfate	37
	Dextran sulfate	85
	Polystyrene sulfonate	37
	Polyvinyl sulfate	37
Lysozyme	Heparin	91
	Hyaluronic acid	92
	Synthetic polyglucose sulfates	82
	RNA	92
	DNA	92
Trypsin	Heparin	93
	Polyglutamic acid	94

TABLE III (*Continued*)
Polyelectrolyte Inhibitors of Enzymes

Enzyme	Polyelectrolyte	Reference
Polyanion inhibitors—continued:		
Pepsin	Heparin	95
	Chondroitin 4-sulfate	95
	Amylose sulfate	96
	Rice-starch sulfate	96
	Carrageenin	96
	Anhydromannuronan	95
	Polyvinyl sulfate	96
Lipoprotein lipase	High concentrations of polyanions	15
Fumarase	Heparin	97
	Chondroitin 4-sulfate	97
	RNA	97
DNase	Galacturonan sulfate	87
Arginase	RNA	98
β-Glucuronidase	Polystyrene sulfonate	99
	Polymethacrylic acid	99
Lactic dehydrogenase	Polystyrene sulfonate	37
	Polyvinyl sulfate	37
Catalase	Polyacrylic acid	100
	Maleic anhydride copolymers	100
Hexokinase	Phosphates of polyphloretin	101
Glyceraldehyde-3- phosphate dehydrogenase	Polyvinyl phosphate	99
RNA Synthesis	Acid polysaccharides	102
Respiratory activities of mitochondria	Dextran sulfate	89
Polycation inhibitors:		
Pepin	Polylysine	103
Trypsin	Polylysine	104
	Polyglutamic acid	105
Phosphorylase a	Protamine, salmine	45
Chymotrypsin	Polyamino acids containing tryosine	106
Lipoprotein lipase	Protamine	9
Trehalose phosphate synthetase	Polyornithine	107
Amino acid incorporation into nuclear protein	Polylysine	31
Photosystem I	Histone	108

and were also present in some serums, appeared to be acid mucopolysaccharides, with hyaluronic acid being a major component. On the
other hand, Toole, Jackson, and Gross (119) found that the addition
of purified hyaluronate blocked the formation of colonies and cartilage
nodules in stationary cultures of cells from embryonic chick somites and
limb buds. These workers proposed that hyaluronate might act as a
regulator or inhibitor of mesenchymal cell aggregation in embryo-
genesis and that its synthesis and removal might be part of the
mechanism of timing of migration, aggregation, and subsequent differentiation. The addition of chrondromucoprotein to a suspension
culture of chondrocytes was shown by Novo and Dorfman (120) to
result in a marked stimulation in the rate of chondromucoprotein synthesis. A number of other polyanionic substances could also stimulate
the rate of synthesis of chondromucoprotein, including chondroitin
sulfate, heparan sulfate, heparin, and dermatan sulfate, which suggests
that these polyanions might interact with the cell surface and somehow
stimulate the synthetic machinery. A mucopolysaccharide produced
by Neurospora has been reported by Reissig and Glasgow (121) to
inhibit its growth and to cause vacuolation and agglutination of the
cells. However, in this case the polysaccharide appears to be polycationic since the complete acetylation of the galactosamine residues
abolished its activity.

Other studies on cell-surface components have implicated glycoproteins or glycolipids, especially those containing sialic acid, in various
interactions. The sialoglycoproteins appear to constitute the outer-
most region of cell surfaces, and the large number of negative charges
are undoubtedly important in various responses. Burnet (122) showed
that treatment of red blood cells with cell-free filtrates of *Vibrio cholerae*
(neuraminidase) destroyed the receptor sites for influenza virus. This
treatment also reduces the net surface negativeity of these cells by more
than 80%, as shown by Ada and Stone (123). Soupart and Clewe (124)
found that removal of sialic acid from the zona pellicuda of ova inhibited the penetration of sperm which indicates that sialic acid also
plays a role in the attachment of sperm to the ovum. Some cells, such as
embryonic chicken muscle cells, do not aggregate normally in the
presence of neuraminidase. Kemp (125) found that the aggregations
of these cells were much smaller and there was a much larger proportion of free cells in the presence of the enzyme. The result of Oppenheimer et al. (126) also suggested that cell adhesion may depend on the

introduction of hexosamine and/or sialic acid into surface components. Roseman (127) has suggested that cell adhesion may involve the binding of surface-bound glycosyl transferases on one cell with their complex carbohydrate acceptors on an adjacent cell. Various other studies dealing with changes in cell-surface composition during virus transformation have also implicated these components in cell adhesion and contact inhibition. Excellent reviews by Ginsburg and Neufeld (128) and Heath (129) should be referred to for more comprehensive coverage.

B. INTERACTION OF POLYANIONS WITH LIPOPROTEINS

Various polyanions interact with serum proteins both *in vivo* and *in vitro*. In fact, such an interaction has been suggested as being involved in the etiology of atherosclerosis. Bernfeld, Donahue, and Berkowitz (130) studied the interaction of human plasma and serum with various polyanions and found that a number of these compounds (e.g., heparin, sulfated pectic acid, sulfated amylose) formed complexes with β-lipoglobulins. In the pH range 7.5–8.6 no other serum proteins interacted with these polyanions. Depending on the nature of the polyanion and its concentration, three types of interaction were observed, ranging from insoluble to soluble and from irreversible to reversible. In further studies on complex formation between β-lipoproteins and various polyanions, Bernfeld et al (131) found that polymers containing a sulfate ester or sulfonic acid formed insoluble complexes with β-lipoproteins provided the polyanion did not have functional groups other than —SO_3H or —OH. The presence of carboxyl groups (as in the uronic acids of glycosaminoglycans) increased the solubility of the complex. The presence of N-acetyl or N-sulfatyl groups not only increased the solubility but also decreased the affinity of polyanion for lipoprotein. Iverius (132) examined the interaction between human-plasma lipoproteins and various glycosaminoglycans by studying the binding of protein to agarose gels containing covalently bound glycosaminoglycans. Very-low-density lipoproteins (VLDL) and low-density lipoproteins (LDL) bound to gels containing heparin, dermatan sulfate, heparan sulfate, and chondroitin sulfate provided the ionic strength was sufficiently low. In contrast, high-density lipoprotein and acetylated VLDL or LDL did not bind. The lipoproteins also did not bind to gels containing covalently bound hyaluronic acid. These studies suggested an electrostatic binding of polyvalent anionic and cationic sites, with the charge density of polysaccharide being an important parameter in the strength of the interaction. Bihari-Varga and Vegh (133) studied the composition of complexes formed by adding aortic mucopolysaccharide to normal and hyperlipemic serum and showed a correlation between the amount of complex formed and the concentration of total lipid, cholesterol, and cholesterol esters.

That these ionic interactions also occur *in vivo* is indicated by the results of various workers. Tracy et al. (134) found evidence of complex formation between low-density lipoproteins and mucopolysaccharide in human aorta intima. They showed that there was a correlation between the amount of lipoprotein sequestered by the aortic intima and the severity of atherosclerosis. The complex demonstrated by these workers was similar to that shown by Bihari-Varga, Gergely, and Giro (135) by adding mucopolysaccharide from human aorta to whole serum. Srinivasan et al. (136,137) extracted complexes of lipoprotein–mucopolysaccharide from fatty streaks of human aorta with isotonic NaCl. Partial characterization of these complexes indicated that they were composed of low- and very-low-density lipoproteins and chondroitin sulfate and/or heparitin sulfate.

C. EFFECT OF POLYANIONS ON ULCER FORMATION

A number of sulfated polysaccharides have been shown to inhibit the activity of pepsin. Levey and Sheinfeld (138) found that heparin, chondroitin sulfate, and Paritol-C (polyhydromannuronic acid sulfate) inhibited the proteolytic action of pepsin on casein, whereas hyaluronic acid and sodium sulfate had no effect. Oral administration of chondroitin sulfate to Shay rats (25 mg/animal) markedly reduced the number of gastric ulcers found in these animals. Chondroitin sulfate also inhibited the action of pepsin *in vivo*. Bianchi and Cook (139) found that a synthetic polysaccharide, amylopectin sulfate, given orally to pyloric-ligated rats or to rats induced to form gastric ulcers prevented or reduced the ulceration. Houck, Bhayana, and Lee (140) reported similar findings with carrageenin, and Piper and Fenton (141) with two other sulfated polysaccharides.

D. OTHER POLYANION EFFECTS

Concanavalin A was shown to interact with heparin by Cifonelli, Montgomery, and Smith (142) and by Doyle and Kan (143), who found that concanavalin A formed precipitinlike complexes with heparin. The interaction was sensitive to pH and salt concentration, and was inhibited by the addition of methyl-α-D-mannopyranoside. Originally it was thought that anionic polysaccharides might be inducers of interferon, but it now seems fairly well established that the induction is caused by double-stranded RNA, DNA, or polynucleotides (144).

Gelman et al (76) studied the interactions between chondroitin sulfate c and poly-L-lysine by circular dichroism. Their data indicated that ionic interactions between chondroitin sulfate and polylysine forced the polypeptide to adopt the α-helical conformation rather than the "charged-coil" form that is expected at neutral pH in the absence of the polysaccharide.

E. EFFECTS OF POLYCATIONS ON CELLS AND TISSUES

Sela and Katchalski (59) have described a number of effects of polyamino acids such as polylysine on various systems. For example, polylysine retards blood clotting apparently by inhibiting thrombin formation; protamine prevents the clearing effect of heparin in alimentary lipemic rats; polylysine inhibits the infectivity of tobacco mosaic virus as well as some animal viruses and bacteriophages; various basic peptides, including polylysine, inhibit the growth of numerous bacteria; polylysine promoted phagocytosis by human white blood cells.

VI. Protein–Protein Interactions

A number of enzyme systems that which require the participation of two proteins in order to have certain catalytic activities have been described (3–5). Although any direct similarity of these enzymes to polyelectrolyte phenomena is obscure at this time, there are certain analogies between these various systems. For example, in the case of the enzyme lactose synthetase, the protein α-lactalbumin has been called a "modifier" or "specifier" since it apparently alters the acceptor specificity of the enzyme. In the presence of lactalbumin lactose synthetase catalyses the transfer of galactose from UDP–galactose to glucose to form lactose, whereas in the absence of lactalbumin the enzyme transfers galactose to N-acetylglucosamine residues (5). Recently Klee and Klee (145) have shown that lactalbumin forms a complex with the lactose synthetase and that this complex is only stable enough to be observed in the presence of one of the substrates.

Trehalose phosphate synthetase and polynucleotide phosphorylase, which were discussed previously, are somewhat analogous to lactose synthetase in that polyelectrolytes are able to alter the substrate specificity of these enzymes. In the presence of various polyanions, such as heparin, trehalose phosphate synthetase is able to utilize both pyrimidine (UDP–glucose, TDP–glucose, CDP–glucose) and purine (ADP–glucose and GDP–glucose) sugar nucleotides as glucosyl donors, whereas in the absence of polyanion the enzyme is only active with the purine sugar nucleotides. Polynucliotide phosphorylase catalyzes the polymerization of ADP, CDP, GDP, and UDP as well as the phosphorolysis of polynucleotides. However, in the presence of polycations, such as polylysine, ADP and GDP polymerization is greatly stimulated while UDP and CDP incorporation is inhibited, as is the phosphorolysis reaction. Thus these two enzymes resemble some of the enzyme systems that require two proteins except that in these cases alterations in substrate specificity are induced by polyelectrolytes.

In regard to the lactose synthetase, it is interesting to note that α-lactalbumin has been shown to have an amino acid sequence quite similar to that of egg-white lysozyme (146). Hill et al. (147) found that lysozyme was not able to replace lactalbumin in the lactose reaction. These workers indicated that although these two proteins have fairly similar amino acid sequences, they have quite different charges at pH 7.0; lactalbumin has an isoelectric point of about 5.0 and therefore would probably be negatively charged, whereas lysozyme has an isoelectric point of about 11 and is positively charged at pH 7.0. Since lysozyme is present in the developing egg while substrates for the enzyme (bacterial cell walls) are probably not present, the authors suggest that lysozymes may play a role in the egg similar to that played by lactalbumin in milk. Grebner and Neufeld (148) found that a xylosyl transferase from mouse-mast-cell ascites tumors is stimulated by the addition of lysozyme to the incubations. Another basic protein, ribonuclease, could duplicate this effect, but protamine, histone, or α-lactalbumin could not. Stimulation only occurred in older preparations of the enzyme that had been stored at 4°C for 17 days or more. However, these findings could suggest a role for lysozyme as an "activator" of glycosyl transferases.

Proteins may be polycationic or polyanionic, depending on the nature of the amino acids at the surface and the pH of the suspending medium. Thus Bernfeld and coworkers (37) found that various proteins could act as polycations and activate certain enzymes. Crestfield, Stein, and Moore (150) found that bovine pancreatic ribonuclease A formed stable aggregates when solutions of the enzyme in 50% acetic acid were lyophilized. These aggregates had the same specific activity as ribonuclease A and could be dissociated back to monomer by heating at 65°C for 10 min. They suggested that the amino-terminal portion of one enzyme molecule combines specifically with the appropriate segment of its neighbor in such a manner as to preserve the appropriate conformation necessary for catalysis. Under conditions of low ionic strength trehalose phosphate synthetase forms aggregates that retain their catalytic activity. However, these aggregates no longer require polyanions in order to utilize UDP–glucose as a glucosyl donor, which implies that the enzyme molecules themselves are active in replacing the polyanion. Thus various protein–protein interactions bear definite resemblances to polyelectrolyte–protein interactions in spite of their definite specificity.

VII. Conclusion

In recent years many studies have shown that polyelectrolytes may have pronounced effects on isolated proteins as well as on cells and tissues. These effects can range from activation or inhibition of enzymatic activity to the stimulation or prevention of certain cellular activities. The fact that polyelectrolytes can bind to or alter protein activity is perhaps not surprising since these compounds are highly charged at physiologic pH values. One question that arises, however, is whether these interactions occur *in vivo* and, if so, what is their significance? There is reasonable evidence to indicate that polyanions, such as sulfated glycosaminoglycans, do interact with cells and proteins in connective tissue and in the circulatory system, and these interactions may have important implications in various disease conditions as well as in normal differentiation and development. However, the question of whether polyanions can control enzymatic activities *in vivo* by activation or inhibition is difficult to assess and cannot be answered at this time. Furthermore, the role of polycations is even less clear since many of these compounds, such as polylysine, are probably not naturally occurring. However, it seems possible that basic compounds such as histones or proteins (i.e., lysozyme) may function as enzyme modulators. It is to be hoped that future experimentation will resolve some of these questions.

With regard to the mechanisms involved in polyelectrolyte–protein interactions, it seems fairly obvious that the two molecules are attracted to each other because of opposing surface changes, and, at least in some cases, a complex between protein and polyelectrolyte results. However, how this leads to activation or inhibition of the enzyme in question is not clear. In some cases the polyelectrolyte appears to stabilize the protein and even to change the conformation of the enzyme to a more active form. In other cases the polyelectrolyte seems to be involved in the formation of the enzyme–substrate complex or in somehow bringing the two into proper alignment. Although most activations do not seem to involve the removal of an inhibitor, this mechanism does seem to be involved in the activation of pepsinogen. In terms of polyelectrolyte inhibitions there is very little information available as to the mechanisms involved, but these inhibitions could be the result of conformational changes in the protein to less active or inactive forms, or to the prevention of enzyme–substrate interactions.

Some of the examples discussed in this chapter involve fairly specific interactions, as, for example, protein–nucleic acid (or polynucleotide) or protein–protein interactions. Whether there are completely different phenomena from other polyelectrolyte interactions or whether they simply reflect differences in the degree of specificity is an open question and in fact may be more a case of semantics. Even the "general" polyelectrolyte–enzyme interactions show degrees of specificity since one polyanion or polycation is generally a better effector than others. That the spacing of charges and charge density are important seems to be well documented. Therefore these specificities may simply be a matter of the polyelectrolyte that best fits the charge on the protein.

Acknowledgments

I am indebted to Dr. Robert Weisman for many helpful suggestions and discussions and to Miss Linda Christian for her very able assistance in preparing this manuscript. The work discussed from the author's laboratory was supported by grants from the National Institute of Allergy and Infectious Diseases and the Robert A. Welch Foundation.

References

1. Bernfeld, P., in *The Amino Sugars*, E. Balazs and R. Jeanloz, Eds., Vol. II, Academic Press, New York, 1966, p. 214.
2. Bernfeld, P., in *Metabolic Inhibitors*, R. M. Hochster and J. H. Quastel, Eds., Vol. II, Academic Press, New York, 1963, p. 437.
3. Frieden, C., *Ann. Rev. Biochem.*, 40, 653 (1971).
4. Ginsburg, A., and Stadtman, E. R., *Ann. Rev. Biochem.*, 39, 429 (1970).
5. Ebner, K. E., *Acc. Chem. Res.*, 3, 41 (1970).
6. Hahn, P. F., *Science*, 98, 19 (1943).
7. Weld, C. B., *Can. Med. Assoc. J.*, 51, 578 (1944).
8. Robinson, D. S., *Adv. Lipid Res.*, 1, 133 (1963).
9. Korn, E. D., *J. Biol. Chem.*, 215, 1 (1955).
10. Korn, E. D., *J. Biol. Chem.*, 215, 15 (1955).
11. Robinson, D. S., *Quant. J. Exp. Physiol.*, 41, 195 (1956).
12. Robinson, D. S., and French, J. E., *Pharmacol. Rev.*, 12, 241 (1960).
13. Reber, K., and Studer, A., *Experientia*, 14, 462 (1958).
14. Korn, E. D., *Methods Biochem. Anal.*, 7, 145 (1959).
15. Bernfeld, P., and Kelley, T. F., *J. Biol. Chem.*, 238, 1236 (1963).
16. Spensley, P. C., and Rogers, H. J., *Nature*, 173, 1190 (1954).
17. Patten, R. L., and Hollenberg, C. H., *J. Lipid Res.*, 10, 374 (1969).
18. Whayne, T. F., Jr., and Felts, J. M., *Circulation Res.*, 27, 941 (1970).
19. Olivecrona, T., and Lindahl, V., *Acta Chem. Scand.*, 23, 3587 (1969).
20. Olivecrona, T., Egelrud, T., Iverius, P., and Lindahl, V., *Biochem. Biophys. Res. Commun.*, 43, 524 (1971).
21. Iverius, P., Lindahl, V., Egelrud, T., and Olivecrona, T., *J. Biol. Chem.*, 247, 6610 (1972).
22. Liu, C., Patterson, B. W., Lapp, D. W., and Elbein, A. D., *J. Biol. Chem.*, 244, 3827 (1969).
23. Lapp, D. W., Patterson, B. W., and Elbein, A. D., *J. Biol. Chem.*, 246, 4567 (1971).
24. Tandon, A., Saxena, K. C., and Krishna Murti, C. R., *Indian J. Chem.*, 7, 29 (1970).
25. Pamer, T., Ph.D. thesis, New York Medical College, 1968.
26. Anderson, W., *J. Pharm. Pharmacol.*, 21, 266 (1969).
27. Horowitz, M. I., Pamer, T., and Glass, G. B. J., *Proc. Soc. Exp. Biol. Med.*, 133, 853 (1970).
28. Pinchot, G. B., *J. Biol. Chem.*, 229, 1 (1957).
29. Pinchot, G. B., *J. Biol. Chem.*, 229, 25 (1957).
30. Shibko, S., and Pinchot, G. B., *Arch. Biochem. Biophys.*, 94, 257 (1961).
31. Allfrey, V. G., and Mirsky, A. E., *Proc. Natl. Acad. Sci. U.S.*, 43, 589 (1957).
32. Allfrey, V. G., and Mirsky, A. E., *Proc. Natl. Acad. Sci. U.S.*, 44, 981 (1958).
33. Sekiguchi, M., and Sibatani, A., *Biochim Biophys. Acta*, 28, 455 (1958).
34. Debreceni, N., Behine, M. T., and Ebisuzaki, K., *Biochem. Biophys. Res. Commun.*, 41, 115 (1970).
35. Bernfeld, P., Bernfeld, H. C., Nesselbaum, J. S., and Fishman, W. H., *J. Am. Chem. Soc.*, 76, 4872 (1954).

36. Bernfeld, P., Tuttle, L. P., and Hubbard, R. W., *Arch. Biochem Biophys.*, *92*, 232 (1961).
37. Bernfeld, P., Berkeley, B. J., and Bieber, R. E., *Arch. Biochem. Biophys.*, *111*. 31 (1965).
38. Jeffree, G., *Biochim. Biophys. Acta*, *20*, 503 (1956).
39. Dolin, M., *Biochem. Biophys. Res. Commun.*, *6*, 11 (1961).
40. Dolin, M., *J. Biol. Chem.*, *237*, 1626 (1962).
41. Dietz, G. W., Jr., and Grunberg-Manago, M., *Biochem. Biophys. Res. Commun.*, *28*, 146 (1967).
42. Fitt, P. S., Dietz, F. W., Jr., and Grunberg-Manago, M., *Biochim. Biophys. Acta*, *151*, 99 (1968).
43. Fitt, P. S., and Wille, H., *Biochem. J.*, *112*, 489 (1969).
44. Fitt, P. S., and Wille, H., *Biochem. J.*, *112*, 497 (1969).
45. Krebs, E. G., *Biochim. Biophys. Acta*, *15*, 508 (1954).
46. Kaufman, B., Basu, S., and Roseman, S., *J. Biol. Chem.*, *243*, 5804 (1968).
47. Goldman, D. S., and Lornitzo, F. A., *J. Biol. Chem.*, *237*, 3332 (1962).
48. Lornitzo, F. A., and Goldman, D. S., *J. Biol. Chem.*, *239*, 2730 (1964).
49. Perlman, G. E., *J. Mol. Biol.*, *6*, 452 (1963).
50. Katchalski, E., Berger, A., and Neumann, H., *Nature*, *173*, 998 (1954).
51. Anai, M., Hirahashi, T., and Takagi, Y., *J. Biol. Chem.*, *245*, 767 (1970).
52. Buttin, G., and Wright, M. R., *Cold Spring Harbor Symp. Quant. Biol.*, *33*, 259 (1968).
53. Barbour, S. D., and Clark, A. J., *Proc. Natl. Acad. Sci. U.S.*, *66*, 955 (1970).
54. Nobrega, F. G., Rola, F. H., Pasetto-Nobrega, M., and Oishi, M., *Proc. Natl. Acad. Sci. U.S.*, *69*, 15 (1972).
55. Yarus, M., *Ann. Rev. Biochem.*, *38*, 841 (1969).
56. DeLange, R. J., and Smith, E. L., *Ann. Rev. Biochem.*, *40*, 229 (1971).
57. Geiduschek, E. P., and Haselkorn, R., *Ann. Rev. Biochem.*, *38*, 647 (1969).
58. Burgess, R. R., *Ann. Rev. Biochem.*, *40*, 711 (1971).
59. Sela, M., and Katchalski, E., *Adv. Protein Chem.*, *14*, 391 (1959).
60. Smolarsky, M., and Tal, M., *Biochim. Biophys. Acta*, *213*, 401 (1970).
61. Calvo, J. M., and Fink, G. R., *Ann. Rev. Biochem.*, *40*, 943 (1971).
62. Sakai, H., *J. Biol. Chem.*, *242*, 1458 (1967).
63. Hofer, H. W., and Pette, D., *Life Sci.*, *4*, 1591 (1965).
64. Hofer, H. W., and Pette, D., *Life Sci.*, *5*, 199 (1966).
65. Pette, D., and Hofer, H. W., in *Control of Energy Metabolism*, B. Chance and R. W. Estabrook, (Eds.) Academic Press, New York, 1965
66. Shapiro, B., *Biochemistry*, *8*, *659* (*1969*).
67. Anderson, W. B., and Stadtman, E. R., *Arch. Biochem. Biophys.*, *143*, 428 (1971).
68. Kovach, J. S., Phang, J. M., Blasi, F., Barton, R. W., Ballesteros-Olmo, A., Goldberger, R. F., *J. Bacteriol.*, *104*, 787 (1970).
69. Hatfield, G. W., and Burns, R. O., *J. Biol. Chem.*, *245*, 787 (1970).
70. Hatfield, G. W., and Burns, R. O., *Proc. Natl. Acad. Sci. U.S.*, *66*, 1027 (1970).
71. Morris, R. W., and Herbert, E., *Biochemistry*, *9*, 4819 (1970).
72. Langerkvist, V., and Waldenström, J., *J. Biol. Chem.*, *240*, PC2264 (1965).
73. Yarus, M., and Berg, P., *J. Mol. Biol.*, *28*, 479 (1967).
74. Makman, M. H., and Cantoni, G. L., *Biochemistry*, *4*, 1434 (1965).

75. Letendre, C. H., Humphreys, J. M., and Grunberg-Manago, M., *Biochim. Biophys. Acta*, *186*, 46 (1969).
76. Gelman, R. A., Rippon, W. B., and Blackwell, J., *Biochem. Biophys. Res. Commun.*, *48*, 708 (1972).
77. Buruiana, L. M., *Naturwissenschaften*, *45*, 293 (1957).
78. Hummel, J. P., Anderson, D. O., and Pattel, C., *J. Biol. Chem.*, *233*, 712 (1958).
79. Rogers, H. J., *Biochem. J.*, *40*, 583 (1946).
80. McLean, D., *Chem. Ind. (London)*, *60*, 219 (1942).
81. Astrup, T., and Alkjaersig, N., *Nature*, *166*, 568 (1950).
82. Mora, P. T., and Young, B. G., *Nature*, *181*, 1402 (1958).
83. Follner, N., and Fellig, J., *Naturwissenschaften*, *39*, 523 (1952).
84. Fellig, J., and Wiley, C. E., *Arch. Biochem. Biophys.*, *85*, 313 (1959).
85. Walton, K. W., *Brt. J. Pharmacol.*, *7*, 370 (1952).
86. Dickman, S. R., *Science*, *127*, 1392 (1958).
87. Roth, J. S., *Arch. Biochem. Biophys.*, *44*, 265 (1953).
88. Vandendriessche, L., *Arch. Biochem. Biophys.*, *65*, 347 (1956)
89. Ogata, E., and Kondo, K., *J. Biochem.*, *71*, 423 (1972).
90. Myrbach, K., and Persson, B., Arkiv Kemi, *5*, 177 (1953).
91. Kerby, G. P., and Eadie, G. S., *Proc. Soc. Exp. Biol. Med.*, *83*, 111 (1953).
92. Sharnes, R. C., and Watson, D. W., *J. Bacteriol.*, *70*, 110 (1955).
93. Horwitt, M. K., *Science*, *92*, 89 (1940).
94. Dellert, E. E., and Stahmann, M. A., *Nature*, *176*, 1028 (1955).
95. Levey, S., and Scheinfeld, S., *Gastroenterology*, *27*, 625 (1954).
96. Ravin, L. J., Baldinus, J. G., and Mazur, M. L., *J. Pharm. Sci.*, *51*, 857 (1962).
97. Fischer, A., and Herrmann, H., *Enzymologia*, *3*, 180 (1937).
98. Moss, S., *Science*, *115*, 69 (1952).
99. Bernfeld, P., Jacobson, S., and Bernfeld, H. C., *Arch. Biochem. Biophys.*, *69*, 198 (1957).
100. Berdick, M., and Morawetz, H., *J. Biol. Chem.*, *206*, 959 (1954).
101. Diczfalusy, E., Ferno, O., Fex, H., Hogberg, B., Linderot, T., and Rosenberg, T., *Acta Chem. Scand.*, *7*, 913 (1953).
102. Aoki, Y., and Koshiwara, H., *Exp. Cell Res.*, *70*, 431 (1971).
103. Katchalski, E., Grossfeld, I., and Frankel, M., *J. Am. Chem. Soc.*, *70*, 2094 (1948).
104. Tsuyuki, E., Tsuyuki, H., and Stahmann, M. A., *J. Biol. Chem.*, *222*, 479 (1956).
105. Dellert, E., and Stahmann, M. A., *Nature*, *176*, 1028 (1955).
106. Rigbi, M., Seliktar, E., and Katchalski, E., *Bull. Res. Council Israel*, *6A*, 313 (1957).
107. Elbein, A. D., and Lapp, D. L., unpublished observations.
108. Brand, J., Baszynski, T., Crane, F. L., and Krogmann, D. W., *J. Biol. Chem.*, *247*, 2814 (1972).
109. Mayer, K., Davidson, E., Linker, A., and Hoffman, P., *Biochim. Biophys. Acta*, *21*, 506 (1956).
110. Kraemer, P. M., in *Biomembranes*, L. Manson, Ed., Vol. I, Plenum Press, New York, 1971.
111. Kraemer, P. M., *Biochemistry*, *10*, 1437 (1971).
112. Kojima, K., and Yamagata, T., *Exp. Cell Res.*, *67*, 142 (1967).

113. Lippmann, M., *Trans. N.Y. Acad. Sci.*, 27, 342 (1965).
114. Mascona, A. A., *Develop. Biol.*, 18, 250 (1968).
115. Kahn, T., and Overton, J., *J. Exp. Zool.*, 171, 161 (1969).
116. Kvist, T. N., and Finnegan, C. V., *J. Exp. Biol.*, 175, 241 (1970).
117. McConnachie, P. R., and Ford, P., *J. Embryol. Exp. Morphol.*, 16, 17 (1966).
118. Pessac, B., and Defendi, V., *Science*, 175, 898 (1972).
119. Toole, B. P., Jackson, G., and Gross, J., *Proc. Natl. Acad. Sci. U.S.*, 69, 1384 (1972).
120. Novo, O., and Dorfman, A., *Proc. Natl. Acad. Sci. U.S.*, 69, 2069 (1972).
121. Reissig, J. L., and Glasgow, J. E., *J. Bacteriol.*, 106, 882 (1971).
122. Burnet, M., *Austr. J. Exp. Biol. Med.*, 26, 37 (1948).
123. Ada, G. L., and Stone, J. D., *Brit. J. Exp. Pathol.*, 31, 263 (1950).
124. Soupart, P., and Clewe, T. H., *Fertility Sterility*, 16, 677 (1965).
125. Kemp, R. B., *Nature*, 218, 1255 (1968).
126. Oppenheimer, S. B., Edidin, M., Orr, C. W., and Roseman, S., *Proc. Natl. Acad. Sci. U.S.*, 63, 1395 (1969).
127. Roseman, S., *Chem. Phys. Lipids*, 5, 270 (1970).
128. Ginsburg, V., and Neufeld, E., *Ann. Rev. Biochem.*, 38, 371 (1969).
129. Heath, E. C., *Ann. Rev. Biochem.*, 40, 29 (1971).
130. Bernfeld, P., Donahue, V. M., and Berkowitz, M. E., *J. Biol. Chem.*, 226, 51, (1957).
131. Bernfeld, P., Nisselbaum, J. S., Berkeley, B. J., and Hanson, R. W., *J. Biol. Chem.*, 235, 2852 (1960).
132. Iverius, P. H., *J. Biol. Chem.*, 247, 2607 (1972).
133. Bihari-Varga, M., and Vegh, M., *Biochim. Biophys. Acta*, 144, 202 (1967).
134. Tracy, R. E., Dzoga, K. R., and Wissler, R. W., *Proc. Soc. Exp. Biol. Med.*, 118, 1095 (1965).
135. Bihari-Varga, M., Gergely, J., and Giro, S., *J. Ather. Res.*, 4, 106 (1964).
136. Srinivasan, S. P., Dolan, P., Radhakrishnamurthy, B., and Berenson, G. S., *Atherosclerosis*, 16, 95 (1972).
137. Srinivasan, S. P., Dolan, P., Radhakrishnamurthy, B., and Berenson, G. S., *Prep. Biochem.*, 2, 83 (1972).
138. Levey, S., and Sheinfeld, S., *Gastroenterology*, 27, 625 (1954).
139. Bianchi, R. G., and Cook, D. L., *Gastroenterology*, 47, 409 (1964).
140. Houch, J. C., Bhayana, J., and Lee, T., *Gastroenterology*, 39, 196 (1960).
141. Piper, D. W., and Fenton, B., *Gastroenterology*, 40, 638 (1961).
142. Cifonelli, J. A., Montgomery, R., and Smith, F., *J. Am. Chem. Soc.*, 78, 2488 (1956).
143. Doyle, R. J., and Kan, T. J., *FEBS Letters*, 20, 22 (1972).
144. Kleinschmidt, W. J., *Ann. Rev. Biochem.*, 41, 517 (1972).
145. Klee, W. A., and Klee, C. B., *J. Biol. Chem.*, 247, 2336 (1972).
146. Brew, K., Vanaman, T. C., and Hill, R. L., *J. Biol. Chem.*, 242, 3747 (1967).
147. Hill, R. L., Brew, K., Vanaman, T. C., Trayer, I. P., and Mattock, P., *Brookhaven Sym. Biol.*, 21, 139 (1968).
148. Grebner, E. E., and Neufeld, E. F., *Biochim. Biophys. Acta*, 192, 347 (1969).
149. Crestfield, A. N., Stein, W. H., and Moore, S., *Arch. Biochem. Biophys.*, Suppl. 1, 217 (1962).

ENZYMES OF ARGININE BIOSYNTHESIS AND THEIR REPRESSIVE CONTROL

By HENRY J. VOGEL and RUTH H. VOGEL, *New York*

CONTENTS

I. Introduction

The arginine biosynthetic system of *Escherichia coli* has, for two decades, been a fertile source of metabolic and genetic findings, some of them unique. The eight-step pathway involved (1–5) proceeds from glutamate (1) via four *N*-acetylated intermediates, *N*-acetylglutamate (1,6), *N*-acetyl-γ-glutamyl phosphate (7), *N*-acetylglutamic γ-semialdehyde (1), and α-*N*-acetylornithine (1), and continues via ornithine (8), citrulline (8), and argininosuccinate (4,9) to yield arginine (Fig. 1). Each of the eight steps is catalyzed by a characteristic enzyme: an acetyltransferase (10), a kinase (7,11,12), a reductase (7,13,14), an aminotransferase (1,2,15,16), a deacetylase (1,17–19), a carbamoyltransferase (20,21), an argininosuccinate-forming enzyme (22), and an argininosuccinate-splitting enzyme (9). The systematic and other names of these enzymes are listed in Table I.

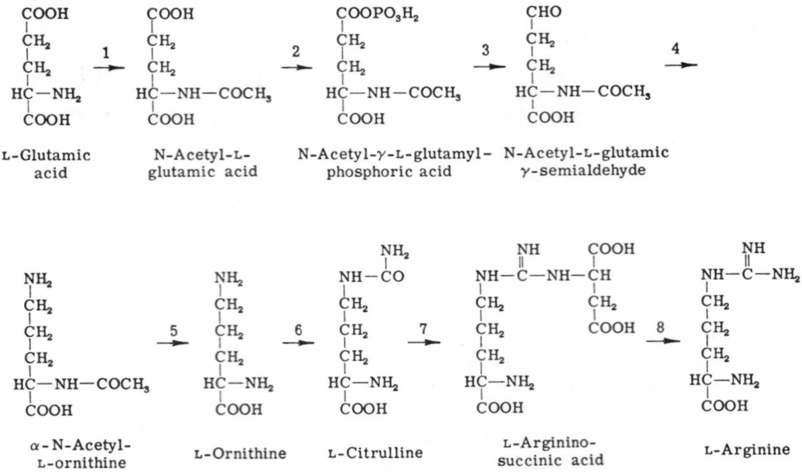

Top row of pathway:

L-Glutamic acid
$COOH$ — CH_2 — CH_2 — $HC-NH_2$ — $COOH$

→ (1) →

N-Acetyl-L-glutamic acid
$COOH$ — CH_2 — CH_2 — $HC-NH-COCH_3$ — $COOH$

→ (2) →

N-Acetyl-γ-L-glutamyl phosphoric acid
$COOPO_3H_2$ — CH_2 — CH_2 — $HC-NH-COCH_3$ — $COOH$

→ (3) →

N-Acetyl-L-glutamic γ-semialdehyde
CHO — CH_2 — CH_2 — $HC-NH-COCH_3$ — $COOH$

→ (4) →

Bottom row of pathway:

α-N-Acetyl-L-ornithine
NH_2 — CH_2 — CH_2 — CH_2 — $HC-NH-COCH_3$ — $COOH$

→ (5) →

L-Ornithine
NH_2 — CH_2 — CH_2 — CH_2 — $HC-NH_2$ — $COOH$

→ (6) →

L-Citrulline
NH_2 / $NH-CO$ — CH_2 — CH_2 — CH_2 — $HC-NH_2$ — $COOH$

→ (7) →

L-Argininosuccinic acid
NH $COOH$ / $NH-C-NH-CH$ — CH_2 CH_2 — CH_2 $COOH$ — CH_2 — $HC-NH_2$ — $COOH$

→ (8) →

L-Arginine
NH / $NH-C-NH_2$ — CH_2 — CH_2 — CH_2 — $HC-NH_2$ — $COOH$

Fig. 1.   Pathway of arginine synthesis in *Escherichia coli*.

The pathway thus accomplishes the conversion of the γ-carboxyl group of glutamate to the guanidinomethyl group of arginine. The acetylation and deacetylation steps can be regarded as furnishing an uncommon protective device that prevents what would otherwise be a free α-amino group from participating in interfering intramolecular reactions (i.e., the formation of pyrrolidonecarboxylate or Δ^1-pyrroline-5-carboxylate). The three nitrogen atoms of the guanidino moiety are assembled via diverse routes: one of the nitrogen atoms is introduced through the transamination of an aldehyde (with glutamate as amino donor); a second is provided by carbamyl phosphate (and hence indirectly by glutamine); and a third is added in a two-step process, with argininosuccinate as intermediate (and with aspartate as nitrogen source). Carbamyl phosphate also supplies the atom that, in the arginine molecule, becomes the amidino carbon (which, therefore, is derived from bicarbonate). The enzymatic mechanism for the formation of carbamyl phosphate from glutamine and bicarbonate (23) has been elucidated.

The biosynthetic enzymes are specified by structural genes (22, 24–26) whose designations (27) are included in Table I. Strain K12 of *E. coli* has two genes, *argF* and *argI*, coding for nonidentical species

TABLE I

Enzymes and Genes of the Arginine Biosynthetic Pathway in *Escherichia coli*

| Step | Enzyme | | Gene |
	Systematic name	Other names	
1	Acetyl-CoA:L-glutamate N-acetyltransferase (EC 2.3.1.1)	Glutamate acetyltransferase, N-acetylglutamate synthetase	*argA*
2	ATP:N-acetyl-L-glutamate 5-phosphotransferase (EC 2.7.2.8)	Acetylglutamate kinase, N-acetyl-γ-glutamokinase	*argB*
3	N-Acetyl-L-glutamate-5- semialdehyde: NADP$^+$ oxidoreductase (phosphorylating) (EC 1.2.1.38)	Acetylglutamylphosphate reductase, N-acetylglutamic γ-semialdehyde dehydrogenase	*argC*
4	N^2-Acetyl-L-ornithine: 2-oxoglutarate aminotransferase (EC 2.6.1.11)	Acetylornithine aminotransferase, acetylornithine δ-transaminase	*argD*
5	N^2-Acetyl-L-ornithine amidohydrolase (EC 3.5.1.16)	Acetylornithine deacetylase, acetylornithinase	*argE*
6	Carbamoylphosphate:L-ornithine carbamoyltransferase (EC 2.1.3.3)	Ornithine carbamoyltransferase, ornithine transcarbamylase	*argF*, *argI*
7	L-Citrulline:L-asparate ligase (AMP-forming) (EC 6.3.4.5)	Argininosuccinate synthetase	*argG*
8	L-Argininosuccinate arginine-lyase (EC 4.3.2.1)	Argininosuccinate lyase, argininosuccinase	*argH*

of ornithine carbamoyltransferase (21,28). Of the structural genes, four are closely linked to form a cluster containing the sequence *argE, argC, argB, argH* (29); the remaining ones are scattered. In addition to the structural genes, there is a regulatory gene, *argR* (22,30), which is involved in the control of the formation of all the enzymes. The linkage relationships among the *arg* genes are illustrated in Figure 2. In the presence of the *argR*$^+$ allele, the enzymes are repressible by arginine (31,32); in the presence of an *argR*$^-$ allele, they are produced at high levels even with an excess of arginine (Figure 3).

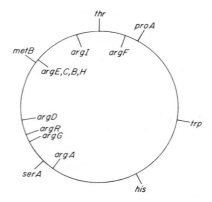

Fig. 2.   Linkage relationships among genes of the arginine biosynthetic system in *Escherichia coli* K12 and reference markers.

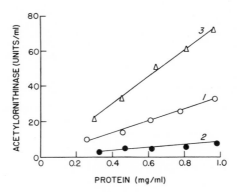

Fig. 3.   Formation of acetylornithine deacetylase in *Escherichia coli* under various regulatory conditions. Plots *1* and *2* represent formation of the enzyme in the wild-type ($argR^+$) strain W cultivated without or with added L-arginine hydrochloride (0.2 mg/ml), respectively. Plot *3* represents the formation of the enzyme in the regulatory-gene ($argR^-$) mutant W2D. Enzyme and total-protein concentrations are expressed per milliliter of extract (1 ml corresponding to 7.5 ml of original culture). The slopes of the plots are the so-called differential rates of enzyme synthesis indicating partial repression (*1*), full repression (*2*), and genetic derepression (*3*). From Vogel et al. (*5*).

II. Arginine Biosynthetic Enzymes

A. GLUTAMATE ACETYLTRANSFERASE

Extracts of *E. coli* contain an enzyme (10) that, in the presence of magnesium ions and a reducing agent such as hydrogen sulfide, catalyzes the reaction

$$\text{L-glutamate} + \text{acetyl CoA} \rightarrow N\text{-acetyl-L-glutamate} + \text{CoA}$$

N-Acetylglutamate formation can be followed by a hydroxylamine procedure, in which this acetylamino acid yields acethydroxamic acid (10). The acetyltransferase proved relatively unstable on extraction. However, by means of cell suspension experiments, it could be shown that this first enzyme of the arginine pathway is susceptible to feedback inhibition of its activity by arginine (33). Recently, a sensitive *in vitro* assay was developed (34), and stabilization of the extracted acetyltransferase, with preservation of inhibitability by arginine, was achieved (35).

B. ACETYLGLUTAMATE KINASE

The phosphorylation of the γ-carboxyl group of *N*-acetylglutamate is carried out by an enzyme that has been extracted from *E. coli* (7,11, 12). This reaction proceeds as follows:

$$N\text{-acetyl-L-glutamate} + \text{ATP} \rightleftharpoons N\text{-acetyl-}\gamma\text{-L-glutamyl phosphate} + \text{ADP}$$

Magnesium ions are required. Acetylglutamate kinase can be assayed by a method in which the ADP produced, in the presence of pyruvate kinase and 2-phosphoenolpyruvate, leads to the formation of pyruvate, which is determined colorimetrically as its 2,4-dinitrophenylhydrazone.

For the preparation of the kinase, an *argR⁻* mutant (strain W2D) is used, which produces this enzyme at derepressed levels. The kinase can be purified 65- to 70-fold by fractionation with streptomycin and with ammonium sulfate, followed by chromatography on hydroxylapatite and on DEAE-cellulose and by gel filtration on Sephadex G-100.

The activity of the enzyme has a relatively broad pH optimum and does not vary appreciably from pH 6.8 to 7.8. *N*-Acetyl- L-glutamate and ATP give K_m values of 6.0 and 1.0 mM, respectively. By sucrose gradient centrifugation, the enzyme exhibits a sedimentation coefficient ($s_{20,w}$) of approximately 4.2S.

C. ACETYLGLUTAMYLPHOSPHATE REDUCTASE

The conversion of N-acetyl-γ-glutamyl phosphate to the corresponding aldehyde can be represented thus:

$$N\text{-acetyl-}\gamma\text{-L-glutamyl phosphate}$$
$$\text{NADPH} + H^+ \downarrow \uparrow \text{NADP}^+$$
$$N\text{-acetyl-L-glutamate }\gamma\text{-semialdehyde} + \text{phosphate}$$

A reductase capable of mediating this reaction has been obtained from $E.\ coli$ (7,13,14). For enzyme assays, the reaction can be followed, in the reverse of the biosynthetic direction, by spectrophotometric measurement of the reduction of NADP$^+$. The semialdehyde substrate is prepared enzymatically (13,14).

An $argR^-$ mutant (strain W2D) that synthesizes the reductase at derepressed levels is a convenient source of this enzyme, which has been purified some 300 times by fractionation with ammonium sulfate and with alumina Cγ, by column chromatography on DEAE-cellulose, and by gel filtration on Sephadex G–200 (13,14).

The enzymatic reaction with the semialdehyde as substrate proceeds optimally at pH 9.3. N-Acetyl-L-glutamate γ-semialdehyde and phosphate show K_m values of 0.6 and 3.0 mM, respectively. Half-maximal velocity is attained at 0.17 mM NADP$^+$. The reaction can be carried out with arsenate in the place of phosphate. In the presence of phosphate, Fe^{2+}, Mn^{2+}, and Ni^{2+}, and particularly Cu^{2+} and Zn^{2+}, are inhibitory, as are sulfhydryl reagents. A sedimentation coefficient ($s_{20,\mathrm{w}}$) of 7.9S has been determined for the reductase.

D. ACETYLORNITHINE AMINOTRANSFERASE

The transamination of N-acetyl-L-glutamate γ-semialdehyde to form α-N-acetyl-L-ornithine occurs as follows:

$$N\text{-acetyl-L-glutamate }\gamma\text{-semialdehyde} + \text{L-glutamate}$$
$$\downarrow \uparrow$$
$$\alpha\text{-}N\text{-acetyl-L-ornithine} + \alpha\text{-ketoglutarate}$$

An aminotransferase catalyzing this reaction, which requires pyridoxal 5-phosphate, has been prepared from $E.\ coli$ (1,2,15,16). For enzyme assays, the reaction is carried out in the reverse of the biosynthetic direction, and the N-acetylglutamic γ-semialdehyde formed is hydrolyzed to glutamic γ-semialdehyde, which cyclizes to Δ^1-pyrroline-5-carboxylic acid and reacts with o-aminobenzaldehyde to yield a yellow dihydroquinazolinium compound (15).

For preparative purposes, the aminotransferase has been extracted from strain W2D and has been purified approximately hundredfold by fractionation with ammonium sulfate, with acetic acid, and with alumina Cγ and by chromatography on DEAE-cellulose (16).

The purified enzyme preparation is inactive unless pyridoxal 5-phosphate is added. The pH optimum for the formation of N-acetyl-L-glutamate γ-semialdehyde is 8.0. α-N-Acetyl-L-ornithine and α-keto-glutarate exhibit K_m values of 0.3 and 1.0 mM, respectively. For pyridoxal 5-phosphate, half-maximal velocity is obtained at 0.0017 mM. Cupric ions, ferrous ions, and sulfhydryl reagents tend to inhibit the aminotransferase; Ni^{2+} is somewhat stimulatory. The sedimentation coefficient ($s_{20,w}$) of the enzyme is 9.2S, as determined in 0.01 M potassium phosphate (pH 7.0) containing 0.2 M KCl.

Strain W2D, which is used as source of the aminotransferase, is genetically derepressed ($argR^-$), as already pointed out. In the parental ($argR^+$) strain W, the aminotransferase, like the other enzymes in its pathway, is repressible by arginine. Following a mutation at a gene termed $argM$, which is not closely linked to any of the other arginine genes (26), strain W derivatives form an acetylornithine aminotransferase that is inducible by arginine (36). The inducibility depends on the presence of a functional $argR^+$ gene (25,37) and differs from the inducibility of certain of the arginine enzymes in $E.\ coli$, strain B (38,39), which has an atypical $argR$ gene (40,41).

E. ACETYLORNITHINE DEACETYLASE

α-N-Acetyl-L-ornithine gives rise to ornithine through the following hydrolytic reaction:

$$\alpha\text{-}N\text{-acetyl-L-ornithine} + H_2O \rightarrow \text{L-ornithine} + \text{acetate}$$

In the presence of Co^{2+} and a sulfhydryl compound such as glutathione, this deacetylation is performed by an enzyme that has been obtained from $E.\ coli$ (1,17–19). An assay method has been devised in which the enzymatically produced ornithine is determined through reaction with ninhydrin at acid pH, followed by colorimetry under alkaline conditions (18).

The enzyme can be extracted from strain W2D and can be purified some hundredfold by a procedure involving fractionations with strep-tomycin and ammonium sulfate, column chromatography on hydroxyl-apatite, and gel filtration on Sephadex G–100 (19).

The purified enzyme is inactive unless a suitable thiol, preferably glutathione, is provided. The Co^{2+} ion is stimulatory. Optimal activity of acetylornithine deacetylase occurs at a pH of approximately 7.0. α-N-Acetyl-L-ornithine gives a K_m value of 2.8 mM. The enzyme deacetylates not only α-N-acetyl-L-ornithine but also α-N-acetyl-L-arginine (42), N-acetyl-L-glutamate γ-semialdehyde (30), α-N-acetyl-L-histidine (43), and the N-acetyl (18) and N-formyl (44,45) derivatives of L-methionine. The Cu^{2+}, Ni^{2+}, and Zn^{2+} ions, ethylenediaminetetraacetate, and sulfhydryl reagents are inhibitory. The isoelectric point of the deacetylase is at pH 4.4. A sedimentation coefficient ($s_{20,w}$) of 4.9S is indicated by sucrose gradient centrifugation (19).

F. ORNITHINE CARBAMOYLTRANSFERASE

With carbamyl phosphate as carbamyl donor, ornithine yields citrulline according to the reaction

$$\text{L-ornithine} + \text{carbamyl phosphate} \rightleftharpoons \text{L-citrulline} + \text{phosphate}$$

A transferase catalyzing this reaction has been extracted from *E. coli*, strain W (20). In strain K12 of *E. coli*, either of the genes *argF* and *argI* specifies a functional transferase that appears to occur in trimeric form; additionally, two kinds of hybrid molecules are produced to constitute a total of four trimeric isoenzymes (21). For transferase assays, citrulline is determined colorimetrically (46).

Escherichia coli ornithine carbamoyltransferase can be purified by treatment at 65°C, ammonium sulfate fractionation, chromatography on DEAE-Sephadex, and gel filtration on Sephadex G–200 (20,21).

The pH optimum of the transcarbamylation is approximately 8.7. L-Ornithine and carbamyl phosphate show K_m values of 1.5 and 0.2 mM, respectively. Sulfhydryl reagents inhibit the enzymatic reaction. A sedimentation coefficient ($s_{20,w}$) of 7.4S has been determined (20).

G. ARGININOSUCCINATE SYNTHETASE

Citrulline is converted to argininosuccinate in the following manner:

$$\text{L-citrulline} + \text{L-aspartate} + \text{ATP} \rightleftharpoons \text{L-argininosuccinate} + \text{AMP} + \text{pyrophosphate}$$

In the presence of magnesium ions, this reaction is mediated by a synthetase, which has been exhibited in extracts of *E. coli* (22) and appears to resemble a corresponding enzyme from steer liver (47). For synthetase assays, AMP formation is coupled to the oxidation of

NADH through lactic dehydrogenase and a suitable ATP-generating system, and the oxidation of NADH is followed spectrophotometrically (47).

H. ARGININOSUCCINATE LYASE

The last step in arginine biosynthesis involves the cleavage of argininosuccinate, as follows:

$$\text{L-argininosuccinate} \rightleftarrows \text{L-arginine} + \text{fumarate}$$

An enzyme catalyzing this reaction has been extracted from *E. coli* (9). For enzyme assays, the reaction is carried out in the presence of arginase (48), and the ornithine formed is determined colorimetrically (9,18). With sufficiently purified enzyme preparations, the formation of fumarate can be followed spectrophotometrically (49).

The argininosuccinate-splitting enzyme from strain W2D has been partially purified by fractionation with ammonium sulfate, acetic acid, and alumina Cγ and by column chromatography on DEAE-cellulose (9).

The optimal pH for the cleavage of argininosuccinate is approximately 7.7. The K_m value for argininosuccinate is 0.15 mM. Sulfhydryl reagents and Zn^{2+} are inhibitory; Co^{2+} and Ni^{2+} give appreciable stimulations. A sedimentation coefficient ($s_{20,w}$) of 9.4S has been found for the lyase (9).

III. Regulation by Repression

A. BACKGROUND

The first clue to what is now known as enzyme repression came from a study of an apparently adaptive phenomenon in the path of arginine biosynthesis (8). Growth rate experiments were performed with arginine and with an arginine precursor, which was later identified (1) as α-N-acetyl-L-ornithine. Cells of an *E. coli* arginine auxotroph grown on a limiting supplement of acetylornithine, when inoculated into medium containing either arginine or acetylornithine, showed almost identical growth rates, without appreciable lag; however, after preliminary growth on limiting arginine, subsequent growth on arginine occurred at approximately the same rate, whereas growth on acetylornithine was preceded by a pronounced lag (8). It was suggested that enzymes of arginine biosynthesis are formed adaptively in response to acetylornithine (8). Alternatively it could have been inferred (which was not done at the time of the brief 1952 report) that excess arginine leads to a decrease in the levels of arginine biosynthetic enzymes. In 1953, when an assay method for acetylornithine deacetylase (1) had become available, enzyme experiments with the arginine auxotroph were described, which seemed to support the adaptive-response interpretation (31). On the other hand, in experiments with wild-type *E. coli*, it was found that cells grown with an arginine supplement yield lower levels of extracted deacetylase than do cells grown on unsupplemented medium (31). Here, arginine, the "end product" of its pathway, was indicated to antagonize the formation of an enzyme participating in arginine biosynthesis. Also in 1953, analogous indications emerged for the pathways leading to methionine (50,51), tryptophan (52), valine (53), and pyrimidines (54).

Further studies on acetylornithine deacetylase supported the view that the "adaptive" response represents the removal of an antagonistic effect of arginine on the synthesis of the enzyme, with acetylornithine functioning as a restrictive source of arginine, and not as an inducer. This view laid the groundwork for the concept of repression, which was advanced at a symposium in 1956 (32). Enzyme repression was defined as a relative decrease, resulting from the exposure of cells to a given substance, in the rate of synthesis of a particular apoenzyme (32). Repression was regarded as a control mechanism that is complementary to enzyme induction; in either case, the cell tends to form enzymes when they are needed and tends not to form enzymes when they are not needed. The picture of repression and induction as control mechanisms was not in harmony with the generalized induction hypothesis (cf. ref. 55), which contemplated the possibility that an induction process is an inherent feature of all enzyme formation. When it became clear that, in the β-galactosidase system, induction constitutes an antagonism of an exogenous inducer to a product specified by the *i* gene (56), the usage of the term "repressor" (32) came to be extended to include the (protein) products of regulatory genes.

In the experiments on acetylornithine deacetylase levels in the arginine auxotroph (31,32) arginine restriction was achieved with the aid of acetylornithine as an exogenous arginine source. The limiting character of acetylornithine results from its uptake via an acetylornithine permeation system, a component of which is itself repressible by arginine (57). An endogenous arginine restriction is obtained on transfer of arginine-grown wild-type cells to an arginine-free medium, in which case a relatively rapid initial rise in ornithine carbamoyltransferase level followed by a drop to a steady-state level can occur (58). In arginine-free media, wild-type cells such as those of strain W form arginine biosynthetic enzymes at partially derepressed levels (see Fig. 3). The release from repression by restricting arginine in any manner is sometimes referred to as physiological derepression.

In a review article in 1960, it was possible to list some two dozen repressible enzymes (59). By that time, enzyme repression was firmly established as a control mechanism impinging on various major areas of biological interest.

B. ARGININE REPRESSOR

The finding of regulatory-gene (argR) mutants (22) for the arginine biosynthetic system suggested the participation of a macromolecular product, now termed "arginine repressor," in repressive control by arginine. Typically, argR⁻ mutants show relatively high levels of arginine biosynthetic enzymes, even on cultivation in the presence of excess arginine. However, various classes of argR mutants are known (60), including a class carrying mutations of a type called argR* (26), which leads to diminished repressibility and diminished derepressibility of the enzymes (5,26). Also, the argR gene of strain B of E. coli (40) can be regarded as a mutant form of the analogous regulatory gene of strain K12.

The enzymes of the linear path of arginine biosynthesis are not the only proteins whose repressibility is governed by argR. For instance, a component of the acetylornithine permeation system is repressible by arginine (57), and argR⁻ mutants of strain W do not show restricted acetylornithine uptake (61) such as is characteristic of argR⁺ organisms (31,32); strain B resembles these argR⁻ mutants in readily taking up acetylornithine (62). An enzyme whose repressibility partially depends on argR is the glutamine-utilizing carbamyl phosphate synthetase (EC 2.7.2.9), with the systematic name ATP:carbamate phosphotransferase (dephosphorylating, amino-transferring). This enzyme catalyzes the formation of carbamyl phosphate via several enzyme-bound intermediates (23), and the carbamyl phosphate so formed is used in the syntheses of citrulline as well as of pyrimidines. Arginine and uracil individually can bring about partial repression of the synthetase and jointly can cause cumulative repression, which is diminished in *argR* mutants (63).

The *argR* gene has certain parallels to *trpR*, which functions in the regulation of tryptophan biosynthetic enzymes (64–66). There is some question, however, whether such genes are generally essential for regulation by repression; for example, the histidine biosynthetic system seems to lack a regulatory gene whose unique function is the specification of a repressor product, although several genes are capable of exerting some kind of effect on regulation (67).

The regulatory gene of the arginine system is known to code for a diffusible product (in the sense that this product has access to various locations in the cell), in view of the pleiotropic *argR⁻* mutations (22, 24,62) and of the results of dominance studies (40,68). The diffusible *argR* product was indicated to be (at least in part) a protein, when an *argR* mutant sensitive to amber suppression was isolated (40). In double-labeling experiments with wild-type and *argR⁻* strains of *E. coli*, a protein fraction thought to contain arginine repressor has been collected (69). An assay method for the functional *argR* product has been devised on the basis of the response of *argR⁻* spheroplasts of a strain K12 derivative to exogenous arginine repressor (70). Recently, *in vitro* activity of arginine repressor was demonstrated: in an acetylornithine-deacetylase-synthesizing system, added extracts from an *argR⁺* strain led to a more pronounced decrease in enzyme yield than did added *argR⁻* extracts (71).

C. REPRESSOR RECOGNITION SITES

The arginine repressor must be able to act at a number of control sites (repressor recognition sites). This is clear from the pleiotropic nature of the *argR⁻* mutations, even without reference to details of the repression mechanism. Presumably, there is at least one control site corresponding to each of the scattered genes of the arginine biosynthetic system, as well as to genes specifying such ancillary proteins as carbamyl phosphate synthetase or acetylornithine permease; the *argECBH* cluster appears to have two control sites. In assessing the total number of repressor recognition sites, we must take into consideration that repression by arginine is both transcriptional and translational, as will be discussed in succeeding sections.

The existence of at least two control sites for the *argECBH* cluster (i.e., its two-operon character) was surmised from the finding that the repression of acetylornithine deacetylase and argininosuccinate lyase is not coordinate (72) and from the isolation of a mutant in which the levels of the enzymes specified by *argCBH* and the repressibility of the lyase are affected, while the level and repressibility of the enzyme specified by *argE* (the deacetylase) are normal (73,74). Strong evidence that *argE* and *argCBH* comprise two separate operons came from an analysis of deletion and nonsense mutations (75). Further genetic studies indicated that the two operons are transcribed divergently, with genetic determinants for a control region located between *argE* and *argC* (76,77).

Relevant to considerations of the number of control sites is the evidence for differential repressor effectiveness. In a strain W mutant with internally restricted arginine formation, full derepression of acetylornithine deacetylase but partial repression of argininosuccinate lyase could be demonstrated; variations in the effectiveness of functional repressor thus seem to occur (78). A similar indication was provided in experiments with a strain B *argR* mutant producing a heat-labile repressor (40). In these two lines of evidence, either arginine or repressor was internally restricted. In experiments with a restrictive external arginine source, namely, α-N-acetyl-L-arginine, a graphic result was obtained (42,79). Acetylarginine is taken up via acetylornithine permease and is deacetylated by acetylornithine deacetylase. In an *argB⁻ argG⁻* mutant of strain W, the supply of arginine from exogenous acetylarginine (which is a restrictive arginine source by itself) can be progressively curtailed by the addition of acetylornithine (42,79). Unexpectedly, the utilization of acetylarginine appears to be antagonized by L-citrulline, which the mutant can accumulate from the added acetylornithine, rather than by the acetylornithine as such; citrulline affects the uptake of acetylarginine. In this organism, when a suitable mixed supplement of acetylarginine and acetylornithine is used, acetylornithine deacetylase is fully derepressed, whereas argininosuccinate lyase is fully repressed (42,79). This finding seems to represent an extreme example of differential repressor effectiveness. The all-or-none effect in the repression of the two enzymes indicates that two separate repressor recognition sites are involved, which is relevant to the previously considered possibility (76,77) that the genetic control elements for *argE* and *argCBH* are partially overlapping.

The repressor recognition sites—be they at the level of the gene or at the level of the ribosome, as discussed below—are thought to participate in repressive complexes, when repression is brought into play (5). In addition to the repressor recognition site, a repressive complex must contain at least arginine repressor plus arginine (or an arginine derivative). Furthermore, there is evidence that finished enzymes of arginine synthesis are components of some, if not all, repressive complexes in the arginine system. Thus in an $argR^+$ derivative of strain K12, but not in a corresponding $argR^-$ derivative, 0.4 M magnesium ions were shown to reduce the activity of acetylornithine aminotransferase (80, cf. 81). This repression-dependent loss of activity was traced to a reversible alteration of the enzyme (Fig. 4), and in view of the

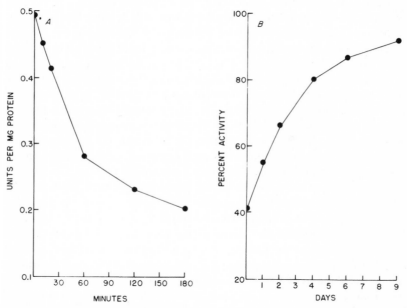

Fig. 4. (A) Time course of the reduction in activity of acetylornithine aminotransferase by treatment of arginine-grown cells of an $argR^+$ derivative of *Escherichia coli* K12 with 0.4 M Mg^{2+} plus L-arginine hydrochloride (0.1 mg/ml), at 37°C with aeration. Aminotransferase activity is determined in extracts. (B) Restoration of acetylornithine aminotransferase activity. An extract, prepared after treatment of cells with Mg^{2+} plus arginine for 2 hr, is kept at 0°C and assayed at intervals. The activity of the extract is plotted as a percentage of that of an untreated-control extract (which remains approximately constant). From Leisinger, Vogel, and Vogel (82).

requirement for repressive conditions, it was inferred that the susceptibility to magnesium arises when the enzyme forms part of the repressive complex (82). Further evidence for the involvement of finished enzymes in repression by arginine was provided by the observation that, on addition of arginine to physiologically derepressed cultures of strain K12 derivatives, there are prompt and sharp reductions in the rates of synthesis of arginine biosynthetic enzymes to values considerably below those of the steady-state repressed rates, followed by gradual increases up to the characteristic repressed values (83). It was suggested that the effects encountered reflect relatively high initial concentrations of arginine biosynthetic enzymes, which are progressively lowered as growth in the presence of arginine proceeds.

D. TRANSCRIPTIONAL REPRESSION

In our first report on the lowering of acetylornithine deacetylase level during the growth of wild-type *E. coli* in the presence of arginine, it was suggested that differences in the sites of enzyme formation with respect to their number per cell or their functioning or both may be involved (31). If in this language, which was written before details of protein synthesis were known, we take sites of enzyme formation to mean deacetylase messages, the suggestion made can be understood as hinting at the possibility of transcriptional repression or translational repression or both. For some time thereafter, our interest was focused on the possibility of translational repression (32,84,85). Other authors also considered translational models for a variety of systems (86–89). Eventually, evidence was obtained that, in *E. coli*, repression by arginine has transcriptional as well as translational components.

Transcriptional repression has mainly been examined with the aid of RNA–DNA hybridization techniques (90–95). One experimental approach is to measure the transcription of messages that is in progress in cultures growing exponentially either under repression or derepression (95). This can be accomplished with the aid of rifampicin, which is known (96) to inhibit the initiation of transcription, but to allow the completion of messages whose formation had been begun at the time of rifampicin addition. In a typical experiment, cells of an $argA^-$ auxotroph, derived from strain W, growing under repression (on arginine) or derepression (on acetylornithine) were treated, at 30°C, with rifampicin and with tritium-labeled uridine (0 time); samples were taken over a 2.5-min interval, and RNA was extracted and hybridized (97) with phage M36 DNA, which carries the *argECBH* gene cluster (95). With this method of determining *argECBH* messages, the extent of their transcription in progress could be ascertained as a function of regulatory conditions. Table II shows that the amount of labeled hybridizable messages increases for 2 min (after which it decreases through message decay); the percent hybridizations corresponding to the peak values are 0.09 and 0.66 for repression and derepression, respectively. The extent of transcription of *argECBH* under derepression, therefore, is indicated to exceed that under repression by a factor of roughly 7 (95). In pulsing experiments with tritium-labeled uridine (in the absence of rifampicin), percent hybridizations of approximately 0.10 and 0.50 were found for steady-state repression and derepression, respectively (95). The thus indicated ratio of about 5 for the derepressed formation rate of the messages to the repressed one appears to be too low to account for the known repressibility of

TABLE II

Extent of Transcription of *argECBH* Messages in Progress
under Repression and Derepression[a]

Regulatory condition	Time of sampling (min)	Specific activity of RNA (10^3 cpm/µg)	Cpm hybridized per 35 µg RNA
Repression	0.5	68	2,200
Repression	1.0	103	4,270
Repression	1.5	144	5,210
Repression	2.0	174	5,280
Repression	2.5	203	5,210
Derepression	0.5	52	8,190
Derepression	1.0	80	15,890
Derepression	1.5	126	22,890
Derepression	2.0	150	34,510
Derepression	2.5	183	34,160

[a] Cells of an *argA⁻* (*argR⁺*) derivative of strain W, growing under repression (L-arginine hydrochloride, 0.1 mg/ml) or depression (α-*N*-acetyl-L-ornithine, 0.75 mg/ml), are treated, at 30°C, with rifampicin, 0.1 mg/ml, plus tritium-labeled uridine (0 time). Samples are taken at the times indicated, and *argECBH* messages are determined by hybridization with DNA from phage M36. From McLellan and Vogel (95).

the enzymes specified by *argECBH*. Therefore, these results would have suggested the existence of translational repression in addition to transcriptional repression, if direct evidence for translational repression had not been available.

E. TRANSLATIONAL REPRESSION

Initially, translational repression by arginine was inferred from the findings that streptomycin sulfate or tetracycline hydrochloride, in *E. coli* under partial growth inhibition, will lower the differential rates of formation of two of the arginine biosynthetic enzymes when the conditions are physiologically derepressive, but not when they are repressive (5,98–100). Typical results for acetylornithine aminotransferase in an *argR*+ strain are shown in Figure 5.

Streptomycin and tetracycline, under the conditions employed, are regarded as translation retarders in that they decrease the rate of peptide chain elongation (101). It is considered that these ribosomal inhibitors will lower the differential rate of formation of a particular arginine biosynthetic enzyme when the decrease in the velocity of

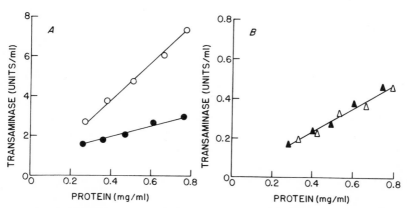

Fig. 5. (*A*) Formation of acetylornithine aminotransferase in an *argA*⁻ (*argR*+) auxotroph of *Escherichia coli* W under physiological derepression (growth on α-*N*-acetyl-ʟ-ornithine, 0.05 mg/ml), in the absence (open circles) or presence (solid circles) of streptomycin sulfate (6 μg/ml). (*B*) Formation of acetylornithine amino-transferase in the same strain under repression (growth on ʟ-arginine hydrochloride, 0.2 mg/ml), in the absence (open triangles) or presence (solid triangles) of strepto-mycin sulfate (6 μg/ml). From Vogel et al. (5); see also Vogel and Vogel (98).

peptide chain elongation caused by the inhibitors sufficiently restricts the amount of enzyme produced per unit time (100). Translational repression is thought to involve the slowing of a phase such as initiation of translation, rather than peptide chain elongation (100). Accordingly, regarding the results in Fig. 5, we can readily account for the drop in differential rate produced by the inhibitor used, under physiological derepression. The failure of the inhibitor to decrease the differential rate of the enzyme under repression indicates that here enzyme formation, in relation to total-protein synthesis, is limited in such a manner as not to be further reduced by some slowing of peptide chain growth (100).

This approach, with tetracycline as translation retarder, was extended to the study of three of the arginine biosynthetic enzymes in the $argR^+$ strain as well as in four different $argR$ mutants (100). Among the results shown in Table III, the case of acetylornithine aminotransferase in the $argR^-$ strains, under arginine excess, seems particularly intriguing. The enzyme is indicated to be under some residual repression (in comparison with the $argR^+$ strain under physiological derepression), with tetracycline not dropping the differential rates. In an $argR^-$ organism (strain 977), under arginine restriction, the residual repression is relieved. From the data for the aminotransferase and the generally similar data for argininosuccinate lyase, it was inferred that the component of repression whose functioning in enzyme formation is brought to light by tetracycline is the translational one (100). The inference is supported by estimates of the transcription rates for the $argECBH$ gene cluster (94,95,100), which contains the gene for the lyase. With respect to strain W2D (Table III), the pronounced residual translational repression indicated for the aminotransferase, but not for the lyase, was attributed to residual activity of the arginine repressor, which does not effect the formation of the two enzymes equally. In the case of acetylornithine deacetylase, the observation that tetracycline lowers the differential rate in every instance given indicates that, for this enzyme, translational repression is impaired in being not as effective as it is for the other two enzymes. However, when the tetracycline concentration was halved, translational repression was in evidence (100), since the repressed differential rate of the $argR^+$ strain was not diminished whereas the derepressed one was (not shown in Table III). In the case of the $argR^*$ strain, tetracycline lowers all differential rates, which is indicative of impaired transla-

TABLE III

Effect of Tetracycline on the Formation of Arginine Biosynthetic Enzymes
under Various Regulatory Conditions (as Relative Differential Rates)[a]

Strain	Regulatory allele	Arginine supply	TC	Enz. 4	Enz. 5	Enz. 8
39A–23R3	$argR^+$	Restrictive	−	100	100	100
39A–23R3	$argR^+$	Restrictive	+	39	28	14
39A–23R3	$argR^+$	Excessive	−	5	8	3
39A–23R3	$argR^+$	Excessive	+	7	5	4
W2D	$argR^-$	Excessive	−	22	79	83
W2D	$argR^-$	Excessive	+	25	19	52
160–37–12	$argR^-$	Excessive	−	24	—	47
160–37–12	$argR^-$	Excessive	+	28	—	50
977	$argR^-$	Restrictive	−	75	94	—
977	$argR^-$	Restrictive	+	34	17	—
977	$argR^-$	Excessive	−	31	53	—
977	$argR^-$	Excessive	+	31	8	—
250	$argR^*$	Restrictive	−	51	46	35
250	$argR^*$	Restrictive	+	23	10	11
250	$argR^*$	Excessive	−	10	17	6
250	$argR^*$	Excessive	+	6	3	2

[a] The strains used are derivatives of strain W, except strain 977, which is derived from strain K12. Abbreviations: TC, tetracycline hydrochloride; Enz. 4, acetyl-ornithine aminotransferase; Enz. 5, acetylornithine deacetylase; Enz. 8, arginino-succinate lyase. The organisms are cultivated at 37°C, with aeration, in a suitably supplemented liquid glucose-salts medium, under arginine excess or arginine restriction, with (+) or without (−) tetracycline hydrochloride (0.25 μg/ml). Arginine restriction is achieved with L-ornithine hydrochloride (0.1 mg/ml) for strain 977 (which is impaired in the conversion of ornithine to arginine) and with α-N-acetyl-L-ornithine (0.05 mg/ml) for the other auxotrophs. The experimental cultures are sampled at intervals. Cell extracts are prepared and assayed for total protein and for the aminotransferase, the deacetylase, and the lyase. From Vogel, Knight, and Vogel (100).

tional repression. In this mutant, low transcriptional derepression (an abnormal repression) appears to occur when arginine is restricted. The narrow repression–derepression range of this mutant, therefore, was inferred to involve translational as well as transcriptional abnormalities (100).

Studies with accumulated messages for the arginine biosynthetic enzymes have provided an approach to translational aspects of repression, in which the translation of the messages is separated from their formation: the messages were accumulated by arginine starvation of suitable auxotrophs; further transcription was inhibited by rifampicin or miracil D; and enzyme formation was measured as a function of (a) arginine excess or restriction and (b) the presence of a wild-type ($argR^+$) or mutant ($argR^-$) regulatory gene in the organisms (102). The strains used had a complete block in the early part of the arginine pathway and had a "leaky" block in argininosuccinate lyase, which permitted limited arginine synthesis from added ornithine. It was thus possible to supply ornithine (arginine restriction), to supply arginine (arginine excess), or to withhold a source of arginine (arginine starvation).

Results of experiments on the translation of accumulated acetylornithine deacetylase messages are presented in Figure 6. The plots in Figures 6A and B correspond to data for $argR^+$ and $argR^-$ strains,

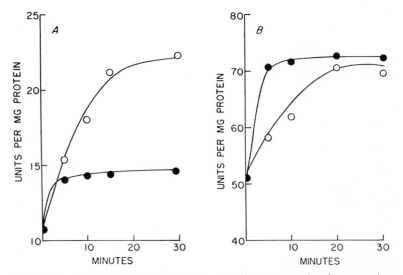

Fig. 6.   Acetylornithine deacetylase formation from accumulated message in an $argR^+$ (A) and $argR^-$ (B) of *Escherichia coli* K12. After accumulation of message by arginine starvation at 37°C for 30 min, rifampicin (60 μg/ml) is added and allowed to act for 2 min. L-Ornithine hydrochloride (open circles) or L-arginine hydrochloride (solid circles) is added (at 0.1 mg/ml) for translation at 37°C; the addition corresponds to 0 time on the abscissa. From McLellan and Vogel (102).

respectively. Open circles indicate translation in the presence of ornithine (arginine restriction), and solid circles indicate translation in the presence of arginine (arginine excess). It is seen that the lowered extent of translation under arginine excess (compared with that under arginine restriction), as found for the $argR^+$ strain, does not occur in the $argR^-$ strain (102). Qualitatively similar results were obtained for ornithine carbamoyltransferase. It was concluded that the repressor specified by $argR$ is required for the reduced translation that occurs in the $argR^+$ strain with an excess of arginine and that the translation of the messages for these arginine biosynthetic enzymes is clearly implicated in the mechanism of repression by arginine (102).

The reduced extent of translation of messages for the arginine biosynthetic enzymes in the $argR^+$ strain, with an excess of arginine, suggests that, under repressive conditions, arginine interferes with translation or promotes the disappearance of translatable messages or both (102). Since there is evidence that translational repression involves the slowing of a phase such as the initiation of translation, any enhanced rate of message degradation under repression compared with that under derepression could be a consequence of a reduction in the efficiency of the initiation of translation (100). A repression-dependent instability of the messages may, therefore, be a secondary, but significant, feature of translational repression. Evidence for this type of instability has indeed been furnished for translatable (102) as well as hybridizable (95) messages.

Data pointing to the disappearance of translatable message were obtained in experiments with the $argR^+$ strain in which the messages were accumulated and translated at 25°C, first in the presence of arginine for periods from 1 to 10 min and then (after rapid filtration to remove arginine) in the presence of ornithine for 50 min. There was progressively less translation with ornithine present after increasing periods of translation in the presence of arginine (102). In analogous experiments on hybridizable messages, the stability of message specified by $argECBH$ was examined in an $argA^-$ derivative of strain W (95). From the results shown in Figure 7, it was calculated that these messages disappear with half-lives of approximately 1.3 and 5.3 min for arginine excess and arginine restriction, respectively. The relative prolongation of the half-life under arginine restriction appears to be specific for these messages of the arginine system, since the half-lives of total message, under arginine excess or restriction, are approximately

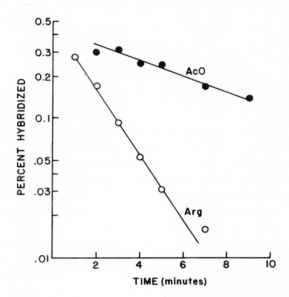

Fig. 7. Degradation of *argECBH* messages under arginine excess or arginine restriction. Cells of an *argA⁻* (*argR⁺*) auxotroph of *Escherichia coli* W are suspended in glucose-salts medium, at 37°C, for arginine starvation (-30 min). At -25 min, tritium-labeled uridine is added, and at -0.5 min, rifampicin (60 µg/ml) and an excess of unlabeled uridine are introduced. At 0 time translation of the accumulated messages is allowed to proceed by the addition of either L-arginine hydrochloride (Arg, arginine excess, repressive conditions) or α-N-acetyl-L-ornithine (AcO, arginine restriction, derepressive conditions). Samples are taken at intervals for the determination of the messages by hybridization with DNA from phage M36. The results are plotted as percentage of RNA hybridized versus time of sampling. From McLellan and Vogel (95).

equal (1.6 min) in the strain used. Similar experiments in a derivative of strain K12, with hybridization of part of the *argECBH* messages, yielded half-lives of 3.0 min (arginine excess) and 7.5 min (arginine restriction). However, in this strain, hybridizable *lac* message decays with a half-life of about 2.2 min, whether arginine is restricted or in excess (95).

Consistent with these findings are the results of experiments of somewhat different design, carried out with the strain W derivative

(103). The organisms growing on α-N-acetyl-L-ornithine (0.5 mg/ml, on which the growth rate of the $argA^-$ strain is only slightly lower than that of the wild type) were pulse-labeled for 3.0 min with tritium-labeled uridine (40 μCi/ml; specific activity, 42 Ci/millimole). A 500-fold excess (on a molar basis) of unlabeled uridine was added, and after this chase, either L-arginine hydrochloride (0.2 mg/ml) was introduced (arginine excess) or no further addition was made (continued arginine restriction). Samples were then taken at intervals, RNA was prepared, hybridizations were performed, and specific-message stability was followed as before (95). Half-life values of 1.6 and 6.6 min were found for the disappearance of message under arginine excess and arginine restriction, respectively (103). These results satisfactorily complement the data from the experiments in which rifampicin was used.

It can thus be seen that the studies with ribosomal inhibitors (5, 98–100) and with accumulated messages (95,102) and also the studies on message stability (95,102,103) represent different approaches supporting the conclusion that translational repression is a significant component of repression by arginine in $E.\ coli$. Further support was provided by an analysis of the onset of repression, from which it was concluded that there is a lack of correspondence between message levels and enzyme levels, such that a translational effect was indicated (83).

With the relatively recent emergence of the evidence for the locales of repression and for the nature of the repressive complex, the time for detailed studies of the molecular mechanism of repression by arginine seems to be at hand. The solution of this problem, which clearly has important implications for the genetic and metabolic regulation of the cell as well as for the process of protein synthesis, constitutes a worthy experimental challenge.

References

1. Vogel, H. J., *Proc. Natl. Acad. Sci. U.S.*, 39, 578 (1953).
2. Vogel, H. J., in *Amino Acid Metabolism*, W. D. McElroy and B. Glass, Eds., Johns Hopkins Press, Baltimore, 1955, p. 335.
3. Vogel, H. J., and Bonner, D. M., in *Encyclopedia of Plant Physiology*, Vol. 11, W. Ruhland, Ed., Springer, Heidelberg, 1959, p. 1.
4. Vogel, H. J., in *Methods in Enzymology*, Vol. 17A, H. Tabor and C. W. Tabor, Eds., Academic Press, New York, 1970, p. 249.

5. Vogel, R. H., McLellan, W. L., Hirvonen, A. P., and Vogel, H. J., in *Metabolic Regulation*, H. J. Vogel, Ed., Academic Press, New York, 1971, p. 463.
6. Vogel, H. J., Abelson, P. H., and Bolton, E. T., *Biochim. Biophys. Acta*, *11*, 584 (1953).
7. Baich, A., and Vogel, H. J., *Biochem. Biophys. Res. Commun.*, *6*, 491 (1962).
8. Vogel, H. J., and Davis, B. D., *Fed. Proc.*, *11*, 485 (1952).
9. Ortigoza-Ferado, J. A., Jones, E. E., and Vogel, H. J., unpublished data.
10. Maas, W. K., Novelli, G. D., and Lipmann, F., *Proc. Natl. Acad. Sci. U.S.*, *39*, 1004 (1953).
11. Sadasiv, E., McLellan, W. L., Baumberg, S., and Vogel, H. J., unpublished.
12. Vogel, H. J., and McLellan, W. L., in *Methods in Enzymology*, Vol. 17A, H. Tabor and C. W. Tabor, Eds., Academic Press, New York, 1970, p. 251.
13. Marschek, W. J., and Vogel, H. J., unpublished data.
14. Vogel, H. J., and McLellan, W. L., in *Methods in Enzymology*, Vol. 17A, H. Tabor and C. W. Tabor, Eds., Academic Press, New York, 1970, p. 255.
15. Albrecht, A. M., and Vogel, H. J., *J. Biol. Chem.*, *239*, 1872 (1964).
16. Vogel, H. J., and Jones, E. E., in *Methods in Enzymology*, Vol. 17A, H. Tabor and C. W. Tabor, Eds., Academic Press, New York, 1970, p. 260.
17. Vogel, H. J., *6th Int. Congr. Microbiol.*, *Rome*, Vol. 1, 267 (1953).
18. Vogel, H. J., and Bonner, D. M., *J. Biol. Chem.*, 218, 97 (1956).
19. Vogel, H. J., and McLellan, W. L., in *Methods in Enzymology*, Vol. 17A, H. Tabor and C. W. Tabor, Eds., Academic Press, New York, 1970, p. 265.
20. Rogers, P., and Novelli, G. D., *Arch. Biochem. Biophys.*, *96*, 398 (1962).
21. Legrain, C., Halleux, P., Stalon, V., and Glansdorff, N., *Eur. J. Biochem.*, *27*, 93 (1972).
22. Maas, W. K., *Cold Sring Harbor Symp. Quant. Biol.*, *26*, 183 (1961).
23. Anderson, P. M., Wellner, V. P., Rosenthal, G. A., and Meister, A., in *Methods in Enzymology*, Vol. 17A, H. Tabor and C. W. Tabor, Eds., Academic Press, New York, 1970, p. 235.
24. Gorini, L., Gundersen, W., and Burger, M., *Cold Spring Harbor Symp. Quant. Biol..*, *26*, 173 (1961).
25. Vogel, H. J., Bacon, D. F., and Baich, A., in *Informational Macromolecules*, H. J. Vogel, V. Bryson, and J. D. Lampen, Eds., Academic Press, New York, 1963, p. 293.
26. Vogel, H. J., and Bacon, D. F., *Proc. Natl. Acad. Sci. U.S.*, *55* 1456(1966).
27. Taylor, A. L., *Bacteriol. Rev.*, *34*, 155 (1970).
28. Glansdorff, N., Sand, G., and Verhoef, C., *Mutation Res.*, *4*, 743 (1967).
29. Glansdorff, N., *Genetics*, *51*, 167 (1965).
30. Itikawa, H., Baumberg, S., and Vogel, H. J., *Biochim. Biophys. Acta*, *159*, 547 (1968).
31. Vogel, H. J., *6th Int. Congr. Microbiol.*, *Rome*, *1*, 269 (1953).
32. Vogel, H. J., in *Chemical Basis of Heredity*, W. D. McElroy and B. Glass, Eds., Johns Hopkins Press, Baltimore, 1957, p. 276.
33. Vyas, S., and Maas, W. K., *Arch. Biochem. Biophys.*, *100*, 542 (1963).
34. Haas, D., Kurer, V., and Leisinger, T., *Eur. J. Biochem.*, *31*, 290 (1972).
35. Haas, D., and Leisinger, T., personal communication.

36. Forsyth, G. W., Theil, E. C., Jones, E. E., and Vogel, H. J., *J. Biol. Chem.*, *245*, 5354 (1970).
37. Bacon, D. F., and Vogel, H. J., *Cold Spring Harbor Symp. Quant. Biol.*, *28*, 437 (1963).
38. Gorini, L., and Gundersen, W., *Proc. Natl. Acad. Sci. U.S.*, *47*, 961 (1961).
39. Vogel, H. J., Albrecht, A. M., and Cocito, C., *Biochem. Biophys. Res. Commun.*, *5*, 115 (1961).
40. Jacoby, G. A., and Gorini, L., *J. Mol. Biol.*, *39*, 73 (1969).
41. Karlström, O., and Gorini, L., *J. Mol. Biol.*, *39*, 89 (1969).
42. Bollon, A. P., Leisinger, T., and Vogel, H. J., *Genetics*, *61*, s6 (1969).
43. Baumberg, S., *Mol. Gen. Genetics*, *106*, 162 (1970).
44. Fry, K. T., and Lamborg, M. R., *J. Mol. Biol.*, *28*, 423 (1967).
45. Adams, J. M., *J. Mol. Biol.*, *33*, 571 (1968).
46. Nakamura, M., and Jones, M. E., in *Methods in Enzymology*, Vol. 17A, H. Tabor and C. W. Tabor, Eds., Academic Press, New York, 1970, p. 286.
47. Ratner, S., in *Methods in Enzymology*, Vol. 17A, H. Tabor and C. W. Tabor, Eds., Academic Press, New York, 1970, p. 298.
48. Ratner, S., Anslow, W. P., Jr., and Petrack, B., *J. Biol. Chem.*, *204*, 115 (1953).
49. Ratner, S., in *Methods in Enzymology*, Vol. 17A, H. Tabor and C. W. Tabor, Eds., Academic Press, New York, 1970, p. 304.
50. Cohn, M., Cohen, G. N., and Monod, J., *C. R. Acad. Sci. Paris*, *236*, 746 (1953).
51. Wijesundera, S., and Woods, D. D., *Biochem. J.* (*London*), *55*, viii (1953).
52. Monod, J., and Cohen-Bazire, G., *C. R. Acad. Sci. Paris*, *236*, 530 (1953).
53. Adelberg, E. A., and Umbarger, H. E., *J. Biol. Chem.*, *205*, 475 (1953).
54. Back, K. J. C., and Woods, D. D., *Biochem. J.* (*London*), *55*, xii (1953).
55. Cohn, M., and Monod, J., *Symp. Soc. Gen. Microbiol.*, *3*, 132 (1953).
56. Pardee, A. B., Jacob, F., and Monod, J., *J. Mol. Biol.*, *1*, 165 (1959).
57. Vogel, H. J., *Proc. Natl. Acad. Sci. U.S.*, *46*, 488 (1960).
58. Gorini, L., and Maas, W. K., *Biochim. Biophys. Acta*, *25*, 208 (1957).
59. Vogel, H. J., in *Control Mechanisms in Cellular Processes*, D. M. Bonner, Ed., Ronald Press, New York, 1961, p. 23.
60. Kadner, R. J., and Maas, W. K., *Mol. Gen. Genetics*, *111*, 1 (1971).
61. Devine, E. A., Vogel, R. H., and Vogel, H. J., unpublished data.
62. Vogel, H. J., *Cold Spring Harbor Symp. Quant. Biol.*, *26*, 163 (1961).
63. Piérard, A., Glansdorff, N., and Yashphe, J., *Mol. Gen. Genetics*, *118*, 235 (1972).
64. Cohen, G., and Jacob, F., *C. R. Acad. Sci. Paris*, *248*, 3490 (1959).
65. Yanofsky, C., *Bacteriol. Rev.*, *24*, 221 (1960).
66. Margolin, P., in *Metabolic Regulation*, H. J. Vogel, Ed., Academic Press, New York, 1971, p. 389.
67. Brenner, M., and Ames, B. N., in *Metabolic Regulation*, H. J. Vogel, Ed., Academic Press, New York, 1971, p. 349.
68. Maas, W. K., and Clark, A. J., *J. Mol. Biol.*, *8*, 365 (1964).
69. Udaka, S., *Nature*, *228*, 336 (1970).
70. Hirvonen, A. P., and Vogel, H. J., *Biochem. Biophys. Res. Commun.*, *41*, 1611 (1970).

71. Urm, E., Kelker, N., Yang, H., Zubay, G., and Maas, W., *Mol. Gen. Genetics*, *121*, 1 (1973).
72. Baumberg, S., Bacon, D. F., and Vogel, H. J., *Proc. Natl. Acad. Sci. U.S.*, *53*, 1029 (1965).
73. Baumberg, S., Bacon, D. F., and Vogel, H. J., *Genetics*, *54*, 322 (1966).
74. Vogel, H. J., Baumberg, S., Bacon, D. F., Jones, E. E., Unger, L., and Vogel, R. H., in *Organizational Biosynthesis*, H. J. Vogel, J. O. Lampen, and V. Bryson, Eds., Academic Press, New York, 1967, p. 223.
75. Cunin, R., Elseviers, D., Sand, G., Freundlich, G., and Glandsorff, N., *Mol. Gen. Genetics*, *106*, 32 (1969).
76. Jakoby, G. A., *Mol. Gen. Genetics*, *117*, 337 (1972).
77. Elseviers, D., Cunin, R., Glansdorff, N., Baumberg, S., and Ashcroft, E., *Mol. Gen. Genetics*, *117*, 349 (1972).
78. Unger, L., Bacon, D. F., and Vogel, H. J., *Genetics*, *63*, 53 (1969).
79. Bollon, A. P., and Vogel, H. J., *J. Bacteriol.*, in press.
80. Vogel, R. H., and Vogel, H. J., *Genetics*, *60*, 233 (1968).
81. Urm, E., Leisinger, T., Vogel, R. H., and Vogel, H. J., *Genetics*, *61*, s59 (1969).
82. Leisinger, T., Vogel, R. H., and Vogel, H. J., *Proc. Natl. Acad. Sci. U.S.*, *64*, 686 (1969).
83. Lavallé, R., *J. Mol. Biol.*, *51*, 449 (1970).
84. Vogel, H. J., *Proc. Natl. Acad. Sci. U.S.*, *43*, 491 (1957).
85. Vogel, H. J., and Vogel, R. H., *Ann. Rev. Biochem.*, *36*, 519 (1967).
86. Halvorson, H. O., *Adv. Enzymol.*, *22*, 99 (1960).
87. Szilard, L., *Proc. Natl. Acad. Sci. U.S.*, *46*, 277 (1960).
88. Cline, A. L., and Bock, R. M., *Cold Spring Harbor Symp. Quant. Biol.*, *31*, 321 (1966).
89. Tomkins, M. G., Gelehrter, T. D., Granner, D., Martin, D., Jr., Samuels, H. H., and Thompson, E. B., *Science*, *166*, 1474 (1969).
90. Cunin, R., and Glansdorff, N., *FEBS Letters*, *18*, 135 (1971).
91. Rogers, P., Krzyzek, R., Kaden, T. M., and Arfman, E., *Biochem. Biophys. Res. Commun.*, *44*, 1220 (1971).
92. Krzyzek, R., and Rogers, P., *J. Bacteriol.*, *110*, 945 (1972).
93. McLellan, W. L., and Vogel, H. J., *Genetics*, *71*, s39 (1972).
94. Knight, G. J., Vogel, R. H., and Vogel, H. J., *Genetics*, *71*, s31 (1972).
95. McLellan, W. L., and Vogel, H. J., *Biochem. Biophys. Res. Commun.*, *48*, 1027 (1972).
96. Mosteller, R. D., and Yanofsky, C., *J. Mol. Biol.*, *48*, 525 (1970).
97. Gillespie, D., and Spiegelman, S., *J. Mol. Biol.*, *12*, 829 (1965).
98. Vogel, R. H., and Vogel, H. J., *Genetics*, *61*, s61 (1969).
99. Vogel, R. H., and Vogel, H. J., *Genetics*, *68*, s70 (1971).
100. Vogel, R. H., Knight, G. J., and Vogel, H. J., *Biochem. Biophys. Res. Commun.*, *48*, 1034 (1972).
101. Pestka, S., *Ann. Rev. Microbiol.*, *25*, 487 (1971).
102. McLellan, W. L., and Vogel, H. J., *Proc. Natl. Acad. Sci. U.S.*, *67*, 1703 (1970).
103. McLellan, W. L., and Vogel, H. J., *Biochem. Biophys. Res. Commun.*, *55*, 1385 (1973).

AMINOACYL-tRNA TRANSFERASES

By RICHARD L. SOFFER*, *New York*

CONTENTS

*Faculty Research Associate of the American Cancer Society.

I. Introduction

This chapter is concerned with a group of incompletely understood enzymes that catalyze the transfer of an aminoacyl residue from tRNA to a specific acceptor molecule. These aminoacyl-tRNA transfer reactions are distinguished from the ribosome-dependent peptidyl transferase reaction, which occurs in *de novo* protein synthesis, by the nature of the acceptors to which the aminoacyl residue is transferred, by the fact that the reactive moiety of the aminoacyl residue is its carboxyl group, and by the lack of requirement for the presence of ribosomes, template nucleic acids, and GTP. The responsible enzymes can be classified on the basis of their acceptor substrates as follows:

1. Aminoacyl-tRNA$\sim$protein transferases are found in bacteria and mammalian tissues, and catalyze the transfer of an aminoacyl residue from tRNA to the NH_2-terminus of specific classes of protein or peptide acceptors.

2. Aminoacyl-tRNA$\sim$phosphatidylglycerol transferases or aminoacyl phosphatidylglycerol synthetases catalyze the formation of aminoacyl esters of phosphatidylglycerol in certain bacteria.

3. Aminoacyl-tRNA$\sim N$-acetylmuramyl-pentapeptide transferases utilize an acceptor whose structure includes this substituted sugar. They catalyze the transfer of an aminoacyl residue from tRNA to the ϵ-amino group of the lysine residue in the peptapeptide and appear to be involved in the biosynthesis of interpeptide bridges in the cell wall of various bacteria.

These enzymes may be profitably considered within a framework that includes the enzyme itself, the donor substrate, the acceptor substrate, and the role of the reaction within the cell. Current knowledge will therefore be summarized with emphasis on the following points:

1. The enzyme. The degree of isolation of these enzymes is of obvious importance in determining their physical properties and the

requirements and kinetic parameters of the reactions they catalyze. It is critical in establishing whether activities for the transfer of different aminoacyl residues are due to the same or different enzyme proteins. It is no less critical in delineating acceptor-substrate specificity, since to do so requires an enzyme preparation free of endogenous acceptors and of activities that might convert acylated substrates to other products and hence obscure the nature of the initial acceptor. If the biological activity of an enzymatically acylated acceptor molecule is to be examined, the enzyme used in its preparation must be sufficiently pure so that the acylated product can be isolated and shown to be otherwise unaltered.

Various complexes among enzyme, aminoacyl-tRNA, and acceptor can be envisioned as possible intermediates in these reactions. An aminoacyl–enzyme thioester intermediate is also plausible in certain cases. Identification of such intermediates would require enzymes whose purity and molecular weight were sufficiently characterized so that experiments involving the stoichiometry of enzyme and substrate binding could be undertaken. No such studies on aminoacyl-tRNA transferases have been reported, and the mechanism of the reactions they catalyze will, therefore, not be further considered.

2. Aminoacyl-tRNA. Overall specificity with respect to aminoacyl residues whose transfer is catalyzed by these enzymes is reflected in studies using different radioactive aminoacyl-tRNAs as substrates. In addition to determining which residues can be transferred, it is desirable to establish whether the basis for this specificity residues in the polyribonucleotide or aminoacyl moiety of the donor. Finally, it is of interest to delineate the functional groups in these structures that contribute to substrate activity and to determine whether any relationship exists between the function of aminoacyl-tRNA in protein synthesis and in the aminoacyl-tRNA transfer reactions.

3. Acceptor. The first criterion that a compound is functioning as acceptor is dependence of the reaction on its presence and proof that an aminoacyl residue is transferred from tRNA to a product not found in its absence. A second criterion is stoichiometric transfer of amino acid at limiting acceptor concentrations. This is particularly important where the possibility exists that an acceptor preparation is not completely pure. Definitive proof that a compound is an acceptor requires isolation of the acylated product and structural studies defining the site of acylation. Knowledge of the structure of the acceptor and the

site at which it is acylated should render feasible delineation of the biochemical determinants of acceptor specificity by the use of related model substrates.

4. Cellular function. The physiologic function of these reactions can be approached indirectly by correlating substrate specificity of the various enzymes with the presence and nature of the acylated products found *in vivo*. In characterizing the factors responsible for the extent and specificity of acylation observed *in vivo* the possible contribution of deacylases should be considered. A more direct approach to understanding the cellular role of aminoacyl-tRNA transferases requires the isolation of mutants in which they are defective. Such mutants should also be lacking acylated endogenous acceptors, and their phenotype should be rationally explicable on the basis of absence of these compounds. A single-step revertant to the parental phenotype should have regained enzymatic activity. If the mutant phenotype can be directly ascribed to the fundamental biochemical lesion, then it may be reversible *in vitro* or *in vivo* by addition, under appropriate conditions, of either the enzyme or of the missing acylated products.

II. Aminoacyl-tRNA∼Protein Transferases

A. THE ENZYMES

Aminoacyl-tRNA∼protein transferases (1) are soluble enzymes that catalyze the transfer of certain aminoacyl residues from tRNA into peptide linkage with specific NH_2-terminal residues of protein or peptide acceptors. The reactions catalyzed by the mammalian and bacterial enzymes are operationally analogous, although their specificities with respect to both the transferred and acceptor aminoacyl residues are entirely different. These reactions represent an enzymatic mechanism for posttranslational modification of the primary structure of certain classes of polypeptides.

1. Arginyl-tRNA∼Protein Transferase

Arginyl-tRNA∼protein transferase (2) is a mammalian enzyme that is found in the soluble fraction of all tissues and of cells in culture. It catalyzes the transfer of arginine into peptide linkage specifically with NH_2-terminal dicarboxylic amino acid residues of acceptor proteins (3) or peptides (4).

Kaji, Novelli, and Kaji (5) were the first to report that extracts of mammalian cells could incorporate arginine into protein in the absence of ribosomes. The reaction they observed with a partially fractionated soluble extract from rat liver required tRNA and an ATP-generating system, was inhibited by pancreatic ribonuclease, and was not dependent on added GTP. Weinstein and Osserman (6) described similar findings with crude supernatant fractions from mouse plasma cell tumors, mouse mammary tumors, and normal rat liver. Our own interest in this reaction began in 1966 when we found that the soluble fraction from sheep thyroid glands could catalyze the direct transfer of [^{14}C]arginine from tRNA to protein (7). Incorporation was not diluted by the presence of [^{12}C]-L-arginine, and the reaction could be distinguished from protein synthesis by the fact that it did not require the presence of ribosomes or magnesium ions and was not inhibited by puromycin. Boiling the extract abolished its activity, which suggested the existence of a specific arginine-transfer enzyme.

Partial isolation of this suspected activity from sheep thyroid glands was accomplished using as an assay the direct transfer of [^{14}C]arginine from tRNA into protein in the absence of magnesium ions (8). The use of arginyl-tRNA as a substrate obviated the requirement for arginyl-tRNA synthetase, and omission of magnesium ions precluded any ribosome-dependent incorporation resulting from contamination of the crude soluble fraction by these particles. During purification a requirement emerged for a heat-stable, nondialyzable cofactor. Bovine albumin or thyroglobulin could fulfill this requirement, whereas several other proteins could not, which suggested the participation of specific protein acceptors in the reaction.

In our subsequent studies the soluble fraction of rabbit liver has been used as the source of arginyl-tRNA∼protein transferase. A new assay was devised because of the lability of arginyl-tRNA at pH 9.0, the optimum for the arginine-transfer reaction. In this assay the transfer reaction is coupled to an arginyl-tRNA-generating system consisting of ATP, magnesium ions, *Escherichia coli* B tRNA and purified *E. coli* B arginyl-tRNA synthetase. The standard protein acceptor is bovine albumin, and puromycin is included to prevent ribosome-dependent incorporation. A procedure was developed in which arginyl-tRNA∼ protein transferase was purified approximately 7000-fold. The method (2) exploits the fact that the properties of the enzyme change dramatically after chromatography on DEAE-cellulose. As a result of this step the enzyme becomes very unstable, its molecular weight diminishes markedly, and it is retained by CM-cellulose (9). The nature of these changes suggests that in crude extracts the enzyme may be bound to an acidic macromolecule, such as tRNA, which is separated during fractionation on DEAE-cellulose. Enzyme purified by this technique was found to catalyze the transfer of about 135 nmoles of arginine onto bovine albumin per minute per milligram at 37°C. Disk gel electrophoresis showed one major band comprising 60–70% of the total protein. If this band represents the enzyme, it may thus be calculated that the pure protein has a specific activity of about 200 nmoles/min mg. The molecular weight of the enzyme determined by gel filtration was approximately 48,000, and a sedimentation constant of 3.6S was estimated by comparison with chicken ovalbumin during centrifugation through a glycerol gradient (9).

The direct transfer of arginine from tRNA to protein catalyzed by the purified enzyme was not inhibited by puromycin or cycloheximide (10) and was completely dependent on the presence of a sulfhydryl compound, a monovalent cation, and a protein acceptor (2,11). When the coupled system was employed and bovine albumin or thyroglobulin was used in limiting concentrations, the reaction was found to stop after the incorporation of 1 mole of arginine per mole of albumin or 2 moles of arginine per mole of thyroglobulin (2). It was found that [14C] arginine incorporated in the presence of both of these acceptors remained reactive with 2,4-dinitrofluorobenzene (DNFB). The site of incorporation onto albumin was established by subjecting the radioactive product to digestion with chymotrypsin (11). A single peptide accounted for most of the radioactivity in the digest. This peptide was purified of nonradioactive peptides, and its structure was shown to be [14C]Arg-Asp-Thr-His. Since Asp-Thr-His was known to be the NH2-terminal tripeptide of bovine albumin (12), these results established that arginine had been transferred into peptide linkage with the NH2-terminal residue.

Similar studies were carried out with peptic digests of the radioactive product made with bovine thyroglobulin (13). Most of the radioactivity was found in two peptides that were isolated and identified as [14C]Arg-Asp and [14C]Arg-Asp-Ile-Phe. These data were consistent with other structural studies demonstrating that bovine thyroglobulin contains two NH2-terminal aspartic acid residues (14) and confirmed that arginine was specifically transferred to the NH2-terminus of the protein acceptor. They also provided new sequence information for this region of the thyroglobulin molecule.

2. Leucyl,Phenylalanyl-tRNA~Protein Transferase

Leucyl,phenylalanyl-tRNA~protein transferase is a soluble enzyme, found in certain enterobacteria, that catalyzes the transfer of these amino acids from tRNA into peptide linkage specifically with NH_2-terminal basic amino acid residues of protein (15) or peptide (16) acceptors.

Kaji, Kaji, and Novelli (17–19) first reported that a partially fractionated extract from *E. coli* could specifically incorporate leucine and phenylalanine into protein in the absence of ribosomes. They demonstrated that aminoacyl-tRNA was an intermediate and that transfer of the aminoacyl residue was abolished by boiling the active fraction. Treatment of this fraction with pancreatic ribonuclease did not alter its activity, indicating that incorporation was not directed by an RNA template. The reaction was inhibited by ribonuclease and puromycin, but not by chloramphenicol. There was no requirement for GTP. When the active fraction was subjected to density-gradient centrifugation, sufficient resolution was achieved to suggest the participation of aminoacyl-tRNA transfer factor, aminoacyl-tRNA synthetase, and protein acceptors in the overall process by which free amino acid was incorporated into protein (20). Incorporated radioactive amino acids were reactive with DNFB and were cleaved by Edman degradation, which indicates the presence of a free α-amino group. Momose and Kaji (20) showed that this "NH_2-group addition" to ribosomal acceptor(s), rather than protein synthesis *de novo*, was responsible for most of the incorporation of radioactive leucine or phenylalanine by a conventional ribosomal system in the absence of added template RNA. Of particular interest was their observation that tryptic digests of products made with the ribosomal system under these conditions contained radioactive peptides that were indistinguishable by paper electrophoresis and chromatography from these obtained by digestion of product made with the soluble fraction alone.

The presence of ribosomal acceptor(s) was also suggested by Otaka and Osawa (21), who found that leucine and phenylalanine were specifically incorporated into *E. coli* "chloramphenicol particles" in a reaction that did not require complete ribosomes or GTP and was not inhibited by chloramphenicol. Incorporated residues sedimented with

the particles and retained a free α-amino group. Aminoacyl-tRNA
was an intermediate in the reaction that required soluble enzyme(s)
as well as the incomplete ribosomal particles.

We have subsequently investigated the enzymatic basis for the transfer of leucine or phenylalanine using an experimental approach similar
to that employed in characterizing the arginine-transfer reaction. We
assumed that E. coli contains a soluble enzyme(s) that catalyzed the
transfer of leucine and phenylalanine from tRNA to acceptor proteins
and reasoned that it could be purified provided an appropriate acceptor was included in the assay system. In our initial studies (22,23)
we used as an assay the direct transfer of radioactive leucine or phenylalanine from tRNA to protein. Reactions were carried out in the absence of magnesium ions, and bovine serum albumin was arbitrarily
selected as a protein acceptor. We developed a purification procedure
(23) that included extraction from protamine sulfate, gel filtration, and
chromatography on DEAE-cellulose. Soluble transfer activities for both
leucine and phenylalanine were copurified several-hundred-fold during
these operations. Their copurification, as well as their identical kinetics
of thermal inactivation, suggested that they were due to the same
enzyme protein. The requirements and inhibitors of the reaction
catalyzed by the partially purified enzyme were identical, regardless
of whether leucyl-or phenylalanyl-tRNA was used as substrate. There
was nearly an absolute dependence on the presence of albumin and
monovalent cation, and a partial requirement for a thiol compound.
Puromycin and p-hydroxymercuribenzoate were inhibitory, as were
various divalent cations, when present at concentrations greater than
10mM. There was no effect of GTP or of a number of antibiotics
(chloramphenicol, streptomycin, pactamycin, sparsomycin) known to
inhibit protein synthesis de novo.

A new method for isolating much more highly purified enzyme has
recently been devised (16). The technique consists of chromatography
on phosphocellulose, precipitation and back extraction with ammonium sulfate; gel filtration on Sephadex G–100; rechromatography
on phosphocellulose; and gel filtration on Sephadex G–75. A peculiar
feature of the enzyme after elution from the first phosphocelluose
column is its insolubility in the absence of salt. The remaining steps are
therefore carried out in the presence of 0.12 M ammonium sulfate.
The procedure exploits the fact that rechromatography on phosphocellulose is associated with a change in the enzyme that makes it much

more retarded during the subsequent gel filtration. The most purified fraction has not been adequately characterized because of its low content of protein. However, preliminary experiments have shown only one faint band on disk gel electrophoresis and have suggested that it catalyzes the transfer of at least 800 nmoles of leucine or phenylalanine per minute per milligram and is purified more than 20,000-fold with respect to the crude soluble fraction. The apparent molecular weight determined by its elution volume from the final Sephadex column was 14,000, whereas the value estimated from glycerol-gradient centrifugation was 25,000. This discrepancy may reflect unusual physical characteristics of the enzyme protein. The fact that enzyme activity in a crude extract can be precipitated with protamine, but also can be retained on phosphocellulose, and the observation that chromatography on this cation-exchange resin causes an apparent decrease in the molecular weight suggest the possibility that the enzyme is a basic protein that is originally bound to tRNA which is removed during the isolation procedure. Leucyl,phenylalanyl-tRNA~protein transferase thus resembles arginyl-tRNA~protein transferase in that the changes in its physical properties during purification are consistent with the removal of a tightly bound acidic macromolecule.

The precise site on the bovine albumin molecule to which leucine or phenylalanine is enzymatically transferred was determined by Leibowitz and Soffer (24). Incorporated radioactive residues were found to be reactive with DNFB and to comigrate with albumin during thin layer electrophoresis. A puzzling observation was made, however, when the transfer reaction was coupled to a phenylalanyl-tRNA-generating system with the objective of preparing sufficient acylated albumin from which to isolate and characterize a radioactive peptide after proteolytic digestion. Even under conditions in which the concentration of albumin was strictly limiting, the incorporation of phenylalanine corresponded to only a minor fraction of the albumin molecules. It was therefore impossible to obtain sufficient product for analysis by classical sequencing techniques. In order to circumvent this problem and to discriminate between addition of an aminoacyl residue at the NH_2-terminus or at an appropriate internal functional group, such as an ϵ-amino group of one or more lysine residues, the following experiment was carried out: A preparation of bovine-serum albumin was acylated with [^{3}H]arginine in the reaction catalyzed by arginyl--tRNA~protein transferase under such conditions that every albumin molecule contained [³H]arginine as its NH₂-terminal residue. This arginylated albumin was used as acceptor for the transfer of either [¹⁴C]leucine or [¹⁴C]phenylalanine. The rationale, which is illustrated in Figure 1, was that if the [¹⁴C]amino acids were specifically added to the NH₂-terminal [³H]arginine residue, tryptic digestion of the product should yield [¹⁴C]leucyi- or [¹⁴C]phenylalanyl [³H]arginine. Edman degradation of this doubly labeled dipeptide would be expected to result in the formation of phenylthiohydantoin [¹⁴C]leucine or phenylthiohydantoin- [¹⁴C]phènylalanine and free [³H]arginine. The structure of these dipeptides could thus be determined using quantities that would be insufficient for standard analytic techniques. The results of the experiment were those predicted for NH₂-terminal addition of the aminoacyl residue. [¹⁴C]Leucyl- and [¹⁴C]phenylalanyl- [³H]arginine were isolated from tryptic digests of the product, and their structures were proved. Unexpectedly, however, arginylated albumin was found to be a much better acceptor than was unmodified albumin, and when present in limiting concentrations, nearly molar equivalents of leucine or phenylalanine could be transferred to it.

The relationship of arginylated albumin to the natural acceptor present in albumin preparations was investigated by comparing arginylalbumin acylated with [³H]leucine and unmodified albumin acylated with [¹⁴C]leucine as schematically summarized in Figure 2. The two labels migrated similarly during electrophoresis on cellulose acetate. After tryptic digestion they were both identified as leucylarginine by paper electrophoresis and paper chromatography. Thus the acceptor in commercial preparations of albumin had the electrophoretic mobility

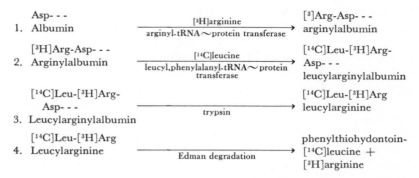

Fig. 1.   Scheme for demonstrating NH₂-terminal addition catalyzed by leucyl, phenylalanyl-tRNA∼protein transferase.

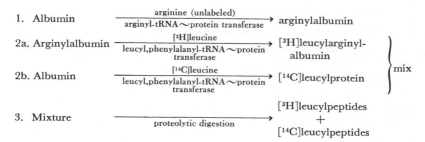

1. Albumin $\xrightarrow[\text{arginyl-tRNA} \sim \text{protein transferase}]{\text{arginine (unlabeled)}}$ arginylalbumin

2a. Arginylalbumin $\xrightarrow[\substack{\text{leucyl,phenylalanyl-tRNA} \sim \text{protein} \\ \text{transferase}}]{[^3\text{H}]\text{leucine}}$ [³H]leucylarginyl-albumin

2b. Albumin $\xrightarrow[\substack{\text{leucyl,phenylalanyl-tRNA} \sim \text{protein} \\ \text{transferase}}]{[^{14}\text{C}]\text{leucine}}$ [¹⁴C]leucylprotein

}mix

3. Mixture $\xrightarrow[\text{proteolytic digestion}]{}$ [³H]leucylpeptides + [¹⁴C]leucylpeptides

Fig. 2. Scheme for comparing acceptor protein in commercial albumin preparations with arginylalbumin.

of albumin, but contained an NH_2-terminal arginine residue to which the leucine was transferred. Furthermore, pepsin and Nagarse digestion of the two acylated protein preparations revealed identical fingerprint patterns for both radioactive isotopes. Similar results were obtained when unmodified albumin was acylated with [¹⁴C]phenylalanine and compared with enzymatically arginylated albumin to which [³H]phenylalanine had been added. These data established that leucyl,phenylalanyl-tRNA~protein transferase catalyzed the transfer of either leucine or phenylalanine into peptide linkage with the NH_2-terminal arginine residue of a fraction (5–10%) of the albumin molecules in a commercial preparation. The most likely explanation for the presence of this unsuspected arginylated fraction of albumin molecules would appear to be posttranslational NH_2-terminal addition catalyzed *in vivo* by arginyl-tRNA~protein transferase. Arginylated albumin thus probably represents one of the first examples of a naturally occurring linear protein containing a residue of its primary structure whose presence and position depend on the specificity of an enzyme, rather than a nucleic acid template.

B. AMINOACYL-tRNA SPECIFICITY

Donor specificity in reactions catalyzed by aminoacyl-tRNA~protein transferases has been studied using as substrate a mixture of 15 [¹⁴C]aminoacyl-tRNAs (1). When highly purified arginyl-tRNA~protein transferase or a crude supernatant fraction from rabbit liver was employed, the transfer of radioactivity from this mixture was completely and specifically eliminated by dilution with [¹²C]arginyl-tRNA. With purified leucyl,phenylalanyl-tRNA~protein transferase or the crude soluble fraction from *E. coli* incorporation was quantitatively abolished by dilution with [^{14}C]leucyl- and [^{12}C]phenylalanyl-tRNA. Thus there is a high degree of donor specificity in these reactions and no evidence of the existence of other transfer activities even in the respective crude soluble fractions.

The aminoacyl moiety appears to contribute more to this specificity than does the polyribonucleotide chain. Arginine, leucine, and phenylalanine can be quantitatively transferred from bulk tRNA of various organisms, which indicates that different isoaccepting species of tRNA can participate in the transfer reactions (2,15). Furthermore, leucyl, phenylalanyl-tRNA~protein transferase was found to catalyze the transfer of [^{14}C]phenylalanine from *E. coli* tRNA^{Val} that had been mischarged using *Neurospora crassa* phenylalanyl-tRNA synthetase, whereas it failed to catalyze the transfer of valine from the same polyribonucleotide donor (15). This result indicates that the nature of the aminoacyl residue can be an absolute determinant of donor specificity. One part of the aminoacyl residue that is critical is its free α-amino group. Thus acetylation of the phenylalanine residue on phenylalanyl-tRNA (15) or the arginine residue on arginyl-tRNA (25) was associated with a loss of substrate activity in the respective reactions. The importance of a free α-amino group in donor-substrate activity indicates that aminoacyl-tRNA~protein transferases cannot function as peptidyl transferases, but does not provide insight into the structural feature(s) that enable the enzymes to discriminate among different aminoacyl residues. It is worth noting in this context that leucyl, phenylalanyl-tRNA~protein transferase can utilize the unnatural substrate *p*-fluorophenylalanyl-tRNA (15).

Although these findings indicate that the aminoacyl residue is a critical determinant of aminoacyl-tRNA donor specificity, there is also evidence of recognition of a site(s) on the prolyribonucleotide moiety. For example, in studies with leucyl,phenylalanyl-tRNA~protein transferase it was found that the rate of transfer of [^{14}C]phenylalanine from *E. coli* tRNA^{Val} was slower than that from *E. coli* tRNA^{Phe} and that the 3'-pentanucleotide fragment of *E. coli* tRNA^{Phe} acylated with [^{14}C]phenylalanine did not function as substrate (15).

C. ACCEPTOR SPECIFICITY

1. Arginyl-tRNA~Protein Transferase

Specificity among protein acceptors was investigated using an assay for acceptance in which the transfer reaction was coupled with an arginyl-tRNA-generating system. Highly purified arginyl-tRNA∼protein transferase containing no endogenous acceptor was employed, and conditions were so selected that incorporation of arginine was stoichiometrically related to the concentration of added protein acceptor. The criterion used to define a true acceptor was that it stimulate the incorporation of molar equivalents of arginine under these conditions. This criterion was found to be extremely critical, since in several instances stimulation by relatively pure protein preparations was found to be due to minor contaminants, rather than to the major protein species. A large number of proteins were examined by this technique (3), and the only acceptors were bovine, human and rabbit albumins; bovine thyroglobulin; two human Bence-Jones proteins; and soybean trypsin inhibitor. These proteins had in common the fact that each was known to contain aspartic or glutamic acid as its NH_2-terminal residue. Further studies were carried out using several homogeneous γG-immunoglobulins that also possessed these residues at their NH_2-terminus (26). No acceptance was found with the native immunoglobulin molecules; however, their isolated light and heavy chains accepted stoichiometric equivalents of arginine. These data thus provided the first evidence that the NH_2-terminal region of light and heavy chains is buried within the native immunoglobulin molecule. In addition, they were consistent with the results obtained with other protein acceptors in suggesting that an accessible NH_2-terminal dicarboxylic acid residue is an important determinant of acceptor protein specificity.

Because only a limited number of pure proteins with defined NH_2-termini are available, it was necessary to develop a more systematic method for determining acceptor specificity. Dipeptides and tripeptides were therefore tested for their ability to inhibit the albumin-dependent enzymatic transfer of [^{14}C]arginine from tRNA into protein, and reaction mixtures were examined by paper electrophoresis for the presence of peptide-dependent radioactive products (4). The rationale of the kinetic experiments was that those low-molecular-weight peptides that were acceptors would compete with albumin and would inhibit incorporation into acid-insoluble material, since their arginylated derivatives would be acid-soluble. Among peptides comprising 17 different NH_2-terminal residues, only those containing aspartic acid, glutamic acid, and, to a much lesser extent, cystine were inhibitory. Of special interest was the fact that the enzyme could discriminate between NH_2-terminal glutamic acid and glutamine, which suggests its potential usefulness in determining the state of amidation of NH_2-terminal dicarboxylic acid residues. The inhibition was kinetically characterized and shown to be strictly competitive with albumin. Furthermore, the results of these experiments correlated perfectly with the presence of peptide-dependent radioactive products in the various reaction mixtures. When Glu-Ala was used as acceptor, the peptide-dependent product was isolated and its structure was established as Arg-Glu-Ala. These results thus provided systematic confirmation then an NH_2-terminal dicarboxylic acid residue to which arginine is transferred represents an absolute determinant of acceptor specificity in the reaction catalyzed by arginyl-tRNA∼protein transferase.

Other aspects of acceptor specificity were delineated by the use of low-molecular-weight model substrates. The L-stereoconfiguration appeared to be an absolute determinant of acceptor activity, since no transfer was observed to D-Glu–D-Glu. It was not determined whether this stringency applies only to the NH_2-terminal residue or to the penultimate one as well. Although all peptides containing NH_2-terminal L-glutamyl or L-aspartyl residues inhibited the albumin-dependent enzymatic transfer of arginine from tRNA to protein, a marked variation in $K_i s$ was observed in both groups, indicating that internal residues can serve as relative determinants of acceptor-substrate activity. Since these peptides inhibit by competing with albumin as acceptor in the reaction, their $K_i s$ values may be considered to provide a rough inverse measure of their affinity for the enzyme. In general, the values were considerably higher than the K_m value (25 μM) for albumin. Among nonpeptide derivatives of the dicarboxylic acids, isoasparagine and isoglutamine were found to be the best acceptors, and substrate activity appeared to be favored by a blocked α-carboxyl and a free β- or γ-carboxyl group.

2. Leucyl,Phenylalanyl-tRNA∼Protein Transferase

Similar techniques have been employed to investigate acceptor-substrate specificity in the reaction catalyzed by leucyl,phenylalanyl-tRNA∼protein transferase. Conditions were established such that true acceptors stimulated the incorporation of molar equivalents of leucine or phenylalanine, and 25 highly purified protein preparations were examined (15). Only three of these, arginylalbumin,α_{S1}-casein, and β-casein A_2, were found to accept stoichiometric quantities of amino acid. Acceptor specificity was identical with respect to the transfer of leucine and phenylalanine. These three proteins are unusual in that each contains an NH_2-terminal arginine residue (24,27,28). Radioactive leucine or phenylalanine incorporated in their presence retained its free α-amino group as judged by the DNFB reaction. Leucylarginine or phenylalanylarginine was identified as the principal radioactive peptide in tryptic digests of arginylalbumin and β-casein A_2. Similar digests of the acylated products made with α_{S1}-casein contained single radioactive peptides that could be clearly distinguished from leucylarginine or phenylalanylarginine by paper chromatography. Since the NH_2-terminal sequence of α_{S1}-casin is Arg-Pro-Lys--- (27), failure to release the dipeptide is consistent with the known resistance of the Arg—Pro bond to scission by trypsin (29), and the unidentified peptides were presumably Leu-Arg-Pro-Lys or Phe-Arg-Pro-Lys. The results of these studies with protein acceptors thus suggested that one determinant of acceptor specificity is an NH_2-terminal arginine residue to which leucine or phenylalanine is transferred.

A more systematic investigation has recently been carried out (16) using a series of dipeptides comprising 19 different NH_2-terminal residues linked to alanine. These were tested for their ability to inhibit the α_{S1}-casein-dependent enzymatic transfer of radioactive leucine or phenylalanine from tRNA into protein, and reaction mixtures containing them were analyzed by paper electrophoresis for the presence of peptide-dependent radioactive products. It was found that Arg-Ala, Lys-Ala, and, to a much lesser extent, His-Ala inhibited the incorporation of leucine or phenylalanine into protein. Kinetic analyses indicated that this inhibition was strictly competitive with α_{S1}-casein. A new radioactive product was demonstrable in reaction mixtures containing each of these peptides. By contrast, dipeptides containing the other 16 NH_2-terminal residues failed to inhibit the α_{S1}-casein-dependent enzymatic transfer of either amino acid to protein, and examination of reaction mixtures containing them did not reveal the presence of peptide-dependent radioactive products. The peptide Ala-Lys was also inactive by both criteria, which suggests that addition to an NH_2-terminal lysine residue occurs at the α-amino, rather than the ϵ-amino, group. Formal chemical proof that phenylalanine could be transferred to an NH_2-terminal lysine residue was obtained with the product of the coupled reaction made using Lys-Ala-Ala as acceptor. The peptide-dependent radioactive compound was purified from a large-scale reaction mixture and identified as Phe-Lys-Ala-Ala by classical sequencing techniques. These studies thus established that an NH_2-terminal basic amino acid residue to which leucine or phenylalanine is transferred represents an absolute determinant of acceptor specificity in the reaction catalyzed by leucyl,phenylalanyl-tRNA~ protein transferase.

A large number of dipeptides containing NH_2-terminal L-arginine or L-lysine linked to different COOH-terminal residues were tested for their ability to inhibit the α_{S1}:casein-dependent transfer of radioactive phenylalanine from tRNA to protein. All were inhibitory, and analyses of reaction mixtures containing them revealed the presence of new radioactive products. However, D-Arg–D-Val was inactive by both criteria, indicating that the L-stereoconfiguration is also an absolute determinant of acceptor specificity. The K_i value of all inhibitory dipeptides was at least 500-times higher than the K_m value for α_{S1}-casein, which was approximately 0.5 μM. In those instances in which the dipeptides differed only in whether the NH_2-terminal residue was lysine or arginine the K_i value observed with the latter was usually slightly lower. The COOH-terminal residue in each series had some influence on the K_i value, which indicates that the penultimate residue may serve as a relative determinant of acceptor-substrate activity.

D. CELLULAR FUNCTION

The biological role of aminoacyl-tRNA~protein transferases is unknown. A striking consequence of their action is the presence of a differently charged NH_2-terminal residue on the acylated acceptor. One working hypothesis is that these enzymes may be part of a system that regulates the activity of those proteins or peptides whose modification they catalyze. This hypothesis leads to the prediction that a specific mechanism, presumably enzymatic, exists for removal of the added aminoacyl residue. An equally plausible hypothesis is that aminoacyl-tRNA~protein transferases perform a biosynthetic, rather than a regulatory, function; that is, they catalyze the last step in the synthesis of specific peptides or proteins of unique function.

The role of these enzymes cannot be understood until their naturally occurring acceptor substrates are characterized. Since the specificity of the enzymes has been delineated, it is possible to predict the classes of molecules that either could serve as acceptors or may have already served. For example, proteins or peptides with an accessible NH_2-

terminal dicarboxylic acid residue could serve as acceptors in the reaction catalyzed by arginyl-tRNA∼protein transferase, whereas those containing the NH_2-terminal sequence Arg-Glu-- or Arg-Asp-- may have already served as acceptor. These considerations suggest that experiments designed to answer the following questions may prove useful in defining the cellular function of aminoacyl-tRNA∼protein transferases:

1. What is the nature and number of acceptors in a crude extract?

2. Are any or all of these acceptors acylated *in vivo*, and if so, is the fractional acylation different for different acceptors, and is it variable during the life cycle of the cell?

3. Is there a potential acceptor whose measurable biological activity is altered by enzymatic acylation, and if so does the acylated acceptor exist *in vivo*, and is there an enzyme that can specifically catalyze removal of the added residue?

4. What are the properties of mutants defective in either the transfer enzymes themselves or in the acceptor substrates whose modification they catalyze?

1. Arginyl-tRNA∼Protein Transferase

Specific protein acceptors have been identified in crude soluble extracts from sheep thyroid glands and from rabbit liver. Thyroid acceptors were studied by allowing the crude supernatant fraction to incorporate radioactive arginine and subjecting the reaction mixture to density-gradient centrifugation (8). Two major radioactive protein peaks were observed. One of these sedimented at 19S with thyroglobulin, the principal protein component of these extracts. The other was found at 4S and was probably arginylated albumin, since albumin is also known to constitute a significant proportion of thyroidal soluble protein (30). A substantial fraction of radioactive arginine incorporated by the thyroid extract into a ribonuclease-resistant product insoluble in cold trichloroacetic acid (TCA) was rendered soluble when the TCA was boiled, suggesting the presence, also, of relatively small peptide acceptors. Because of arginase activity rabbit-liver supernate could only be used in the coupled reaction at low protein concentrations. Its acceptor activity was therefore fractionated using as an assay the ability to stimulate the incorporation of arginine into a product insoluble in hot TCA in the presence of excess highly purified arginyl-tRNA∼

protein transferase and arginyl-tRNA synthetase (3). Acceptance by the crude soluble proteins was 0.13 nmole of arginine per milligram. If the average molecular weight of these polypeptide chains is assumed to be 40,000, this suggests that approximately 0.5% of them can accept arginine. Such a value is probably a low estimate, since inhibitors and degradative enzymes in the crude extract may prevent stoichiometric acylation and since low-molecular-weight polypeptide acceptors are soluble in hot TCA. A single protein accounting for more than 50% of the total acceptance was purified to homogeneity and identified as albumin. Experiments have also been reported (31) in which soluble extracts from various uninfected and virus-infected cells in tissue culture have been incubated with radioactive arginine and then analyzed by disk gel electrophoresis in sodium dodecylsulfate. In each instance the most heavily labeled protein migrated with a mobility corresponding to a molecular weight of 65,000. At least one other acceptor also appeared to be common to all the cell types tested.

Evidence has been obtained that both albumin and thyroglobulin are fractionally arginylated *in vivo*. It is worth noting in this context that the arginine-transfer reaction appears to be essentially irreversible and that attempts to detect the presence in rabbit liver of a soluble deacylating enzyme active on [^{14}C]arginylalbumin have so far been unsuccessful (9). As already described, it has been found that approximately 5-10% of the molecules of commercial preparations of bovine-serum albumin contain arginine, rather than aspartic acid, as their NH$_2$-terminal residue (24). Detection of this low level of acylation would not be possible by conventional techniques and was rendered feasible only by the specificity of leucyl,phenylalanyl-tRNA~protein transferase for basic NH$_2$-terminal residues. This fortunate circumstance has made the bacterial enzyme an important analytic reagent in estimating the percentage of potential acceptor molecules that are actually acylated in the reaction catalyzed by arginyl-tRNA~protein transferase. The technique has also been applied to commercial preparations of pure bovine thyroglobulin (32), and about 0.2 mole of basic termini was observed per mole of protein. Since thyroglobulin contains two NH$_2$-terminal aspartic acid residues, both of which can be acylated *in vitro* (13), this value suggests a fractional acylation *in vivo* of 10%.

The effect of enzymatic arginylation has been examined on two potential peptide substrates with measurable biological activities. These peptides were subjected to incubation with radioactive arginine using the transfer reaction coupled to an arginyl-tRNA-generating system under conditions developed for stoichiometric acylation of acceptors. Porcine β-melanocyte-stimulating hormone, an octadecapeptide containing NH_2-terminal aspartic acid, inhibited the transfer of [^{14}C]arginine from tRNA to albumin with a K_i of 18 μM. The product obtained when it was used as acceptor in the coupled reaction was purified and identified as the expected nonadecapeptide containing 1 mole of arginine at the NH_2-terminus. When this compound was compared with the unmodified peptide in the frog-epithelium bioassay, the two were found to possess identical potencies (33). Similar experiments were carried out with angiotensin II, a vasopressor octapeptide containing the NH_2-terminal sequence Asp-Arg— (1). This compound was a competitive inhibitor of the transfer of arginine to albumin with a K_i of about 0.5 μM, which suggests that it possesses a very high affinity for arginyl-tRNA~protein transferase. The peptide-dependent radioactive product resulting when it was stoichiometrically arginylated in the coupled reaction was purified and identified as [^{14}C]Arg-Asp. Since free aspartic acid is a very poor acceptor in this reaction, a plausible explanation for this unexpected result is that arginylation of angiotensin II renders its Asp—Arg bond susceptible to hydrolysis by a protease contaminating either the preparation of transfer enzyme or of arginyl-tRNA synthetase. In view of the fact that many important biologically active peptides arise by proteolysis of precursors, a modification of this type that renders a protein or peptide susceptible to specific enzymatic cleavage at another site may be of considerable interest.

2. Leucyl,Phenylalanyl-tRNA~Protein Transferase

As could be anticipated from the work of others (20,21), ribosomes that had been washed with 1 M NH_4Cl were found to accept substantial quantities of radioactive leucine or phenylalanine in the reaction catalyzed by purified leucyl, phenylalanyl-tRNA~protein transferase (1,34). With ribosomes from E. coli B this acceptance appeared to be largely due to a single protein on the 30S subunit whose molecular weight in the reduced denatured form was estimated to be 12,000 by disk gel electrophoresis in sodium dodecylsulfate (34).

Recently a mutant lacking leucyl,phenylalanyl-tRNA~protein transferase has been isolated (35). Since this mutant presumably contains no acceptors that have been acylated *in vivo*, a comparison of acceptance by fractions derived from it and from the wild type may provide information as to which proteins are acylated *in vivo*. Approximately 0.5 nmole of phenylalanine per milligram of protein was incorporated into a ribonuclease-resistant product insoluble in cold TCA when the mutant soluble proteins were used as acceptor in the coupled system. Thus at least 2% of the polypeptide chains are potential acceptors if 40,000 is taken as an average molecular weight for these chains. Acceptance by the wild-type soluble proteins was usually about one-third lower, although some variability was observed with these crude extracts. If the difference is due solely to acylation with leucine or phenylalanine *in vivo* (i.e., if the fraction of total protein that can be or has been acylated is identical in the two strains), the results suggest that a significant percentage of the potential soluble acceptor molecules are acylated *in vivo*. When these labeled soluble fractions were analyzed by disk gel electrophoresis, it was found that the proportion of incorporated radioactivity in the radioactive proteins was somewhat different in the two extracts. Although the wild-type extract was not completely lacking in any of the labeled proteins, several of them were less radioactive relative to the mutant proteins than were the others. This finding suggests that specific potential acceptors may be preferentially acylated *in vivo*. In contrast to the results with the soluble fraction, intact ribosomes from the mutant and wild type accepted similar quantities of phenylalanine although considerable variation (0.4–1.1 nmoles/mg protein) was found among different preparations from each strain. The existence of the mutant makes feasible a study of the effect of enzymatic acylation of the 30S particle on its function in protein synthesis *in vitro*, since in such a study it is critical that the initial degree of acylation of the particle be known and that it be low relative to the increment obtained as a result of the enzymatic reaction.

The mutant lacking leucyl,phenylalanyl-tRNA~protein transferase is a K12 F⁻ strain that was isolated after mutagenesis with nitrosoguanidine (35). It contains no detectable activity ($<0.3\%$ wild type) for the transfer of either leucine or phenylalanine which, confirms that a single protein is responsible for both activities. It is viable at 37°C, which indicates that the enzyme is not required for the survival of the organism. Preliminary mapping of the mutation has been carried out using *18 F'* strains whose episomes encompassed the entire *E. coli* genome. Only one strain, that carrying episome F142, was found to transduce the ability to produce the enzyme. There was cotransduction of the two activities. Genetic material carried by this episome is located between markers *tyrA* and *dsdA* on the circular *E. coli* chromosome (36). The episome is known to be unstable, and transductants tended to lose their capacity for enzyme production.

One important phenotypic difference from the wild type has been observed in the mutant. When grown into stationary phase in minimal medium and then resuspended in fresh medium, it exhibits a lag period of as much as 8 hr before resuming growth. The mechanism responsible for this delay is not currently understood. Although the mutant and wild type double at the same rate, the selective pressure against the mutant because of the lag has been used to isolate revertants that have the same growth characteristics as the wild type. These revertants were found to have regained activities for the transfer of leucine and phenylalanine, which suggests that the enzyme may be involved in a fundamental regulatory process. Further work on the physiologic defect in the mutant is required to determine its biochemical basis and to establish the biological function of leucyl, phenylalanyl-tRNA~protein transferase in *E. coli*.

III. Aminoacyl-tRNA~Phosphatidylglycerol Transferases

A. THE ENZYMES

The occurrence of aminoacyl derivatives of phosphatidylglycerol in bacteria was first reported by Macfarlane in 1962 (37). Lennarz and collaborators have investigated the biosynthesis of these compounds and have demonstrated the presence of enzymes that catalyze the transfer of lysyl (38) and alanyl (39) residues from tRNA into ester linkage with the glyceryl moiety of phosphatidylglycerol. These investigators have called the enzymes lysyl (40) and alanyl (41) phosphatidylglycerol synthetase, respectively. They can also be considered as aminoacyl-tRNA~phosphatidylglycerol transferases in accord with the system of nomenclature used in this chapter.

1. Lysyl-tRNA~Phosphatidylglycerol Transferase

In 1966 Lennarz, Nesbitt, and Rice (38) provided evidence that extracts of *Staphylococcus aureus* could carry out the synthesis of lysyl-phosphatidylglycerol, an aminoacyl phospholipid known to occur in this organism (37), and that lysyl-tRNA was an intermediate in this enzymatic process. They showed that a particulate fraction sedimenting between 15,000 and 105,000 g catalyzed the almost quantitative transfer of [^{14}C]lysine from either *S. aureus* or *E. coli* tRNA into material that could be extracted with a 2:1 mixture of $CHCl_3$ and CH_3OH. The reaction appeared to be specific with respect to the transfer of a lysyl residue, since lysine was the only radioactive amino acid found after acid hydrolysis of the extracted product, which had been made using a nixture of 14 radioactive aminoacyl tRNAs as substrate (39). The radioactive product comigrated with lysylphosphatidylglycerol during column chromatography on DEAE-cellulose, during thin-layer chromatography in two solvent systems, and during paper chromatography. Radioactive lysine was generated from it by mild alkaline, as well as acid, hydrolysis, and it gave rise to [^{14}C]lysine hydroxamate when subjected to treatment with hydroxylamine. On the basis of these observations it was identified as O-L-lysylphosphatidylglycerol (Fig. 3).

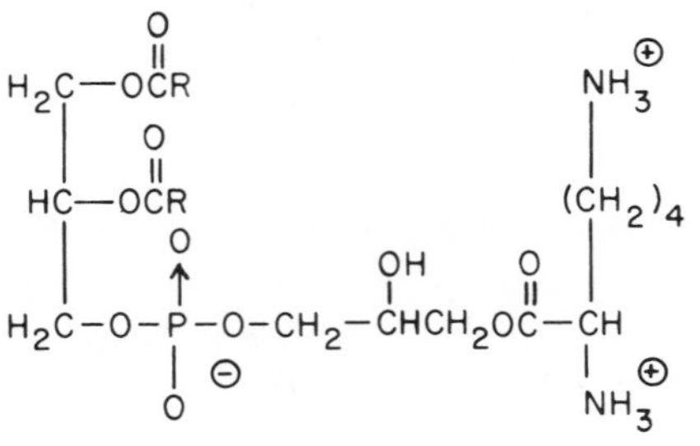

Fig. 3. Structure of O-L-lysylphosphatidylglycerol.

When the active particles were extracted with organic solvents, a marked dependence of the reaction on added phosphatidylglycerol could be demonstrated and there was little or no requirement for anionic surfactants, as compared with the unextracted enzyme preparation (42). The optimal pH for the transfer reaction was 6.9, and the ionic strength of the reaction mixture was found to be critical, although not ion-specific (40). There was no specific requirement for magnesium ions or added sulfhydryl-containing compounds; however, sulfhydryl-reactive reagents inhibited the reaction, and dithiotheritol reversed this inhibition (40). The transfer of [14C]lysine from tRNA to phosphatidylglycerol was not affected by the presence of chloramphenicol or puromycin (40), nor was it diluted by excess [12C]-L-lysine. Preincubation with pancreatic ribonuclease abolished the substrate activity of lysyl-tRNA, and this enzyme also prevented incorporation from free [14C] lysine, which occurred when ATP, magnesium ions, and crude soluble fraction containing lysyl-tRNA synthetase were included in the reaction mixture (38). These results thus established that lysyl-tRNA, in addition to its role in protein synthesis, could also serve as aminoacyl donor in the enzymatic formation of the naturally occurring O-L-lysyl ester of phosphatidylglycerol.

2. Alanyl-tRNA∼Phosphatidylglycerol Transferase

A similar particulate fraction from *Clostridium welchii* was found to catalyze the transfer of [14C]lysine and [14C]alanine from their respective tRNAs into material that was extractable with CHCl$_3$-CH$_3$OH (39,41). When a mixture of 14 radioactive aminoacyl-tRNAs was used as substrate, lysine and alanine were the only radioactive amino acids found in the acid hydrolysate of the extracted product. This specificity was not unexpected, since both lysyl and alanyl phosphatidylglycerol had been reported as constituents of this organism (37). Because of the particulate nature of the enzyme preparation, it was not established whether the two catalytic activities were due to the same or different enzyme proteins. The product made using radioactive alanyl-tRNA as substrate was identified as L-alanylphosphatidylglycerol by the criteria that it migrated identically with the synthetically prepared compound in three different thin-layer-chromatography systems, that it gave rise to alanine in the L-configuration after acid hydrolysis, and that it yielded radioactive alanine hydroxamate after treatment with hydroxylamine. The enzymatic transfer of alanine from tRNA to phosphatidylglycerol proceeded optimally at pH 5.7 and 0.125 M KCl. There was a partial requirement for added phosphatidylglycerol. No transfer occurred if the particles were boiled or if the alanyl-tRNA was preincubated with pancreatic ribonuclease. Preincubation of the enzyme preparation with ribonuclease, which was subsequently removed by treatment with bentonite, did not result in a loss of activity, which suggests that neither ribosomes nor an RNA template participated in the reaction.

Gould and Lennarz (39) also examined particulate fractions from various other bacteria for their ability to catalyze the transfer of radioactive aminoacyl residues from tRNA into material that could be extracted with $CHCl_3$–CH_3OH. Reactions were carried out at pH 7.0 in the presence of phosphatidylglycerol. No transfer of radioactivity from either lysyl-tRNA or a mixture of 14 aminoacyl-tRNAs was observed with extracts from *E. coli*, *Micrococcus leisodeikticus* or *Sarcina lutea*. Particles obtained from *Bacillus megaterium* and *Bacillus cereus* were active with either substrate, and acid hydrolysates of the lipid product made in the presence of the aminoacyl-tRNA mixture contained only radioactive lysine. A preparation from *Streptococcus faecalis* catalyzed the transfer of radioactivity from the aminoacyl-tRNA mixture into a lipid product from which radioactive arginine and lysine were generated by acid hydrolysis. At a lower pH (5.0–5.5) small amounts of two other radioactive aminoacyl lipids could be detected, and the aminoacyl residue of one of these was tentatively identified as alanine. The formation of arginyl lipid also appeared to be favored at the lower pH and could be demonstrated using arginyl-tRNA as the sole radioactive substrate.

B. AMINOACYL-tRNA SPECIFICITY

As already noted, aminoacyl-tRNA~phosphatidylglycerol transferases were found to be highly specific with respect to the residues whose transfer they catalyzed from a mixture of aminoacyl-tRNAs (39). This specificity appears to depend on both the aminoacyl and polyribonucleotide moieties of the donor substrate.

1. *Lysyl-tRNA~Phosphatidylglycerol Transferase*

Donor specificity in the reaction catalyzed by the particulate enzyme from *S. aureus* was investigated by exploiting the properties of the lysine analog *S*-β-aminoethylcysteine (43). This compound was prepared chemically from [^{14}C]cystine and then used with bulk *E. coli* B tRNA for the enzymatic synthesis of *S*-β-aminoethyl-[^{14}C]cysteinyl-tRNA^{Lys}. *S*-β-Aminoethyl-[^{14}C]cysteinyl-tRNA^{Cys} was made by chemical amino-ethylation of *E. coli* B [^{14}C]cysteinyl-tRNA. These aminoacyl-tRNAs were compared with *E.coli* B [^{14}C]lysyl-tRNA^{Lys} as donors in the enzymatic formation of aminoacyl phosphatidylglycerol (40). The most striking finding was that *S*-β-aminoethylcysteine could be enzymatically transferred to lipid from tRNA^{Lys}, but not from tRNA^{Cys}. Appropriate control experiments indicated that the *S*-β-aminoethylcysteinyl-tRNA^{Cys} preparation did not contain an inhibitor and that its structure had not been altered in such a way as to affect its recognition by *E. coli* cysteinyl-tRNA synthetase or its ability to participate in cysteine codon-dependent polypeptide synthesis catalyzed by an extract of *E. coli*. These results thus indicated the existence of an absolute determinant of donor-substrate specificity that was present in the polyribonucleotide chain of *E. coli* B tRNA^{Lys}, but absent from that of *E. coli* B tRNA^{Cys}.

Both lysine and *S*-β-aminoethylcysteine could be almost quantitatively transferred from *E. Coli* tRNA^{Lys} to lipid. The two substrates exhibited similar K_m values in the reaction, and the product made with *S*-β-aminoethylcysteinyl-tRNA^{Lys} was found to cochromatograph with endogenous lysylphosphatidylglycerol and to yield only *S*-β-aminoethylcysteine after acid hydrolysis. One difference was noted, however, in the utilization of the two substrates. The V_{max} observed for the transfer of lysine was approximately double that found for *S*-β-aminoethylcysteine. This observation suggests that, in addition to recognizing a specific polyribonucleotide chain, the enzyme can also distinguish the difference between lysine and *S*-β-aminoethylcysteine esterified to this chain.

2. Alanyl-tRNA∼Phosphatidylglycerol Transferase

Gould et al (41) studied donor specificity in the alanyl-tRNA∼ phosphatidylglycerol transfer reaction catalyzed by particles extracted from *C. welchii* and provided evidence of recognition by the enzyme of both the aminoacyl and polyribonucleotide moieties.

Specificity for the alanine residue was shown by the fact that phenyl-alanyl-tRNA^{Ala}, prepared with *E. coli* tRNA^{Ala} and phenylalanyl-tRNA synthetase from *N. crassa* (44), was not a substrate in the reaction. The effect of blocking the α-amino group of the alanine residue was investigated by using [¹⁴C]alanyl-tRNA that has been acetylated to varying extents with *N*-acetoxysuccinimide. Reactions were carried out in the presence of excess enzyme, so that the transfer of alanine was essentially quantitative. The formation of radioactive lipid from the various preparations was found to be inversely proportional to the percentage of their alanine residues that had been acylated. Only radioactive alanine was found when the lipid products were hydrolyzed under conditions that had been established not to cause scission of the acetyl–alanine linkage. Radioactive lactyl-tRNA prepared by the action of sodium nitrite on alanyl-tRNA was also found to be completely inactive as a donor. That its lack of activity was due to alteration of the aminoacyl residue was indicated by the fact that it could be deacylated, recharged with [¹⁴C]alanine, and shown to participate in the reaction. These data established that the nature of the aminoacyl residue and the presence of a free α-amino group are absolute determinants of donor specificity in the reaction catalyzed by alanyl-tRNA$\sim$phsophatidylglycerol transferase.

The contribution of the ribonucleotide chain to donor specificity was demonstrated (41) by using [¹⁴C]alanyl-tRNA^{Cys} made from [¹⁴C] cysteinyl-tRNA by reduction with Raney nickel. Preparations containing as much as 50% of their radioactivity as alanine were inactive as donor under conditions in which 90% of [¹⁴C]alanine from alanyl-tRNA^{Ala} could be transferred to lipid. Control experiments indicated that these inactive preparations did not contain an inhibitor and could function as an alanyl donor in the cysteine codon-dependent polypeptide synthesis catalyzed by an *E. coli* extract. Additional evidence for recognition by the enzyme of the polyribonucleotide moiety was obtained when [¹⁴C]alanyl-tRNA was subjected to limited digestion with T_1 ribonuclease. The treated tRNA retained its ability to function as substrate, but this activity was markedly heat-labile when compared with the undigested control. Furthermore, when the partially digested preparation was subjected to gel filtration on Sephadex G–75, virtually all of the [¹⁴C]alanine was eluted in a fragment estimated to be about one-half the molecular weight of alanyl-tRNA, and radioactivity could not be enzymatically transferred from this molecule to lipid.

C. ACCEPTOR SPECIFICITY

Specificity among potential lipid acceptor substrates was investigated (42) with the particulate enzyme fraction from *S. aureus* after extraction with organic solvents to remove endogenous lipids. The criterion of acceptance was stimulation by the added lipid of the enzymatic transfer of radioactivity from [^{14}C]lysyl-tRNA into a product that was extractable with $CHCl_3$–CH_3OH. Compounds tested as acceptors were present in more than a thousandfold molar excess, compared with [^{14}C]lysine as lysyl-tRNA. Under these conditions the best substrate, phosphatidylglycerol isolated from *S. aureus*, stimulated the transfer of lysine about tenfold above the endogenous value. This transfer represented a substantial fraction of the lysyl residues, but corresponded to acylation of less than 0.05% of the added lipid molecules. Stimulatory activity was highly specific for phosphatidylglycerol since phosphatidic acid, phosphatidylethanolamine, phosphatidylcholine, phosphatidylethyleneglycol, and cardiolipin were completely inactive. Several derivatives of phosphatidylglycerol also failed to stimulate the transfer of lysine to lipid. These included phosphatidyl-(rac)-1'–glycero-3'–phosphate, diphosphatidylglycerol, the isopropylidene derivative, and lysylphosphatidylglycerol itself. The most significant finding was that phosphatidyl-(2'-deoxy)glycerol showed considerable activity, whereas phosphatidyl-(3'-deoxy)glycerol did not. These data with defined lipids indicated that acceptor-substrate specificity is related to the presence of a free primary hydroxyl group in phosphatidylglycerol and therefore suggested that the position of enzymatic esterification of phosphatidylglycerol is the 3'-hydroxyl, rather than the 2'-hydroxyl, group.

D. CELLULAR FUNCTION

The fact that particles from a number of organisms, including *S. aureus*, *C. welchii*, and *B. megaterium*, specifically catalyze (39) the transfer of the aminoacyl residues known to be present in their aminoacyl phospholipids (37,45) from a mixture of aminoacyl-tRNAs to phosphatidylglycerol strongly implicates the responsible enzymes in the biosynthesis of aminoacylphosphatidylglycerol.

The function of aminoacylphosphatidylglycerol is not known. In *S. aureus* phosphatidylglycerol and its lysyl derivative are the two major lipids. The proportion of phosphatidylglycerol containing lysine has been found to be dependent on the pH of the growth medium (46,47) due to a relative decrease of unacylated phosphatidylglycerol under more acidic environmental conditions (47). It has been suggested that the resulting more positively charged membrane lipids might facilitate growth at lower pH by minimizing proton entry into the cell (46). However, factors other than pH have been found to be associated with alterations of the ratio of unmodified and acylated phosphatidylglycerol in *S. aureus* (47,48), and a similar pH-dependent shift in this ratio was not observed in all bacterial species (49).

The fact that the aminoacyl residue is linked to phosphatidylglycerol by a relatively high-energy bond has suggested the possibility that aminoacylphosphatidylglycerol might serve as an aminoacyl-group donor in a biosynthetic reaction. However, studies with *S. aureus in vivo* (47,48), though demonstrating a metabolic turnover of the lysyl group, have indicated that most of it enters the free lysine pool. The observed rate of turnover of lysylphosphatidylglycerol was much less than the rate of entry of lysine into the cell (47), thus rendering unlikely also the possibility that lysylphosphatidylglycerol serves as a carrier of lysine during its transport into the cell. The enzymatic basis for the turnover of aminoacyl residues of aminoacylphosphatidylglycerol is unknown, and the possibility exists that specific aminoacylphosphatidylglycerol deacylases may participate in regulating the extent of phosphatidylglycerolacylation.

No mutants have been described that have specifically lost the ability to acylate phosphatidylglycerol. Isolation and characterization of such mutants would appear to be a prerequisite in delineating the physiological function of aminoacyl phospholipids.

IV. Aminoacyl-tRNA~N—Acetylmuramyl–Pentapeptide Transferases*

Elucidation of the structure and biosynthesis of the bacterial cell wall, though still incomplete, is clearly one of the major accomplishments of contemporary biochemistry. This important subject has been reviewed elsewhere (52,53). This chapter is concerned only with the tRNA-dependent biosynthesis of interpeptide bridges, the amino acid

---

*The abbreviations used in this section are as follows: UDP–GlcNAc, uridine diphospho-N-acetylglucosamine; UDP–MurNAc–pentapeptide, uridine diphospho-N-acetylmuramyl-L-alanyl-D-glutamyl-L-lysyl-D-alanyl-D-alanine; MurNAc–pentapeptide, N-actylmuramyl-L-alanyl-D-glutamyl-L-lysyl-D-alanyl-D-alanine; Glc-NAc–MurNAc(–pentapeptide)–P-P-phospholipid, disaccharide-pentapeptide lipid intermediate (50). MurNAc(–pentapeptide)–P-P-phospholipid, monosaccharide–pentapeptide lipid intermediate (50). The C$_{55}$ isoprenylpyrophosphate lipid moiety in these compounds has been partially characterized (51).

sequences that link the ε-amino group of the L-lysine residue in one tetrapeptide to the carboxyl group of the D-alanine residue in a tetrapeptide located on another strand of peptidoglycan (Fig. 4).

In 1964 Chatterjee and Park (54) obtained a particulate fraction from *S. aureus* that catalyzed the incorporation of glycine into an alkali-resistant, perchloric acid-insoluble product. The reaction depended on the presence of UDP–GlcNAc and UDP–MurNAc–pentapeptide, the nucleotide precursors of peptidoglycan. Pancreatic ribonuclease inhibited the incorporation of glycine, but not of MurNAc-pentapeptide. These critical observations suggested that a species of RNA participated in the biosynthesis of the oligoglycine interpeptide bridge known to exist in this organism.

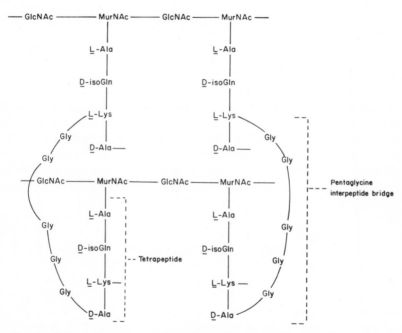

Fig. 4. Schematic representation of two linear strands of peptidoglycan in *Staphylococcus aureus*. Amidation of D-glutamic acid to give an isoglutaminyl residue and transpeptidation with elimination of the COOH-terminal D-alanine residue from the MurNAc–pentapeptide moiety of nascent peptidoglycan are not considered in this chapter. Adapted from Strominger (53).

This reaction (55) and similar reactions in other bacteria (56–59) have subsequently been extensively investigated primarily by Strominger and his collaborators. It has become apparent that the process by which a sequence of aminoacyl residues is specified in the biosynthesis of the interpeptide bridge differs from that occurring in protein synthesis. The concept has emerged that these structures are synthesized by the sequential enzymatic transfer of aminoacyl residues from tRNA starting with acylation of the ε-amino group of the lysine residue in a MurNAc-pentapeptide-containing precursor of peptidoglycan and continuing by addition at the NH$_2$-terminus of the growing peptide chain. The specificity of the responsible transfer enzymes is thought to be important in establishing the sequence of aminoacyl residues, and there is no evidence of the participation of a nucleic acid template in the mechanism by which this is achieved.

A. THE ENZYMES

1. Particulate Aminoacyl-tRNA~GlcNAc–MurNAc(–pentapeptide)–P–P–Phospholipid Transferases

In most of the bacterial species for which evidence exists that aminoacyl-tRNA is an intermediate in biosynthesis of interpeptide bridges, the responsible transferase(s) has been found in a fraction sedimenting between 15,000 and 100,000 g (56–58,60). This fraction also contains all the activities and cofactors required to generate peptidoglycan from added uridine nucleotide precursors, as summarized in Figure 5. In studies on amino acid incorporation catalyzed by these particles in the presence of added nucleotide substrates, reaction mixtures have been analyzed by paper chromatography in a 5:3 mixture of isobutyric acid and 1 N NH$_4$OH, which resolves nucleotide substrate, lipid intermediate, and peptidoglycan (60). The rate, extent, and distribution of incorporated radioactivity under these circumstances is not a function of the activity of a single enzyme. In certain cases extracted GlcNAc–MurNAc(–pentapeptide)–P–P–lipid has been used as acceptor, and transfer of amino acids to this compound has been studied directly in the absence of added nucleotide substrate. Even under these conditions, however, the transfer of an aminoacyl residue from tRNA to acceptor is not an isolated reaction since the particulate enzyme preparation contains activities that convert both the unacylated and acylated acceptor to peptidoglycan.

1. UDP–MurNAc–pentapeptide + P-phospholipid $\rightleftharpoons$ UMP + MurNAc(–penta-peptide)–P-P-phospholipid

2. UDP–GlcNAc + MurNAc(–pentapeptide)-P-P-phospholipid $\rightleftharpoons$ UDP + Glc–MurNAc(–pentapeptide)–P-P-phospholipid

3. GlcNAc–MurNAc(–pentapeptide)–P-P-phospholipid + aceptor → GlcNAc–MurNAc(–pentapeptide)–acceptor (nascent peptidoglycan) + P-phospholipid + P_i

Fig. 5. Sequence of reactions catalyzed by the particulate fraction of *Staphylococcus aureus* in the biosynthesis of peptidoglycan from uridine-nucleotide precursors. From Anderson et al. (50).

Most studies on particulate transferases have utilized only the amino acid(s) known to be present in the interpeptide bridge of the organism under investigation. Since specificity with respect to the transfer of residues from a mixture of aminoacyl-tRNAs has not been reported, these activities will be considered in the context of the bacterium from which they were obtained.

a. *Staphylococcus aureus*. Matsuhashi, Dietrich, and Strominger (55,60) investigated the enzymatic basis for the participation of RNA in the biosynthesis of the pentaglycine interpeptide bridge (61) in *S. aureus*. They studied the incorporation of radioactivity from [³H] glycine and UDP–MurNAc–pentapeptide labeled with either [¹⁴C]-L-lysine or [¹⁴C]-D-alanine. The presence of UDP–GlcNAc was necessary for the incorporation of both isotopes into peptidoglycan, but less so for that into lipid intermediate, reflecting the fact that only the complete disaccharide–pentapeptide unit (Glc-NAc–MurNAc–pentapeptide) can be transferred from the phospholipid carrier (50) to peptidoglycan primer. The incorporation of glycine into lipid intermediate and peptidoglycan required UDP–MurNAc–pentapeptide and a glycyl-tRNA-generating system or glycyl-tRNA itself. It was inhibited by pancreatic ribonuclease, but was unaffected by the presence of chloramphenicol, puromycin, the glycyl analog of puromycin, or GTP. Particles washed in the absence of magnesium ions retained their catalytic activity, which suggests that ribosomes were not a necessary component of the enzyme preparation. These results indicated that glycyl-tRNA was an intermediate in the biosynthesis of the interpeptide bridge by a mechanism distinct from that occurring in protein synthesis.

The effect of various antibiotics known to inhibit cell-wall synthesis *in vivo* on the simultaneous enzymatic incorporation of [^{3}H]glycine and MurNAc–[^{14}C]pentapeptide was also examined (60). Ristocetin and vancomycin were found to inhibit incorporation of both isotopes into peptidoglycan and at much higher concentrations into lipid intermediate as well. The incorporation of glycine into lipid intermediate was more sensitive to these agents than was that of MurNAc–pentapeptide. However, the demonstration of a direct effect of the antibiotics on the isolated enzymatic transfer of glycine to acceptor was not feasible with this unfractionated system.

Analysis of doubly labeled peptidoglycan containing radioactive glycine and MurNAc–pentapeptide complemented the biosynthetic studies in suggesting that the two compounds had been incorporated into the same product. Both isotopes were rendered chromatographically mobile by the action of egg-white lysozyme, and their ratio in fragments obtained as a result of treatment with this enzyme was approximately the same as that found in the undigested peptidoglycan.

[^{14}C]Glycine and UDP–MurNAc–pentapeptide containing a [^{14}C]-L-lysine residue were used as substrates for the enzymatic synthesis of peptidoglycan, which was then analyzed by the DNFB technique (60). A fraction of the ε-amino groups of lysine became substituted in the absence of added glycine, presumably due to endogenous glycine in the enzyme preparation. When a correction was made for this, it was determined that 1 mole of lysine ε-amino group became substituted per mole of added DNFB-reactive glycine. A rough estimate of six glycine residues per chain was obtained from the ratio of total incorporated glycine to glycine that remained reactive with DNFB. No evidence was obtained of the formation of closed polyglycine bridges of the type found *in vivo*. Similar studies were carried out on a degradation product of the labeled lipid intermediate, which could be eluted from paper chromatograms after heating to 100°C for 10 min at pH 4. An average chain length of four to six glycine residues was calculated from the molar ratio of total incorporated glycine to lysine that became substituted as a result of the presence of glycine in the reaction mixture. The presence of radioactive oligoglycine sequences in both the peptidoglycan and lipid intermediate was also demonstrated by using a peptidase from *Myxobacterium* known to cleave the interpeptide bridge in such a way as to release triglycine and tetraglycine (62). Treatment of either enzymatic product yielded the expected peptides.

These studies established that the site of initiation of synthesis of the oligoglycine interpeptide bridge was the ε-amino group of the lysine residue in MurNAc–pentapeptide. Since various attempts to demonstrate glycine peptides linked to tRNA were unsuccessful, the most likely mechanism of synthesis appeared to be the sequential transfer of single glycine residues from tRNA. Because of the insoluble nature of the enzyme preparation, it could not be determined whether more than one catalytic transfer activity was required in this process.

More direct evidence for the sequential transfer of glycine residues from tRNA was obtained independently by Thorndike and Park (63) and by Kamiryo and Matsuhashi (64). Thorndike and Park incubated a particulate fraction with UDP–MurNAc–pentapeptide and UDP–GlcNAc under conditions in which the principal product was GlcNAc-MurNAc(–pentapeptide)–P-P-phospholipid. The particles were then resiolated by centrifugation and used as a source of enzyme and acceptor for the transfer of [^{14}C]glycine in the presence of a glycyl-tRNA-generating system. After incubation of 15°C for various times, they extracted the lipid intermediate, hydrolyzed it in dilute acid to release the disaccharide–pentapeptide to which the glycine residues were attached, and determined the average chain length of the oligoglycine from the ratio of total glycine to that which was reactive with DNFB. They observed an increase in average chain length as a function of the time of incubation.

Kamiryo and Matsuhashi (64) studied the incorporation of [^{3}H]-glycine and [^{14}C]glycine into purified GlcNAc–MurNAc(–pentapeptide)–P-P-phospholipid catalyzed by the particulate enzyme and a glycyl-tRNA-generating system. They analyzed the disaccharide-pentapeptide-oligoglycine isolated from lipid intermediate after prolonged incubation with one isotope, followed by pulse-labeling with the second. Sequential Edman degradation showed that the isotope used for pulse labeling was preferentially located at the NH$_2$-terminal end of the polyglycine chains. Therefore they concluded that elongation occurs at this end, that is, the direction of elongation is precisely the opposite of that in protein synthesis. They also analyzed products containing [^{14}C]glycine that had been incorporated during short (10 min) or long (180 min) incubations. The average chain lengths, as determined by the ratio of total glycine to DNFB-reactive glycine, were, respectively, 3.0 and 5.2. The frequency of different chain lengths in the two preparations was estimated by performing six cycles of Edman degradation.

Triglycine was found to predominate in the product synthesized during the short incubation, whereas that made over a prolonged interval consisted mainly of pentaglycine and tetraglycine. These data thus indicated that the smaller chains were intermediates in the biosynthesis of longer chains that resulted from subsequent addition of glycine residues. Further evidence that chain growth occurred sequentially, starting at the ϵ-amino group of the lysine residue, was obtained by analysis of a product made using [^{3}H]glycine and GlcNAc–MurNAc (–pentapeptide)–P–P-phospholipid containing [^{14}C]lysine. After incorporation of glycine the lysine residue became unreactive with Edman reagent. The frequency with which it was rendered reactive during repeated cycles of Edman degradation was consistent with the distribution of chain lengths estimated by release of phenylthiohydantoin-[^{3}H]glycine during these cycles.

Kamiryo and Matsuhashi (65) have also recently begun to purify glycyl-tRNA~GlcNAc–MurNAc(–pentapeptide)–P–P-lipid transferase from *S. aureus*. They used as an assay either the direct transfer of [^{14}C]glycine from tRNA to purified disaccharide-pentapeptide lipid intermediate or incorporation from free glycine in the presence of a glycyl-tRNA-generating system. Activity was solubilized from particles by treatment with 0.6 M ammonium sulfate and further resolved by extraction from calcium phosphate gel. The RNA content of the soluble preparation was only 1%, and the molecular weight of its activity was tentatively estimated to be about 200,000 by sucrose-gradient centrifugation. These results suggest that ribosomes, which constitute a substantial part of crude particulate fractions, are not actually involved in the synthesis of the oligoglycine bridge. The time course for the incorporation of glycine into lipid intermediate in the presence of soluble enzyme showed the same kinetics and plateau level as had been found for the reaction catalyzed by the particulate fraction, which did not stop until an average chain length of 5.2 glycine residues had been synthesized. Thus no evidence was obtained at this stage of fractionation of the participation of more than a single catalytic activity in the sequential transfer of multiple residues.

b. *Micrococcus roseus*. Roberts, Strominger, and Söll (56) have studied the biosynthesis of the interpeptide bridge in *Micrococcus roseus*. They used a strain in which the sequence of the bridge had been established as (D-Ala)–L-Ala–L-Ala–L-Ala–L-Thr (N$^\epsilon$-L-Lys) (66). A particulate fraction was found to catalyze the incorporation of radio-active threonine into lipid intermediate and peptidoglycan. Threonyl-tRNA could replace tRNA, ATP,·and a source of threonyl-tRNA syn-thetase as substrate. Both UDP–MurNAc and UPD–GlcNAc were required for incorporation into peptidoglycan, which was inhibited by pancreatic ribonuclease, but not by puromycin. By the use of [^{14}C]threo-nine and UDP–MurNAc–pentapeptide containing [^{14}C]lysine as sub-strates, it was estimated that 0.3 residue of threonine was incorporated into peptidoglycan per residue of MurNAc–pentapeptide. That both compounds were present in the same structure was indicated by the fact that the two major lysozyme degradation products obtained from it yielded both radioactive amino acids after acid hydrolysis. Of con-siderable interest was the finding that a similar particulate preparation from a strain of *M. roseus* lacking threonine in its interpeptide bridge failed to catalyze the incorporation of this amino acid.

A series of reactions was carried out in which L-threonine and UDP–MurNAc-pentapeptide were present as radioactive substrates. Lipid intermediates were extracted from the product, and after mild acid hydrolysis the disaccharide derivative with its attached peptide was isolated by paper electrophoresis and analyzed by the DNFB technique. When the reaction was carried out in the presence of L-threonine, all of the lysine residues became unreactive with DNFB, whereas two-thirds of the threonine residues remained reactive. When unlabeled L-alanine was also included in the reaction mixture, all of the radioactive threo-nine became unreactive, suggesting that its amino groups were sub-stituted by the alanine. The question of whether L-alanine could be transferred to lipid intermediate and peptidoglycan, and if so, whether its incorporation required prior acylation of the ε-amino group of the L-lysine residue with L-threonine, was not resolved because an extensive ATP, tRNA-dependent incorporation of [^{14}C]L-alanine occurred in the absence of added uridine nucleotide substrates. The product of this reaction was not solubilized by treatment with lysozyme, and its nature has not been established

c. *Staphylococcus epidermis.* Petit, Strominger, and Söll (57) examined a strain of *S. epidermis*, the cell wall of which contained 1.4 serine and 3.1 glycine residues per residue of glutamic acid. It had been established that no serine residues were attached directly to the ε-amino group of lysine, nor to another serine residue (67). A particulate fraction was found to catalyze the nucleotide substrate, tRNA-dependent incorporation of both amino acids into lipid intermediate and peptidoglycan. When UDP–GlcNAc was omitted from reaction mixtures, large amounts of the two amino acids were found in a product thought to be acylated UDP–MurNAc–pentapeptide. It was not established whether the different aminoacyl residues were attached to the same or different molecules of nucleotide substrate. The acylated nucleotide was thought to derive from reversal of the reaction in which UDP–MurNAc-pentapeptide is converted to MurNAc(–pentapeptide)–P-P-lipid, rather than from direct acylation of UDP–MurNAc–pentapeptide.

Incorporation of radioactivity from the free amino acids was studied using aminoacyl-tRNA-generating systems containing synthetases that had been resolved free of each other. The relative incorporation of the two amino acids was variable, depending on the amount of the respective synthetases. When a crude supernatant fraction was used, approximately equimolar amounts were incorporated. In the presence of both radioactive amino acids omission of glycyl-tRNA synthetase caused a marked reduction in the incorporation not only of glycine but also of serine. In contrast, omission of seryl-tRNA synthetase had little effect on the incorporation of glycine. This observation provided indirect evidence that glycine must first be incorporated onto the ε-amino group of the lysine residue in the pentapeptide before incorporation of serine could occur. Such an interpretation is difficult to reconcile with other experiments in which it was shown that radioactivity could be transferred from [^{14}C]seryl-tRNA into lipid intermediate and peptidoglycan in the absence of added glycine. It is possible, however, that some of the endogenous lipid intermediates were already acylated with glycine and that it was these molecules that accepted serine in the transfer reaction

Analysis with DNFB was carried out on fragments obtained from the peptidoglycan products by digestion with lysozyme. When the biosynthesis reaction mixture contained nonradioactive serine, glycine, and UDP–MurNAc–pentapeptide labeled with [^{14}C]lysine, an amino-acid-dependent blockage of the ε-amino group was found. It was not established whether this was due specifically to glycine or serine. When the reaction was carried out with unlabeled UDP–MurNAc–pentapeptide and either [^{14}C]glycine and [^{12}C]serine or [^{14}C]serine and [^{12}C]glycine, only 13% of the radioactive glycine was reactive with DNFB, compared with 48% of the radioactive serine. Since equimolar amounts of the two amino acids were incorporated, a statistical characterization of the average chain synthesized *in vitro* was that it contained four residues with serine at the NH$_2$-terminus. In the cell wall of this organism, however, at least half the interpeptide bridges have glycine in this position (67).

Direct proof that both amino acids could be incorporated into the same peptide chain was obtained by analysis of peptidoglycan containing incorporated [^{14}C]serine and [^{3}H]glycine in approximately equimolar amounts. This material was subjected to digestion with lysozyme, followed by mild acid hydrolysis. Two major radioactive products were found after these operations. The structure of one of these, comprising about 30% of the incorporated radioactivity, was established as [^{14}C]Ser-[^{3}H]Gly.

d. *Arthrobacter crystallopoietes*. The interpeptide bridge of this organism contains a single alanine residue (68). A particulate fraction was found to catalyze the nucleotide substrate-dependent transfer of radioactive alanine from tRNA to lipid intermediate or peptidoglycan (58). In studying incorporation from the free amino acid better dependence on the presence of added uridine nucleotide substrates was observed with supernatant fraction and tRNA from *S. epidermis* as an alanyl-tRNA-generating system than with homologous preparations. Both [^{14}C]-L-Alanine and UDP–MurNAc–pentapeptide labeled with [^{14}C]-L-lysine were incubated under these conditions, and the disaccharide containing attached peptide was separated by paper electrophoresis after mild acid hydrolysis of the isolated lipid intermediate. As expected, the presence of alanine in the reaction mixture rendered the ε-amino group of lysine unreactive with DNFB. Of special interest was the finding that all the incorporated alanine retained its free α-amino group. Thus the alanylated lipid intermediate synthesized *in vitro* had the same length restriction as that found in the cell wall.

2. *Alanyl-tRNA~UDP-N-Acetylmuramyl-Pentapeptide Transferase*

Lactobacillus viridescens, which contains an interpeptide bridge of the structure (D-Ala)–L-Ser–L-Ala ($N^ε$-L-Lys) (69), is unique among the organisms that have been surveyed in that it possesses a soluble aminoacyl-tRNA transferase (59,70). This enzyme has been found to catalyze the transfer of L-alanine, L-serine, and, to a lesser extent, of certain

other residues from tRNA into peptide linkage with the ε-amino group
of the lysine residue in free UDP–MurNAc–pentapeptide. The available evidence suggests that it catalyzes the transfer of only a single
residue to the acceptor and that the addition of a second residue may
be catalyzed by another enzyme in the particulate fraction of the cell
(59).

Alanyl-tRNA∼UDP–MurNAc–pentapeptide transferase was puri-
fied approximately 500-fold from the soluble fraction of *L. viridescens* by
a procedure that included ammonium sulfate fractionation, chromatography on anion-exchange resins and hydroxylapatite, extraction from
alumina Cγ gel, and gel filtration on Sephadex G–200 (70). Activity was
assayed either in a coupled system containing the free radioactive amino
acid, tRNA, ATP, the appropriate aminoacyl-tRNA synthetase and
UDP–MurNAc–pentapeptide, or with radioactive aminoacyl-tRNA,
in which case nucleotide substrate was the only other required component of the reaction mixture. In the first assay system the product
was detected as radioactivity which was soluble in 5% TCA and retained by charcoal at low pH, but extracted from it with 50% ethanol
containing 0.1 N ammonia. In the second assay system, the formation
of acylated nucleotide was determined by paper chromatography. The
purified enzyme migrated as a single band during disk gel electrophoresis in sodium dodecylsulfate with a mobility corresponding to a
molecular weight of 40,000 for the reduced, denatured protein. It
catalyzed the transfer of 276 nmoles of alanine per minute per milligram at 37°C. During the last step of the purification procedure, gel
filtration on Sephadex G–200, the activities for the transfer of L-alanine
and L-serine were found to coelute, indicating that they were due to the
same enzyme protein.

After labeling with [^{14}C]lysine, UDP–MurNAc–pentapeptide was
acylated with nonradioactive L-alanine or L-serine in reaction mixtures containing the nucleotide substrate in approximately a sixfold
molar excess with respect to the amino acids. Acylated product was
isolated from unreacted nucleotide by thin-layer or column chromatography, and the lysine residue was shown to have lost its DNFB-
reactive ε-amino group. A single cycle of Edman degradation rendered
most of it again reactive with DNFB, with only a slight increment after
a second Edman cycle (59). Similarly, when a 25-fold molar excess of
nucleotide-substrate containing [^{14}C]lysine was used as acceptor for
the transfer of [^{14}C]alanine and the acylated product was separated

from unreacted substrate, it was found to contain equimolar quantities of the two radioactive amino acids after acid hydrolysis (70). These findings indicated that, at least under conditions in which the concentration of acceptor was high relative to that of amino acid, the soluble enzyme catalyzed the transfer of only one residue to the nucleotide substrate. Further evidence for this conclusion was the failure of the crude soluble function to catalyze the transfer of either [^{14}C]-L-alanine or [^{14}C]-L-serine to enzymatically prepared unlabeled UDP–MurNAc–pentapeptide–L–Ala or UDP–MurNAc–pentapeptide–L–Ser (59). It is noteworthy in this context that when UDP–MurNAc–hexapeptides were prepared with [^{14}C]alanine or [^{14}C]serine, radioactivity from them could be incorporated into lipid intermediate or peptidoglycan in the presence of particles from L. viridescens, although the incorporation into peptidoglycan appeared to be considerably less efficient than that of radioactivity from UDP–MurNAc–pentapeptide (59).

Since the soluble enzyme was not found to catalyze the transfer of more than a single aminoacyl residue to nucleotide acceptor, the possibility was examined that the particles might contain an activity(s) that specifically catalyzed the transfer of an additional residue(s) to MurNAc–hexapeptide (59). For these experiments washed particles were used which contained no activity for the transfer of alanine or serine to UDP–MurNAc–pentapeptide. The particles were incubated with either UDP–MurNAc–pentapeptide or UDP–MurNAc–pentapeptide–L–Ala to generate MurNAc(–pentapeptide or hexapeptide)–P-P-phospholipid and centrifuged to remove soluble unreacted substrates. They were then incubated with UDP–GlcNAc to allow formation of peptidoglycan and with either [^{14}C]alanyl or [^{14}C]seryl-tRNA. A radioactive, chromatographically immobile spot typical of peptidoglycan was found in reaction mixtures containing either radioactive amino acid and particles that had been preincubated with UDP–MurNAc–pentapeptide–L–Ala, but was virtually absent in those in which the initial preincubation had been carried out with UDP–MurNAc–pentapeptide. The product into which [^{14}C]alanine had been transferred was almost completely solubilized by treatment with lysozyme, whereas that containing [^{14}C]serine was less sensitive to the action of this enzyme. It was proposed that this difference might be due to a serine-dependent crosslinking reaction, since little chromatographically immobile radioactivity was found if the incubation with

[^{14}C]seryl-tRNA was carried out in the presence of penicillin. These experiments suggest that particles from *L. viridescens* may possess an activity that catalyzes the transfer of L-alanine or L-serine from tRNA to MurNAc–pentapeptide–L-Ala at the level of the monosaccharide– or disaccharide–lipid intermediate or peptidoglycan itself.

B. AMINOACYL-tRNA SPECIFICITY

1. *Aminoacyl-tRNA~GlcNAc–MurNAc(–pentapeptide)–P-P- Phospholipid Transferases*

Most studies on the specificity of aminoacyl-tRNA in reactions catalyzed by these particulate enzymes have dealt with the relative ability of isoaccepting species to participate in peptidoglycan and protein biosynthesis.

Crude tRNA from *S. aureus* was fractionated by reverse-phase chromatography on Chromosorb W (71). Three peaks of glycine-acceptor activity were separated, and each was found to support the incorporation of radioactive glycine into peptidoglycan catalyzed by the particulate enzyme system from this organism in the presence of uridine-nucleotide substrates and a glycyl-tRNA-generating system. Each species was acylated with [^{14}C]glycine and tested for its ability to participate in protein synthesis. Two species were found to be active and one inactive, as judged by the following criteria: (*a*) binding to *E. coli* ribosomes dependent on the presence of known glycine codons; (*b*) transfer of glycine into polypeptide dependent on a 5:1 copolymer of uridylate and guanylate, and catalyzed by extracts of either *E. coli* or *S. aureus*; (*c*) transfer of glycine into polypeptide catalyzed by extracts of rabbit reticulocytes. These investigations thus provided no evidence of specific recognition of the polyribonucleotide chain by glycyl-tRNA~MurNAc(–pentapeptide)–P-P-phospholipid transferase. That a specific recognition site may exist, however, is suggested by the fact that a species of glycyl-tRNA had previously been obtained that participated in protein synthesis, but not in that of peptidoglycan (55,60). This species of glycyl-tRNA has only been found in one batch of *S. aureus*, however, and its significance is unclear (71).

Two species of tRNA^{Thr} have been isolated from bulk tRNA of *M. roseus* (56). [^{14}C]Threonyl-tRNA prepared from each of these was found to participate in both peptidoglycan and protein synthesis.

Four species of seryl-acceptor tRNAs were separated from bulk tRNA of *S. epidermis* (57). After acylation with [^{14}C]serine, each could function as substrate for the transfer of [^{14}C]serine into lipid intermediate or peptidoglycan. However, one of the four was found to be inactive in protein synthesis, as judged by the criteria described above. More recently seven species of serine-accepting tRNA have been obtained (72). One of these did not function in the biosynthesis of peptidoglycan, which suggests the presence on the polyribonucleotide chain of a determinant for donor specificity in the transfer reaction catalyzed by particles from this organism.

Glycyl-acceptor tRNA from *S. epidermis* has also been investigated (73). Four isoaccepting species were resolved, and [^{14}C]glycine could be transferred from each into peptidoglycan. However, one of these failed to serve as substrate for protein synthesis *in vitro* and was not bound to *E. coli* ribosomes in the presence of any of the known glycine codons. This species accounted for 40% of the total tRNA^{Gly} and was subsequently (74) found to consist of two distinct components, called tRNA$_{IA}^{Gly}$ and tRNA$_{IB}^{Gly}$, whose sequences were determined. They were found to differ from each other at seven points and from previously characterized species of tRNA in possessing only one modified base, a 4-thiouridine residue, and in lacking a GC sequence commonly found in the dihydrouridine loop. Of special interest was the finding that they contained the sequence GUGC in place of the usual GTψC. This region is thought to be a ribosomal binding site (75); its alteration may account for the failure of these glycyl-tRNAs to function in protein synthesis since their anticodon loops contained the sequence UCC in the proper position to accommodate the glycine codons GGG and GGA.

Evidence has been obtained of recognition of the polyribonucleotide moiety of aminoacyl-tRNA in the alanine-transfer reaction catalyzed by the particulate fraction of *A. crystallopoietes* (58). Alanyl-tRNA^{Cys} prepared by the reduction of *E. coli* cysteinyl-tRNA^{Cys} could not substitute for *E. coli* alanyl-tRNA^{Ala} as an alanine donor in the uridine-nucleotide-dependent acylation of lipid intermediate and peptidoglycan. Appropriate experiments excluded the presence of an inhibitor in the inactive preparation and demonstrated that in response to cysteine codons it could contribute alanine residues for polypeptide synthesis catalyzed by an *E. coli* extract.

2. *Alanyl-tRNA~UDP-N-Acetylmuramyl–Pentapeptide Transferase*

With crude soluble fraction from *L. viridescens* as a source of transferase and of aminoacyl-tRNA synthetases, and with homologous
tRNA, Plapp and Strominger (59) observed the incorporation of radioactivity into UDP–MurNAc–pentapeptide from L-serine, L-alanine,
and, to a lesser extent, from glycine, but not from L-theonine, L-lysine
or L-aspartic acid. They also found (70) that the purified enzyme
catalyzed the transfer of L-alanine, L-serine, and L-cysteine from their
respective tRNAs and of L-alanine from alanyl-tRNA^{Cys}. Thus there
is apparently broad specificity with respect to the aminoacyl residue
whose transfer is catalyzed by this enzyme, despite the fact that
L-alanine is the only residue found linked directly to L-lysine in the
cell wall of this organism (69).

C. ACCEPTOR SPECIFICITY

1. *Aminoacyl-tRNA~GlcNAc–MurNAc(–pentapeptide)–P-P-Phospholipid Transferases*

It is clear that the ε-amino group of the lysine residue in MurNAc–
pentapeptide is the site to which aminoacyl residues are transferred
from tRNA in reactions catalyzed by these particulate enzymes. However, the presence in the particles of the activities and cofactors required to generate peptidoglycan from uridine-nucleotide substrates
renders it difficult to establish directly the level at which acylation
initially occurs. Matsuhashi et al. (60) have studied the uridine-nucleotide-dependent transfer of radioactive glycine from a glycyl-tRNA-
generating system into UDP–MurNAc–pentapeptide, lipid intermediate, and peptidoglycan in reactions catalyzed by the particulate
fraction of *S. aureus*. From the kinetics of incorporation into these components under various conditions they have inferred that the initial
acceptor in this system is GlcNAc–MurNAc(–pentapeptide)–P-P-
phospholipid.

The possibility was first considered that UDP–MurNAc–pentapeptide might be the acceptor, and a compound tentatively identified
as UDP–MurNAc–pentapeptide–oligoglycine was detected in reaction
mixtures from which UDP–GlcNAc had been omitted. This compound
was considered to have arisen from the glycylation of MurNAc(–pentapeptide)–P-P-phospholipid followed by a UMP-dependent reversal of
the reaction by which MurNAc–pentapeptide is incorporated into the

lipid carrier (Fig. 5, reaction 1), rather than by direct glycylation of the nucleotide. The evidence for this conclusion was that accumulation of radioactive glycine in the nucleotide lagged behind that in the lipid intermediate and that, if UMP was added after incorporation into lipid had occurred, radioactivity rapidly disappeared from the lipid intermediate with a corresponding increase in the amount of radioactive nucleotide. After this rapid transfer there was no further incorporation of glycine, presumably because lipid intermediate could not be formed in the presence of UMP, and the nucleotide substrate itself was not an acceptor. The finding that radioactive nucleotide appeared much diminished in the presence of UDP–GlcNAc was attributed to the irreversibility of the reaction by which MurNAc(–pentapeptide)–P-P-phospholipid is converted to GlcNAc–MurNAc(–pentapeptide)–P-P-phospholipid (Fig. 5, reaction 2). These data can thus be interpreted as indicating that UDP–MurNAc–pentapeptide is not the primary acceptor to which glycine is transferred. Such an interpretation also suggests that under appropriate circumstances (i.e., in the absence of the disaccharide–pentapeptide lipid intermediate) the monosaccharide derivative, MurNAc(–pentapeptide)–P-P-phospholipid, can serve as acceptor.

Additional evidence indirectly suggesting that the nucleotide is not the initial acceptor was obtained by following the incorporation of radioactivity from a mixture of UDP–MurNAc–pentapeptide–oligo-[^{14}C]glycine and UDP–MurNAc–pentapeptide containing [^{3}H]-D-alanine. The velocity of transfer of ^{3}H to lipid intermediate and peptidoglycan was greater than that of ^{14}C, indicating that the glycylated nucleotide is a relatively poor substrate for the enzymes catalyzing peptidoglycan biosynthesis in *S. aureus*.

That GlcNAc–MurNAc(–pentapeptide)–P-P-phospholipid could function as acceptor was indicated by the following experiment. The complete system, except for glycine, was incubated in the presence of a concentration of vancomycin that inhibited the synthesis of peptidoglycan, but allowed accumulation of lipid intermediate. Radioactive glycine was added after the accumulation had occurred. A rapid incorporation into lipid intermediate was observed during the second incubation, but there was little formation of radioactive nucleotide. Had the acceptor of [^{14}C]glycine been MurNAc(–pentapeptide)–P-P-phospholipid, then presumably UDP–MurNAc–pentapeptide–oligo-[^{14}C]glycine would have been produced from it as decsribed above.

Failure to generate the glycylated nucleotide thus suggested that transfer had occurred at the level of the disaccharide derivative.

A similar experiment indicated that peptidoglycan is probably not an acceptor. The complete system, except for glycine, was incubated to allow formation not only of lipid intermediate but also of nascent peptidoglycan. Vancomycin was then added together with radioactive glycine; incorporation into lipid intermediate, but not peptidoglycan, was observed. Incorporation into both products was found in a control reaction not containing antibiotic. Therefore either newly formed peptidoglycan could not function as acceptor or the transfer of glycine to it was inhibited by a concentration of vancomycin that did not affect the transfer to lipid intermediate.

The kinetics of incorporation of glycine into various derivatives of MurNAc-pentapeptide in the uridine-nucleotide-dependent reactions catalyzed by particles from *S. aureus* thus implied that the two lipid intermediates could function as acceptor, whereas the nucleotide derivative and peptidoglycan itself could not. A preferential utilization of GlcNAc–MurNAc(–pentapeptide)–P-P-phospholipid as acceptor substrate was suggested by the fact that the rate of accumulation of [^{14}C]glycine in lipid intermediate was 4 times higher in the presence of UDP–GlcNAc than in its absence.

Kamiryo and Matsuhashi provided direct evidence that the disaccharide–pentapeptide lipid intermediate could act as acceptor (64,65). They studied the incorporation of glycine into lipid carrier catalyzed by the particulate fraction in the absence of both uridine-nucleotide substrates. The reaction was shown to be completely dependent on the addition of purified GlcNAc–MurNAc(–pentapeptide)–P-P-phospholipid, and the radioactive lipid product was isolated and characterized as GlcNAc–MurNAc(–pentapeptide–oligoglycine)–P-P-phospholipid. The acceptor properties of MurNAc(–pentapeptide)–P-P-phospholipid were not examined under these conditions.

2. Alanyl-tRNA∼UDP-N-Acetylmuramyl–Pentapeptide Transferase

Acceptor specificity of alanyl-tRNA∼UDP–MurNAc–pentapeptide transferase is more readily accessible to direct experimental investigation than is that of the particulate transferases, since the enzyme preparation is presumably free of contaminating activities that can generate acceptor or convert it to other products. Plapp and Strominger (70) estimated a K_m of 2×10^{-7} M for UDP–MurNAc–pentapeptide in the alanine-transfer reaction using alanyl-tRNA and the purified enzyme. To compare acceptance of radioactive alanine or serine by various substrates, they coupled the transfer reaction to an aminoacyl--tRNA-generating system that included *L. viridescens* tRNA and a crude tRNA-free supernatant fraction from *S. faecalis* that lacks transfer activity. Potential acceptors were present at 5×10^{-4} M, and reaction mixtures were analyzed by paper chromatography for the presence of acceptor-dependent radioactive products. Similar results were obtained with both amino acids. Under the conditions they employed, UDP–MurNAc–pentapeptide appeared to be the best acceptor, although only a small fraction of the molecules was acylated during the reaction. Phosphoacetylmuramyl–pentapeptide prepared by treating the nucleotide with venom phosphodiesterase and MurNAc–pentapeptide obtained from it by incubation with diesterase and alkaline phosphomonoesterase were found to accept less than 2% of the amount of amino acid that could be incorporated into the undegraded nucleotide. It was found that UDP–MurNAc–L-Ala–D-Glu–L-Lys and UDP–MurNAc–L-Ala–D-Glu–*meso*-Dap–D-Ala–D-Ala accepted, respectively, about 7 and 12% as much amino acid as UDP–MurNAc–pentapeptide. It was not established whether GlcNAc–MurNAc(–pentapeptide)–P-P-phospholipid or UDP–MurNAc–hexapeptide could function as acceptor for the purified enzyme. However, no acceptance by the latter was demonstrable when a crude supernatant fraction of *L. viridescens* was used as a source of transferase (59).

These findings indicate that both the nucleotide moiety and the peptide chain contain determinants of acceptor specificity in the reaction catalyzed by alanyl-tRNA∼UDP–MurNAc–pentapeptide transferase. Recognition of the UMP residue is of particular interest, since this part of the molecule is lacking in the lipid intermediates that appear to function as acceptor in reactions catalyzed by the particulate transferases.

D. CELLULAR FUNCTION

Aminoacyl-tRNA transferases appear to play an important role in the biosynthesis of interpeptide bridges attached to the ε-amino group of lysine in the tetrapeptide of bacterial peptidoglycan. A substantial fraction of the appropriate aminoacyl-tRNA probably participates in the reactions catalyzed by these enzymes, since it has been estimated that the amount of glycine and serine in the bridge of *S. epidermis* is at least equal to that found in the total cellular protein (76). It is there-fore, not unlikely that certain isoaccepting species of tRNA that are inactive in protein synthesis may have evolved specifically for this purpose (73). The evidence is most compelling that bridge formation is initiated by the enzymatic transfer of an aminoacyl residue from tRNA to the ϵ-amino group of lysine in an acceptor-containing MurNAc–pentapeptide and that elongation occurs at the NH_2-ter-minus of the nascent bridge by the sequential transfer of additional aminoacyl residues from tRNA. Although it is clear that specificity with respect to sequence and length of the bridge is achieved by a mechanism distinct from that operative in protein synthesis, the extent to which aminoacyl-tRNA transferases are responsible for this speci-ficity has not been completely established. In some instances a good correlation has been found to exist between the reaction observed *in vitro* and the known structure of the cell wall. Examples of this are the dependence of serine incorporation on the prior incorporation of glycine in the reactions catalyzed by the particulate fraction of *S. epidermis* (57), the fact that only a single alanine residue can be trans-ferred to the pentapeptide with the particulate system from *A. crystal-lopoietes* (58), and the finding that an enzyme preparation from a strain of *M. roseus* lacking threonine in its cell wall failed to catalyze the incor-poration of this amino acid (56). There appear to be other instances, however, in which the specificity of the transferase cannot entirely account for the naturally occurring sequence. For example, in *L. viridescens* alanine is the amino acid linked to the lysine residue of the tetrapeptide in the cell wall, but the purified soluble enzyme has been found to catalyze the transfer of several different residues to UDP–MurNAc–pentapeptide (70).

A major obstacle in defining the enzymatic basis for the biosynthesis of specific bridge structures has been that most of the responsible amino-acyl-tRNA transferases are located in the same particulate fraction as the other enzymes and lipid intermediates involved in the synthesis of peptidoglycan. Studies with these complex fractions possess the ad-vantage that they may more correctly mimic the circumstances *in vivo* than do single reactions catalyzed by isolated enzymes. However, such studies cannot directly establish whether single or multiple activities exist for the transfer of different aminoacyl residues and whether the transfer enzyme(s) discriminate among acceptor substrates containing lysine residues that are acylated to varying extents with specific residues. It is not unlikely that sequence and length may derive in part from acceptor specificity of different transferases—that is, that the different enzymes may recognize only the bridge structure whose synthesis has been catalyzed by the preceding enzyme in the biosynthesis scheme. This situation probably exists in *L.viridescens*, where the first enzyme, a soluble alanyl-tRNA~UDP–MurNAc–pentapeptide transferase, appears to recognize the free lysine residue of the pentapeptide, but not the hexapeptide in which it is acylated, whereas the particulate transferase apparently preferentially utilizes the hexapeptide derivative (59). Alternatively it is also possible that a single enzyme might distinguish between a free and acylated lysine residue and in so doing might be sterically altered in such a way as to recognize an appropriate aminoacyl-tRNA and hence catalyze the transfer of a specific aminoacyl residue to the correct acceptor.

Resolution of these questions will require solubilization and isolation of the particulate activities and of the differentially acylated potential acceptors. Selection of mutants with altered or defective interpeptide bridges should be useful in establishing the role and number of transfer enzymes, and such mutants might provide a source of acceptor substrates specific for the enzyme they are lacking. Genetic techniques may also be valuable in exploring the possible contribution of specific peptidases to the process by which the sequence and length of this important structure are determined.

Acknowledgments

Unpublished investigations reported in this chapter were supported by research grants from the National Institutes of Health (GM 11301; AM 12395) and the National Science Foundation (GB 35203).

References

1. Soffer, R. L., Horinishi, H., and Leibowitz, M. J., *Cold Spring Harbor Symp. Quant. Biol.*, *34*, 529 (1969).
2. Soffer, R. L., *J. Biol. Chem.*, *245*, 731 (1970).
3. Soffer, R. L., *J. Biol. Chem.*, *246*, 1602 (1971).
4. Soffer, R. L., *J. Biol. Chem.*, *248*, 2918 (1973).
5. Kaji, H., Novelli, G. D., and Kaji, A., *Biochim. Biophys. Acta*, *76*, 477 (1963).
6. Weinstein, I. B., and Osserman, E. F., *Acta Union Int. Contre Cancer*, *20*, 932 (1964).

7. Soffer, R. L., and Mendelsohn, N., *Biochem. Biophys. Res. Commun.*, *23*, 252 (1966).
8. Soffer, R. L., *Biochim. Biophys. Acta*, *155*, 228 (1968).
9. Soffer, R. L., *Trans. N.Y. Acad. Sci.*, *32*, 974 (1970).
10. Soffer, R. L., and Horinishi, H., *Fed. Proc.*, *27*, 778 (1968).
11. Soffer, R. L., and Horinishi, H., *J. Mol. Biol.*, *43*, 163 (1969).
12. Shearer, W. T., Bradshaw, R. A., Gurd, F. R. N., and Peters, T., Jr., *J. Biol. Chem.*, *242*, 5451 (1967).
13. Soffer, R. L., *J. Biol. Chem.*, *246*, 1481 (1971).
14. Spiro, M. J., *J. Biol. Chem.*, *245*, 5820 (1970).
15. Leibowitz, M. J., and Soffer, R. L., *J. Biol. Chem.*, *246*, 5207 (1971).
16. Soffer, R. L., *J. Biol. Chem.*, *248*, in press (1973).
17. Kaji, A., Kaji, H., and Novelli, G. D., *Biochem. Biophys. Res. Commun.*, *10*, 406 (1963).
18. Kaji, A., Kaji, H., and Novelli, G. D., *J. Biol. Chem.*, *240*, 1185 (1965).
19. Kaji, A., Kaji, H., and Novelli, G. D., *J. Biol. Chem.*, *240*, 1192 (1965).
20. Momose, K., and Kaji, A., *J. Biol. Chem.*, *241*, 3294 (1966).
21. Otaka, E., and Osawa, S., *Biochim. Biophys. Acta*, *119*, 146 (1966).
22. Leibowitz, M. J., and Soffer, R. L., *Biochem. Biophys. Res. Commun. 36*, 47 (1969).
23. Leibowitz, M. J., and Soffer, R. L., *J. Biol. Chem.*, *245*, 2066 (1970).
24. Leibowitz, M. J., and Soffer, R. L., *J. Biol. Chem.*, *246*, 4431 (1971).
25. Soffer, R. L., and Pestka, S., unpublished data.
26. Soffer, R. L., and Capra, J. D., *Nature New Biol.*, *36*, 44 (1971).
27. Grosclaude, F., Mercier, J.-C., and Ribadeau-Dumas, B., *Eur. J. Biochem.*, *16*, 447 (1970).
28. Ribadeau-Dumas, B., Brignon, G., Grosclaude, F., and Mercier, J.-C., *Eur. J. Biochem.*, *20*, 264 (1971).
29. Smythe, P. G., in *Methods in Enzymology*, Vol. XI, C. H. W. Hirs, Ed., Academic Press, New York, 1967, p. 216.
30. Jonckheer, H. M., and Karcher, D. M., *J. Clin. Endocrinol. Metab.*, *32*, 7 (1971).
31. Goz, B., and Voytek, P., *J. Biol. Chem.*, *247*, 5892 (1972).
32. Soffer, R. L., and Leibowitz, M. J., unpublished data.
33. Soffer, R. L., and Lerner, A., unpublished data.
34. Leibowitz, M. J., and Soffer, R. L., *Proc. Natl. Acad. Sci. U.S.*, *68*, 1866 (1971).
35. Soffer, R. L., and Savage, M., *Proc. Natl. Acad. Sci. U.S.*, *71*, in press (1974).
36. Taylor, A. L., and Trotter, C. D., *Bacteriol. Rev.*, *31*, 332 (1967).
37. Macfarlane, M. G., *Nature*, *196*, 136 (1962).
38. Lennarz, W. J., Nesbitt, J. A., III, and Reiss, J., *Proc. Natl. Acad. Sci. U.S.*, *55*, 934 (1966).
39. Gould, R. M., and Lennarz, W. J., *Biochem. Biophys. Res. Commun.*, *26*, 510 (1967).
40. Nesbitt, J. A., III, and Lennarz, W. J., *J. Biol. Chem.*, *243*, 3088 (1968).
41. Gould, R. M., Thornton, M. P., Liepkalns, V., and Lennarz, W. J., *J. Biol. Chem.*, *243*, 3096 (1968).
42. Lennarz, W. J., Bonsen, P. P. M., and Van Deenen, L. L. M., *Biochemistry*, *6*, 2307 (1967).
43. Stern, R., and Mehler, A. H., *Biochem. Z.*, *342*, 400 (1965).

44. Barnett, W. E., and Epler, J. C., *Cold Spring Harbor Symp. Quant. Biol.*, *31*, 549 (1966).
45. Op den Kamp, J. A. F., Houtsmuller, V. M. T., and Van Deenen, L. L. M., *Biochim. Biophys. Acta*, *106*, 438 (1965).
46. Houtsmuller, U. M. T., and Van Deenen, L. L. M., *Biochim. Biophys. Acta*, *84*, 96 (1964).
47. Gould, R. M., and Lennarz, W. J., *J. Bacteriol*, *104*, 1135 (1970).
48. Lennarz, W. J., *Acc. Chem. Res.*, *5*, 361 (1972).
49. Houtsmuller, V. M. T., and Van Deenen, L. L. M., *Biochim. Biophys. Acta*, *106*, 564 (1965).
50. Anderson, J. S., Matsuhashi, M., Hasken, M. A., and Strominger, J. L., *J. Biol. Chem.*, *242*, 3180 (1967).
51. Stone, J. K., and Strominger, J. L., *J. Biol. Chem.*, *247*, 5107 (1972).
52. Ghuysen, J.-M., *Bacteriol. Rev.*, *32*, 425 (1968).
53. Strominger, J. L., *Harvey Lect.*, *64*, 179 (1970).
54. Chatterjee, A. N., and Park, J. T. *Proc. Natl. Acad. Sci, U.S.*, *51*, 9 (1964).
55. Matsuhashi, M., Dietrich, C. P., and Strominger, J. L., *Proc. Natl. Acad. Sci. U.S.*, *54*, 587 (1965).
56. Roberts, W. S. L., Strominger, J. L., and Söll, D., *J. Biol. Chem.*, *243*, 749 (1968).
57. Petit, J.-F., Strominger, J. L., and Söll, D., *J. Biol. Chem.*, *243*, 757 (1968).
58. Roberts, W. S. L., Petit, J.-F., and Strominger, J. L., *J. Biol. Chem.*, *243*, 768 (1968).
59. Plapp, R., and Strominger, J. L., *J. Biol. Chem.*, *245*, 3667 (1970).
60. Matsuhashi, M., Dietrich, C. P., and Strominger, J. L., *J. Biol. Chem.*, *242*, 3191 (1967).
61. Mandelstam, M. H., and Strominger, J. L., *Biochem. Biophys. Res. Commun.*, *5*, 446 (1961).
62. Tipper, D. J., Strominger, J. L., and Ensign, J. C., *Biochemistry*, *6*, 906 (1967).
63. Thorndike, J., and Park, J. T., *Biochem. Biophys. Res. Commun.*, *35*, 642 (1969).
64. Kamiryo, T., and Matsuhashi, M., *Biochem. Biophys. Res. Commun.*, *36*, 215 (1969).
65. Kamirgo, T., and Matsuhashi, M., personal communication.
66. Petit, J.-F., Munoz, E., and Ghuysen, J. M., *Biochemistry*, *5*, 2764 (1966).
67. Tipper, D. J., *Fed. Proc.*, *27*, 294 (1968).
68. Krulwich, T. A., Ensign, J. C., Tipper, D. J., and Strominger, J. L., *J. Bacteriol.*, *94*, 734 (1967).
69. Kandler, O., Plapp, R., and Holzapfel, W., *Biochim. Biophys. Acta*, *147*, 252 (1967).
70. Plapp, R., and Strominger, J. L., *J. Biol. Chem.*, *245*, 3675 (1970).
71. Bumsted, R. M., Dahl, J. L., Söll, D., and Strominger, J. L., *J. Biol. Chem.*, *245*, 779 (1968).
72. Hilderman, R. H., and Riggs, H. G., Jr., *Biochem. Biophys. Res. Commun.*, *50*, 1095 (1973).
73. Stewart, T. S., Roberts, R. J., and Strominger, J. L., *Nature*, *230*, 36 (1971).
74. Roberts, R. J., *Nature New Biol.*, *237*, 44 (1972).
75. Ofengand, J., and Henes, C., *J. Biol. Chem.*, *244*, 6241 (1969).
76. Gale, E. G., Shepherd, C. J., and Folkes, J. P., *Nature*, *182*, 592 (1958).

AMINOACYL-tRNA SYNTHETASES: SOME RECENT RESULTS AND ACHIEVEMENTS

By LEV L. KISSELEV and OL'GA O. FAVOROVA,

Moscow, U.S.S.R.

CONTENTS

I. Introduction

Aminoacyl-tRNA synthetases are key enzymes of cell metabolism because their activity leads to the utilization of amino acids for protein synthesis. The increasing interest in this group of enzymes is stimulated by several circumstances. The synthetases interact with three chemically very different substrates—tRNA, amino acid, and ATP—and they catalyze the formation of three products—aminoacyl-tRNA, AMP, and pyrophosphate. The synthetases belong to the group of enzymes that can recognize combinations of nucleotides in nucleic acids; containing no nucleotide cofactors, they can be regarded as true protein partners in selective protein–nucleic acid interactions. A characteristic feature of protein biosynthesis is the nearly absolute specificity with which amino acids are incorporated into each position of the polypeptide chain, as determined genetically. This means that at least the same degree of specificity exists in the ability of synthetases to discriminate both among at least 20 amino acids and among at least 60 species of tRNA within each cell. Therefore the fidelity of synthetases in catalyzing aminoacyl-tRNA formation may be regarded as one of the main factors involved in the correct reproduction of the genetically determined primary structure of proteins.

The current state of investigation of this group of enzymes is reflected by numerous recent reviews of the field (1–6). This area of research is considered also in reviews concerned with general aspects of protein synthesis (7–9), with recognition mechanisms (10), and with tRNA structure and function (11–15). The earlier literature is covered by Zamecnik and Novelli (16,17). The preparation of aminoacyl-tRNA synthetases from bacterial sources has been discussed by Muench (18). A separate review was devoted recently to aminoacyl-tRNA synthetases and tRNAs from plants (18a).

However, we believe that, in spite of the many excellent reviews, a survey of the experimental evidence obtained mainly between 1970 and 1973 would be useful because many new lines of work on the synthetases developed during this period, such as preparation of cognate and noncognate enzyme–substrate complexes and their chemical investigation, genetically altered synthetases, crystallization, and studies on the order of substrate addition and tertiary structure. This chapter is an attempt to survey these recent developments; whenever possible the aspects discussed thoroughly by previous reviewers were avoided. To obtain a more comprehensive knowledge of the present state of the art, the reader is referred to the reviews by Mehler and Chakraburtty (3), by Loftfield (4,5), and by Favorova, Parin, and Lavrik (6).

II. Structure

A. PRIMARY, SECONDARY, AND TERTIARY STRUCTURES

The amino acid composition of a number of synthetases, mainly of those isolated from unicellular organisms, has been determined. Surveys of these data have been already published (3,6,9) and can be now extended by new data for tryptophanyl- (19), glutaminyl- (20), glutamyl- (21) and isoleucyl- (33c) tRNA synthetases from *Escherichia coli* and the valine enzyme from yeast (22). The amino acid composition of the lysine enzyme of *E. coli* was examined in two preparations obtained in somewhat different ways (22,23). Whereas Stern and Peterkofsky (23) reported 0.4 residue of proline and 3.2 residues of methionine per enzyme molecule with a molecular weight of approximately 100,000, Rymo, Lundvik, and Lagerkvist (22) found 35 proline residues and 31 methionine residues per molecule with a molecular weight of 104,000. The reason for these considerable discrepancies is unknown, but it is very unusual for large globular proteins to contain practically no proline and less than two methionine residues per subunit. In the case of lysine enzyme from yeast (22,24), tyrosine enzyme from *E. coli* (24a,24b) and tryptophan enzyme from beef pancreas (25,26) the amino acid compositions of two preparations determined in two laboratories are similar. Characteristic of the amino acid composition is the predominance of dicarboxylic acids and the relatively small content of basic amino acids. Thus the synthetases are acidic proteins and in electrophoresis migrate to the anode in neutral media (27).

At present nobody has yet succeeded in determining the complete primary structure of a synthetase. Presumably this is not due to any essential difficulties in comparison with other proteins; the absence of such information must be rather explained by the fact that it was only in 1964–1966 that first synthetases were obtained in a pure state in analytical amounts, whereas preparative amounts became available only 2–3 years ago. However, very recently great progress was achieved in Cambridge with sequence analysis of the *E. coli* methionyl-tRNA synthetase (50), an enzyme that was isolated in a pure state rather long ago (24d).

Of considerable interest are partial studies of the primary structure of synthetases near the active center. The polypeptide fragments from this region are obtained in modern enzymology by the affinity-labeling techniques, that is, by the use of substrate analogs retaining the affinity to enzyme and bearing a functional grouping that reacts with the protein to form a covalent bond. This approach to synthetase studies was used for the first time by Bruton and Hartley (28), who added a *p*-nitrophenyl-chloroformate grouping to the α-NH$_2$-group of the methionine residue of methionyl-tRNA. Within the enzyme–substrate complex the analog formed a covalent bond with the α-NH$_2$-group of the methionyl-tRNA synthetase lysine residue. By ribonuclease and protease digestion, the peptide containing the [^{14}C]methionine residue formerly belonging to methionyl-tRNA has been isolated, and its amino acid sequence was found to be Phe-Thr-Tyr-Gly-Lys-Leu-His-Asn.

This approach was extended by Santi, Marchant, and Yarus (33) with *E. coli* isoleucyl-tRNA synthetase. *N*-Bromoacetyl-Ile-tRNA was used as an alkylating agent (it is known that bromoacetamides react with histidine, cysteine, methionine, and lysine residues of proteins). The results demonstrate that *N*-bromoacetyl-Ile-tRNA binds reversibly to the enzyme by noncovalent interactions with the tRNA moiety, and then forms a covalent bond between the alkylating group and the amino acid residue of the enzyme.

Evidently the applicability of this affinity-labeling method depends on the presence of a grouping that can be alkylated by the chloroformate derivative (e.g., of a lysine, a cysteine, or a histidine residue) near the aminoacyl residue of the analog of aminoacyl-tRNA. An attempt to apply this method to another enzyme, namely, bovine-liver valyl-tRNA synthetase, was a failure: no covalent bond was formed (29). Another type of affinity labeling based on the amino acid analogs containing an active chlorine atom has been proposed very recently (30). It was shown that chloromethyl ketone, analog of L-valine, formed a covalent bond with valyl-tRNA synthetase, presumably in the active center of the enzyme.

Unfortunately the information obtained by the affinity-labeling techniques is of somewhat limited value: when primary and spacial structures are unknown, it remains uncertain whether the alkylated peptide participates directly in the active-center formation or just occupies a nearby position; on the other hand, some of the amino acid residues of the peptide, if involved in the active center, may not participate in catalytic or binding functions. However, wider application of the affinity-labeling techniques to synthetases of different specificity as well as to synthetases of the same amino acid specificity isolated from different organisms will most probably help to overcome these difficulties.

The secondary structure of synthetases has not been studied to any considerable extent. A circular dichroism spectrum has been reported for bakers' yeast tyrosyl-tRNA synthetase, but the purity of the enzyme was not documented (31). The α-helix content of the pancreatic tryptophanyl-tRNA synthetase is about 30–40% (25,32).

The general shape and the dimensions of synthetases in solutions are unknown, although the sedimentation constants and the Stokes radii have been determined for a number of synthetases [for references see Favorova et al. (6)]. Small-angle X-ray scattering study of dilute solutions of lysyl-tRNA synthetase from yeast was undertaken (49) in 0.1 M phosphate buffer at pH 7.0, 21°. The radius of gyration R (37.5 Å), the molecular weight M (114,000), and the volume V (295,000 Å³) were determined. The shape of the molecule can be represented by an oblate ellipsoid with the semiaxes A = 62.7, B = 50.1, and C = 23.5 Å.

Yeast leucyl-tRNA synthetase, which also consists of two subunits and has a molecular weight of 120,000, has been studied by electron microscopy (34). Its plane dimensions obtained for negatively stained samples are 77 × 120 Å. The third dimension ("height") determined by the fixed-angle shadowing method was about 27 Å. Thus the shape of the molecule is a slightly asymmetric flattened ellipsoid.

A very interesting observation has been made by Cassio and Waller (35) with *E. coli* K12 methionyl-tRNA synthetase. Treatment of the individual synthetase (mol. wt. 173,000) with proteolytic enzymes results in an enzymatically active form of molecular weight 64,000. Detailed analysis revealed that the tetramer consisting initially of four subunits with molecular weights of 43,000 is transformed by the limited proteolysis into a dimer composed of subunits with molecular weights of 32,000. Thus removal of about 20% of the protein leads to dissociation of the tetramer into dimers, but affects neither the activating nor the acylating function. Moreover, the Michaelis constants as well as the specificity toward the substrate amino acid remain practically the same.

In view of these studies, the structure of each of the *E. coli* methionyl-tRNA synthetase subunits can be described as a dense "nucleus" resistant to proteases surrounded by a loose polypeptide unimportant for the enzymatic activity. This interpretation by Cassio and Waller is in line with the data of Waller et al. (36) concerning the crystallization of the enzyme. It was found that the protease-modified enzyme readily crystallizes, whereas all attempts to crystallize native methionyl-tRNA synthetase were unsuccessful, presumably because of the presence of the above-mentioned loose external polypeptides.

It is quite probable that this peculiarity of methionyl-tRNA synthetase is not a unique feature. For instance, when lysyl-tRNA synthetase was isolated from yeast in the presence of protease inhibitor the molecular weight was found to be 140,000; in the absence of inhibitor it drops to 117,000. The molecular weight of each subunit was 70,000 and 50,000–60,000, respectively. The same effect may be obtained with limited tryptic digestion. In all cases the oligomers were active in both reactions (37). The beef-pancreas tryptophanyl-tRNA synthetase composed of two identical subunits has a molecular weight of 108,000 (25)–120,000 (26) daltons when isolated in the presence of protease inhibitors. After isolation without inhibitors an 80,000–85,000 form appears active in both functions (33a). The same pattern of hydrolysis was observed with trypsin although no activity was detected (33a, 33b).

Fluorescence studies (38) with some individual synthetases revealed that a large fraction of the hydrophobic tryptophan residues of the enzyme is exposed (i.e., situated on the surface of the protein). This peculiarity of the tertiary structure may be due to possible role of the tryptophan residues in the binding of aminoacyl-tRNA synthetases to tRNAs (39,40).

B. QUATERNARY STRUCTURE

As summarized in Table I, many types of subunit structure are possible (α_4, $\alpha_2\beta_2$, α_2, $\alpha_1\beta_1$, α_1). It is noteworthy, however, that the enzymes classified as belonging to the α_2- or α_4-group in most cases have been so categorized only on the basis of data indicating apparent identity of subunits. As for the enzymes classified as belonging to the $\alpha\beta$-type, their structures seem to be established by virtue of the difference in subunit molecular weights.

The molecular weights of synthetases are scattered around 72,000–120,000 on the one hand and around 175,000–260,000 on the other. The majority of enzymes in the first group have molecular weights of about 90,000–110,000. In some cases enzymes of similar specificity have similar molecular weights and probably subunit structures [e.g., methionyl-tRNA synthetases of E. coli (63), Bacillus brevis (64), and reticulocytes (65) or phenylalanine-tRNA synthetases of yeast (61), *E. coli* (59), and *Drosophila melanogaster* (66)]. However, in the case of tryptophanyl-tRNA synthetases from *E. coli* and beef pancreas both enzymes belong to the same α_2-group, but their molecular weights differ (19,25,26). The same is true of lysyl-tRNA synthetases from *E. coli* and yeast (22).

The enzymes referred to the α_1-group do not change in molecular weight after treatment with guanidinium chloride, urea, sodium dodecylsulfate, 2-mercaptoethanol, maleic anhydride, or oxidants. However, at least one of the enzymes of this group, leucyl-tRNA synthetase of *E. coli*, consists of two "masked" subunits, as reported in the interesting series of publications by Rouget and Chapeville (45,67). At first it was found that two forms of leucine-activating enzyme can be isolated from *E. coli* B: one form was not capable of catalyzing leucyl-tRNA formation, but catalyzed ATP–PP$_i$ exchange, whereas the other catalyzed both reactions. The molecular weights of the two forms were indistinguishable (104,000). Interconversion of the two forms was induced by a factor present in the 105,000-g supernatant of *E. coli* extract. More detailed studies showed that the conversion of the completely active enzyme E_2 into the partially active enzyme E_1 is catalyzed by a protease with an approximate molecular weight of 20,000. The reverse transformation is induced by a peptide with a molecular weight of 3000.

Mild tryptic hydrolysis of the pure E_2 enzyme affords a peptide that is able to reassociate with E_1, forming an enzyme indistinguishable from E_2; the only difference is that native E_2 is stable in 8 M urea and in 6 M guanidinium chloride, whereas E_1 and the E_1 "repaired" by the polypeptide decompose into two subunits.

On the basis of this evidence, Rouget and Chapeville (67) proposed a scheme for the structure of leucyl-tRNA synthetase according to which the enzyme consists of two presumably identical polypeptide chains ("subunits") covalently bound by a peptide of molecular weight 3000. The subunits are relatively resistant to proteolysis and can be imagined as rather compact globules connected by the peptide fragment. The presence of the covalent bonds between the "subunits" and the fragment is unimportant for the "subunit" association as well as for the catalytic activity, but removal of the fragment results in a change in the specificity to the amino acid and in abolition of the function of esterification of tRNA with the amino acid.

TABLE I
Subunit Structure of Some Purified Synthetases[a]

Synthetase for	Source	Molecular weight	Number and molecular weight of subunits	Reference
		α_1-*Structure*		
Arginine	*Bacillus stearother-mophilus*	78,000	1 × 78,000	156
Aspartate	Bakers' yeast	100,000	1 × 100,000	41
Glutamine	*E. coli*	69,000	1 × 69,000	20
Isoleucine	*E. coli* B	110,000	1 × 110,000	42, 43
	E. coli MRE600	102,000	1 × 102,000	33c
Leucine	*E. coli*	105,000	1 × 105,000	44, 45
Tyrosine[d]	Bakers' yeast	40,000–46,000	1 × 40,000	46
Valine	*E. coli*	110,000	1 × 110,000	43
	Yeast	122,000	1 × 122,000	22
		α_2-*Structure*		
Alanine	*E. coli*	160,000	2 × 80,000 (?)	47
Arginine[b]	*E. coli*	72,000	2 × 40,000 (?)	48
Histidine	*Salmonella typhimunium*	80,000	2 × 39,000	50c
Leucine	Yeast	120,000	2 × 60,000	50a
Lysine	*E. coli*	104,000	2 × 52,000	22
	Yeast	138,000	2 × 70,000	22
Methionine	*E. coli*	180,000	2 × 90,000	50
Proline	*E. coli*	94,000	2 × 47,000	51
Serine	*E. coli*	95,000	2 × 48,000	52
	Hens liver	120,000	2 × 58,000	50b
	Yeast	120,000	2 × 60,000	53
		95,000	2 × 48,000	100e
Tryptophan	*E. coli*	74,000	2 × 37,000	54
	Beef pancreas[e]	120,000	2 × 60,000	26, 55, 56
	Human placenta	118,000	2 × 58,000	59b
Tyrosine	*E. coli*	97,000	2 × 48,000	24b, 24c
		$\alpha\beta$-*Structure*		
Glutamate	*E. coli*	102,000	1 × 56,000 1 × 46,000	21, 57

TABLE I (*Continued*)

Synthetase for	Source	Molecular weight	Number and molecular weight of subunits	Reference
		$\alpha_4{}^e$-*Structure*		
Glutamine	*Micrococcus cryophilus*	190,000	$4 \times 53,000$	58
Phenylalanine	*E. coli*	181,000	$4 \times 43,000$	59
Tyrosine	*Saccharomyces cerevisiae*	116,000	$4 \times 31,500$	59a
		$\alpha_2\beta_2$-*Structure*		
Glycine	*E. coli*	220,000	$2 \times 33,000$ $+ 2 \times 80,000$	60
Phenyla-lanine	Bakers' yeast	220,000	$2 \times 50,000$ $+ 2 \times 60,000$	61
		262,000	$2 \times 61,000$ $+ 2 \times 70,000$	62
	Rat liver		$2 \times 69,000$ $+ 2 \times 74,500$	33d

ᵃ This table includes only those enzymes for which information regarding the subunit structure has been published. The references give molecular weights of the subunits, and not necessarily the purification of the enzyme or the determination of its molecular weight.

ᵇ Hirshfield and Bloemers (314) found no dissociation into subunits after poly-acrylamide-gel electrophoresis in the presence of dodecylsulfate.

ᶜ Preddie (69) suggested an $\alpha_2\beta_2$-structure for this enzyme, but this seems to be incorrect (55, 56).

ᵈ Discrepancy between molecular weight estimations and the number of sub-units (α_1-α_4) found for yeast tyrosyl-tRNA synthetase isolated from commercial yeast (46) and from pure strain (59a) is remarkable. It is possible that this enzyme is composed of two subunits and the molecular weight observed (46) reflects the dissociation of the α_2-type protein (cf. 24b). On the other hand, hidden breaks caused by protease action may convert α_2-type enzyme into α_4-type (59a, cf. 50, 56).

ᵉ The real existence of this group of aminoacyl-tRNA synthetase is not evident. Appearance of hidden breaks due to protease contaminations were not fully excluded and if gene duplication for some synthetases occurs (50d), it may explain the two-fold dimunitation in molecular weight of each subunit (cf. 45), that is, $\alpha_2 \rightarrow \alpha_4$ conversion.

It follows from the studies cited that the uninterrupted polypeptide chain in synthetases of the α_1-type is not incompatible with a structure composed of "physical" subunits.

It seems timely to make similar studies of other synthetases of the α_1-type; the close similarity in the properties of these enzymes in other respects suggests the possibility that they also consist of "physical" subunits.

"Physical" subunit structure is characteristic of α_2-type enzymes as well as of those of the α_1-type. It was believed earlier that E. coli methionyl-tRNA synthetase is a tetramer (63) composed of subunits with molecular weights of 43,000–48,000. More detailed studies revealed that the enzyme consists of two presumably identical subunits (50). In the course of isolation each of the two subunits can be cleaved into two polypeptide chains of equal length without loss of enzymatic activity. The lability of the bond between the two globules that form the subunit as well as the equal lengths of the two fragments obtained are very reminiscent of the behavior of E. coli leucyl-tRNA synthetase.

All these data make one cautious about the claims of α_2- and α_4-structure because the lability of the interglobular bond to protease action may lead to cleavage(s) and to doubling of the number of chains. The existence of enzymes composed of two similar or even identical halves within one polypeptide chain suggests that this phenomenon could be due to duplication of the structural gene and to subsequent elimination of the intergene termination signal. Other explanations are also possible.

Internal duplication was not observed with the enzyme from the multicellular organism, beef-pancreas tryptophanyl-tRNA synthetase. Subunits of this enzyme can be converted by limited proteolysis from a 60,000 to 25,000 molecular weight, thus resembling the first sight splitting of the subunit into two halves as described for methionyl- (50) and leucyl- (45) tRNA synthetases. However, quantitative analysis showed that each subunit lost half of its amino acids and that the peptide fingerprint was profoundly different before and after limited proteolysis (33e). In other words each polypeptide chain was not scissed in the middle, but half of each chain was hydrolyzed.

Glutamyl-tRNA synthetase from E. coli seems to be the first enzyme among aminoacyl-tRNA synthetases that belongs to the $\alpha\beta$-type (21, 57). The 56K subunit is catalytically active in both reactions, whereas 46S protein is fully inactive in a separate state. In the aminoacylation reaction the 46S protein decreases the K_m values of the 56K enzyme for glutamate and ATP by a factor of 17 and 6, respectively, and increases the K_m value for tRNA^{Glu} by a factor of 2. Like the *E. coli* DNA-dependent RNA polymerase (see ref. 70), this synthetase seems to be composed of a "core" enzyme and of a "factor" protein playing a regulatory role.

C. CRYSTALLIZATION

As with other proteins, crystallization of synthetases seems to be a necessary prerequisite for the determination of the three-dimensional structure.

Formation of crystals in aminoacyl-tRNA synthetase preparations was reported long ago (71,72), but these first observations were poorly documented, and the investigations were not continued.

The first well-characterized crystals of a synthetase were obtained by Rymo et al. (73,74), who obtained 0.3–1-mm-long monocrystals of yeast lysyl-tRNA synthetase in phosphate buffer. Regular trigonal or ditrigonal bipyramids were formed in the presence of KCl, sodium arsenate, or a GMP–CMP mixture on dialysis against phosphate buffer. The crystallization technique was subsequently improved (75), which made it possible to obtain crystals of regular form in the presence of different ions and buffers. Both the old and the improved techniques give the same $P3_121$ space group and the same dimensions of the trigonal unit cell ($a = 118$, $c = 190$ Å). The high content of water (65–78%) was presumably the reason for the rapid destruction of the crystals under X-ray beams, leading to poor diffraction patterns even with crystals of good macroscopic form.

Crystals suitable for high-resolution X-ray diffraction studies have been obtained by Waller *et al.* (36) from methionyl-tRNA synthetase. The enzyme was pretreated with trypsin to remove the functionally unessential part of the molecule. The content of water in these crystals was 48%, and they survived 100 hr under X-ray irradiation without loss of activity. The asymmetric unit of the crystals consisted of a single molecule. It was possible to obtain a heavy-atom derivative by soaking the crystals with $Pt(CN)_4^{2-}$ and $UO_2F_5^{3-}$. In the latter case it appears that there is probably only one site of high occupancy per asymmetric unit for the heavy ion (75a).

A native enzyme—yeast leucyl-tRNA synthetase—was crystallized (77) by simple dialysis of a concentrated enzyme solution against 50%

saturated ammonium sulfate at 4°C and pH 4.6–6.5. The dimensions of the orthorhombic crystals obtained were 0.4–0.5 × 0.4 × 0.5 × 0.2 × 0.3 mm; the unit-cell dimensions were $a = 75.5$, $b = 110.7$, and $c = 124.0$ Å; the space group was 1_{222} or $1_{2_12_12_1}$. Unfortunately degradation of crystals occurred in 24 hr under X-ray irradiation. However, the distinct character of the diffraction patterns encourages further studies of these crystals; such work may provide information on the tertiary structure of the enzyme.

Variable pH and ammonium sulfate concentrations were studied as conditions for the crystallization of tyrosyl-tRNA synthetase from *Bacillus stearothermophilus* (75b) and found to be optimal around 55% saturation at 0° and pH 6–7. The crystals can be produced up to sizes of 0.32 × 0.32 × 0.12 mm and are well-ordered, diffracting strongly out to 2.7 Å. The crystal system is trigonal and the space group is P3$_1$21. The unit cell is $a = b = 64.4$ Å, $c = 238$ Å. Soaking the crystals in 10^{-5} M MeHg produced a clean single-site isomorphous MeHg derivative from which the Patterson difference was derived.

Two-dimensional crystals of pancreatic tryptophanyl-tRNA synthetase have been obtained by back extraction of the ammonium sulfate precipitate of the pure enzyme in the presence of tryptophan (78). Rodlike particles of approximately 1 μ in length and 125 ± 10 Å in diameter were seen in the electron microscope. The particles tended to associate into "stacks" of 10–15 rods packed side by side. Along the axes of the rods 40–50 Å density periods are visible. Dissociation of the "rods" gives rise to "washers" with central holes. Optical diffraction data suggest that the rods are multistranded helices. Further investigation of these paracrystals may provide information on the tertiary structure of the enzyme.

Thus the tertiary and quaternary structures of synthetases are now being studied using crystalline and paracrystalline preparations. It is hoped that these direct approaches will lead to an elucidation of the three-dimensional structure of the enzymes, which is necessary for understanding the mechanism of their actions.

III. Enzyme–Substrate Complexes

Isolation of enzyme–substrate (ES) complexes and investigation of their properties is a necessary stage in the studies of any enzyme. With the aminoacyl-tRNA synthetases this stage is especially important because these enzymes have three chemically very different substrates: amino acid, nucleoside triphosphate, and polynucleotide. Studies on enzyme-substrate complexes, make it possible to distinguish between the catalytic functions (activation and acylation) and the process of substrate binding.

A. DETECTION AND ISOLATION

Various methods have been applied to the study of the ES complexes of the synthetases. On the one hand, there are methods that do not involve the isolation of complexes; for instance, studies of the effect of substrates on the thermostability of the synthetases (79–81), on resistance to proteolysis (81), or the action of other denaturing factors (82) or studies of the effect of substrates on the fluorescence of the enzyme (28,38,83–87). Fluorescence techniques are used to investigate such characteristics of the enzyme as quantum yield, position and half-width of the fluorescence maximum of the aromatic amino acid residues of the enzyme protein (mainly tryptophan) (28,38,83), or the kinetics of the fluorescence quenching in the presence of denaturing agents (85). Both approaches—inactivation studies and fluorescence techniques—have been applied to study the binding of all three substrates. A direct method for the detection of ES complexes is equilibrium dialysis. In the case of the synthetases it is applicable, obviously, only to the low-molecular-weight substrates [i.e., amino acids and ATP (22,28,43,88)]. Such techniques have made it possible to determine the substrate constant (K_s) values—the true dissociation constants of the ES complexes, which are an adequate measure of the affinity of the enzymes for the substrates.

Also, ES complexes have been studied by methods involving their isolation. In this approach, the incubation mixture containing the enzyme and the substrate is subjected to fractionation by gel filtration, gel chromatography, sucrose-gradient centrifugation, or electrophoresis (see Table II). The latter three techniques are used mainly to isolate the complexes of the enzymes with tRNA and provide separation from not only excess substrate but also from the unbound enzyme.

The use of immobilized systems is a very important approach to the investigation of the synthetase ES complexes. Affinity chromatography of synthetases on tRNA-containing columns (89,90) makes it possible to detect the ES complexes, to study the conditions of their association and dissociation, and to purify the synthetases corresponding to the immobilized tRNAs. It was found (91,91a) that valyl- and lysyl-tRNA synthetases of rat liver form complexes with tRNA^{Val} and tRNA^{Lys} bound covalently to columns of polyacryl hydrazide agar gel after oxidation of the terminal ribose residues with sodium periodate. The absorption of synthetases by the columns takes place at low ionic strength, and elution is achieved by increasing the latter. The procedure provides thirtyfold to ninetyfold enrichment. The same approach has been applied (92) to the isoleucyl-tRNA synthetase of *E. coli*. The immobilization was achieved by binding isoleucyl-tRNA^{Ile} to bromo-acetamidobutyl-Sepharose; the enrichment factor was 27.5. Remi, Birmele, and Ebel (93) used periodate-oxidized tRNA^{Phe} bound to hydrazide-Sepharose with the bond subsequently stabilized by NaBH$_4$ reduction. The column was successfully used to obtain highly purified yeast phenylalanyl-tRNA synthetase. To decrease the effect of the nonspecific adsorption of proteins, the crude extract was passed at first through a similar column with bulk tRNA containing no tRNA^{Phe}.

With polyuridylic acid instead of tRNA, a column is obtained that selectively adsorbs the lysyl-tRNA synthetase of *E. coli* (94,95). Polyuridylic acid is an inhibitor of this enzyme and forms complexes with it (96,97); thus adsorption on the column is due to enzyme–inhibitor complex formation. The purity of the enzyme obtained was about the same as of that obtained by conventional purification procedures. The method can be applied both to the investigation of the enzyme–inhibitor complex and, due to ready availability of polyuridylic acid, to preparative-scale isolation of the purified enzyme.

An inverse technique—that is, binding of a synthetase to a Sepharose column and adsorption of the specific tRNA—provides highly purified tRNA; this approach has been explored with the isoleucine (98) and serine (98a) ES pair of *E. coli*.

Studies on all the aminoacyl-tRNA synthetases examined indicate that a free COOH group plays little part in the association of the amino acid with the enzyme (1, 5, 6). This finding provides the basis for an affinity column in which the amino acid with a free NH$_2$ group essential for binding with enzyme is linked to an insoluble support via the carboxyl end.

Based on these considerations a method was developed (100b, 100c) using activated Sepharose coupled with methionine or phenylalanine. In both cases highly purified preparations were obtained, although some nonspecific retention of the proteins on the columns (possibly due to ion-exchange effects) were noticed. When Sepharose was coupled even with aminoalkyl instead of aminoacyl residues some aminoacyl-tRNA synthetases were still bound to the column with a mechanism that probably differs from that of the ordinary ion exchange (98b).

Affinity chromatography was modified in such a way as to avoid the necessity of the chemical coupling procedure (100d). Crude yeast synthetase may be loosely retained on the phosphocellulose column and selectively eluted with purified tRNA specific for the given amino acid. Affinity elution looks very attractive as a convenient way for enzyme purification, although at the moment this procedure has not yet been repeated in another laboratory.

An interesting way of investigating the E–tRNA complexes was suggested by Yaniv and Gros (105), Befort et al. (114), and Lagerkvist and Rymo (115). The sensitivity of the tRNA to nuclease digestion was compared in the free state and in the E–tRNA complex. In the *E. coli* valyl-tRNA synthetase–tRNA^{Val} complex the region adjacent to position 8 was preferentially protected in comparison with the rest of the molecule (105) after T1-RNase digestion. Similarly yeast tRNA^{Val} in a complex with cognate enzyme retained amino-acid-acceptor activity after incubation with T1-RNase under conditions where tRNA^{Val} alone was completely inactivated (115,116). Yeast tRNA^{Phe} of the T1-RNase-treated E–tRNA complex sedimented with phenylalanyl-tRNA synthetase in a sucrose gradient and probably was an intact tRNA^{Phe} (114).

A detailed study was undertaken by Hörz and Zachau (128) with complexes of yeast tRNA^{Phe} and tRNA^{Ser} with their cognate amino-acyl-tRNA synthetases. It was found that tRNA^{Phe} was strongly protected by the cognate enzyme at both nuclease-sensitive sites, the dihydrouridine loop and the anticodon; tRNA^{Ser} was shielded by the cognate synthetase only at one of the two nuclease targets, the GCCA terminus, whereas splitting at the other target, the anticodon, was not prevented by the synthetase. In addition it was shown that the phenyl-alanine enzyme protected 1 mole of tRNA^{Phe} and the serine enzyme protected 2 moles of tRNA^{Ser} against nuclease attack. This approach seems to be more sensitive than the purely physical ones. For example, attempts to find changes in the kinetics of release of slowly exchanging hydrogens from Ile-tRNA (*E. coli*) previously equilibrated with ^3H_2O and then allowed to interact with cognate enzyme were unsuccessful (129).

Table II surveys the methods used to detect, to isolate, and to study the ES complexes. In addition to the methods listed, electrophoresis (103,115,122), nonequilibrium dialysis (104), influence of the substrates on the chemical modification of synthetases (82) have been also applied. Isolation of the complexes by means of gel chromatography and gel filtration is not included in the table since data on many of them have been already compiled by Allende and Allende (123). In addition to their table other recent publications may also be cited here (103,104,111–113,116–121).

As a rule complexes of synthetases with amino acids cannot be isolated by gel filtration.* However, the complexes of synthetases with ATP are stable enough and can be easily isolated by gel filtration. It is considered important to demonstrate that ATP remains unchanged after binding to the enzyme. With threonyl-tRNA synthetase this was done (124) using ATP labeled with ^3H at the adenosine moiety and with ^{32}P at the γ-phosphate moiety. The complex contained ^3H and ^{32}P in a molar ratio of $1:1$. In other cases ATP isolated from the complex after its decomposition was identified by electrophoresis (125) or ion-exchange chromatography (118).

The binding of radioactive ATP to the enzyme in the presence of Mg^{2+} in the absence of added amino acids has been observed also by other workers (126,127), but the possibility has not been ruled out that the binding was due to the presence of endogenous amino acids.

Methods involving the isolation of stable complexes result in separation of one or both unreacted components and thus can be used for purifying the complexes and for their subsequent investigation. On the other hand, the preparation of the complexes under equilibrium conditions does not afford the ES pair in a pure state, but makes it possible to determine the quantitative characteristics of the binding, such as the number of binding sites, affinity constant, or related constants.

B. PROPERTIES OF THE COMPLEXES

Table III surveys some quantitative data characterizing the affinity of synthetases for substrates. These values correspond to equilibrium

---

*Only one exception is known: Dorizzi, Labouesse, and Labouesse (120) described a complex of tryptophan with tryptophanyl-tRNA synthetase from beef pancreas that could be isolated by gel filtration. These data contradict the results obtained earlier with the same enzyme in another laboratory: no E–Trp complex formation was noted by means of the same technique (117,118).

conditions. Two of the methods of the association-constant determination—that based on estimation of the protection constant in thermal inactivation, and that based on fluorescence-intensity measurement in the presence of substrate—are in principle general for all the substrates if these substrates induce the corresponding changes in the enzymes.

The literature contains also a great deal of data characterizing the ES affinity by nonequilibrium methods. The constants thus obtained are sometimes in a good accord with the true (equilibrium) ones, suggesting thermodynamic stability of the corresponding complexes (i.e. their long lifetime compared with the time of the isolation).

The method of filtration through nitrocellulose filters, though rapid and convenient, needs allowance for some additional factors and is applied most often in comparative studies where relative, rather than absolute, values are of importance.

The duration of the experiment is several hours in the majority of the gel-filtration and sucrose-gradient-centrifugation procedures. The complexes obtained can therefore be used for different studies, but application of the method to determine the affinity constants is unreliable. The alternative may consist in equilibrating the gel-filtration column with free substrate and in estimating the "lack" of substrate after the peak of the ES complex, as made for the tRNA^{Leu}–leucyl-tRNA synthetase pair of *E. coli* (111,112). This procedure makes it possible to find the equilibrium dissociation constant.

Remarkable is the wide range of the absolute K_s values (from 10^3 to 10^8, see Table III). The affinity of tRNA for synthetases is usually higher than that of the smaller substrates. The affinities of ATP and amino acids are in many cases of the same order of magnitude. However, it still remains unknown, with a few exceptions (38,109,111), to what extent these quantitative relationships are retained when the ES complex is formed in the presence of one or two of the other substrates.

Studies of the isolated ES complexes reveal in fact the potential properties of the partners, but the superposition of the properties of these complexes is not a direct description of the enzymatic reaction itself. For this reason studies of ES complexes must be aided by first of all kinetic investigations because only a combination of at least two approaches may lead to an understanding of the mechanism of the enzyme action.

In general, there is a correlation between the number of subunits and the number of substrate-binding centers for each of the substrates.

TABLE II
Enzyme–Substrate Complexes

Synthetase for	Source	Substrate	Reference
	Detected by changes in thermostability		
Arginine	*E. coli*	Arginine	99, 100
	Bacillus stearothermophilus	tRNA	100a
		tRNA Arginine	
Glutamate	*E. coli*	tRNA	101
Isoleucine	Lupine seeds	tRNA	102
	Bacillus stearothermophilus	tRNA	103
Lysine	Rat liver	Lysine	82
Proline	Rat liver	Proline	80
Phenylalanine	*E. coli*	Phenylalanine	104
		ATP	
		tRNA	
	Rat liver	Phenylalanine	33d
		ATP	
		tRNA	
Tryptophan	Beef pancreas	Tryptophan	81
Valine	Rat liver	Valine	82
		tRNA	79
	E. coli	tRNA	105
	Detected by changes in fluorescence properties		
Glutamate	*E. coli*	tRNA	101
Isoleucine	*E. coli*	ATP	85
		tRNA	
Methionine	*E. coli*	$tRNA_F^{Met}$	28
		$tRNA_M^{Met}$	
		Methionyladenylate	88
		Methioninyladenylate	
Phenylalanine	*E. coli*	tRNA	86
Serine	Yeast	tRNA	87, 106
Tryptophan	Beef pancreas	tRNA	107
Valine	*E. coli*	$tRNA_{1,2}^{Val}$	38
		Valine	
		ATP	
	Detected by equilibrium dialysis in solution		
Isoleucine	*E. coli*	Isoleucine	43
		Isoleucyladenylate	

158

TABLE II (*Continued*)

Synthetase for	Source	Substrate	Reference
Lysine	*E. coli*	Lysine ATP	22
	Yeast	ATP Lysine	
Methionine	*E. coli*	Methionine ATP	28, 88
Valine	Yeast	Valine ATP	22
	E. coli	Valyladenylate	43

Detected by equilibrium dialysis on the column

Leucine	*E. coli*	ATP tRNA$_1^{Leu}$	111 112

Detected by fixation on nitrocellulose filters

Isoleucine	*E. coli*	tRNA Isoleucyladenylate	108–110
	B. stearothermophilus	tRNA	103
Leucine	*E. coli*	tRNA	111
Serine	*E. coli*	tRNA	44

Detected by sucrose-density-gradient centrifugation

Glutamate	*E. coli*	tRNA	101
Glutamine	*E. coli*	tRNA	20
Leucine	*E. coli*	tRNA	113
Phenylalanine	Yeast	tRNA	114
	E. coli		86, 104
Serine	*E. coli*	tRNA	113
Tryptophan	*E. coli*	tRNA	54
Valine	Yeast	tRNA	115, 116
	E. coli		105

Detected by affinity chromatography

Isoleucine (enzyme fixed)	*E. coli*	tRNA	98
Isoleucine	*E. coli*	tRNA fixed	92
Lysine	Rat liver	tRNA fixed	90, 91
Phenylalanine	Yeast	tRNA fixed	93
Valine	Rat liver	tRNA fixed	90, 91

159

TABLE III

Enzymes–Substrate Complexes: Some Characteristics

Synthetase for	Molar E/S ratio[a]	Substrate	κ^b	Conditions of measurement	Method of ES complex detection	References
Glutamate, E. coli	1:1	tRNA^{Glu}	$\kappa = 3.6 \times 10^{-7}$	40°C, 0.05 Na–Mes buffer, pH 6.0 + 0.02 M ME + 10% glycerol + 0.005 M MgCl$_2$	Thermal inactivation	101
Isoleucine, E. coli	1:1	Ile-tRNA	$\kappa_a = 1.7 \times 10^8$	17°C, 0.05 M K-phosphate, pH, 5.5, + 0.01 M MgCl$_2$ + 0.01 M ME	Adsorption on nitrocellulose filters	108, 109
	1:1	Isoleucine	$\kappa_a = 0.95 \times 10^5$		Equilibrium dialysis	43
	1:1		$\kappa_s = (7 \pm 2) \times 10^{-6}$	10°C, 0.05 M K-phosphate, pH 8.2, + 0.01 M ME	Fluorimetry with a reporter group	157
			$\kappa_s = (7.7 \pm 0.8) \times 10^{-6}$	10°C, 0.05 M K-phosphate, pH 8.2, + 0.01 M ME + ATP	Fluorimetry with a reporter group	157

160

	1:1	ATP^{-4}	$\kappa_s = (1.4 \pm 0.4) \times 10^{-4}$	10°C, 0.05 M K-phosphate, pH 8.2, + 0.01 M ME	Fluorimetry with a reporter group	157
			$\kappa_s = (1.0 \pm 0.3) \times 10^{-4}$	10°C, 0.05 M K-phosphate, pH 8.2, + 0.01 M ME, + isoleucine	Fluorimetry with a reporter group	157
Isoleucine, *Bacillus Stearothermophilus*	1:1	tRNA	$\kappa_a = 0.4 \times 10^9$	15°C, 0.05 M K-phosphate, pH 5.5, + 0.01 M MgCl$_2$ + 0.01 M ME	Adsorption on nitrocellulose filters	103
			$\kappa = 0.11 \times 10^{-9}$	70, 5°C, 0.01M tris–HCl, pH 7.5	Thermal inactivation	
Lysine, *E. coli*	1:2	Lysine[d]	$\kappa_a = 6.4 \times 10^3$ $\kappa_a = 3.6 \times 10^3$	4°C, 0.1 M K-phosphate, pH 7.0	Equilibrium dialysis	22
	1:2	ATP	$\kappa_a = 0.3 \times 10^3$	4°C, 0.1 M K-phosphate, pH 7.0	Equilibrium dialysis	22
Lysine, yeast	1:2	Lysine[d]	$\kappa_a = 6.2 \times 10^3$ $\kappa_a = 5.8 \times 10^3$	4°C, 0.1 M K-phosphate, pH 7.0	Equilibrium dialysis	22
	1:2	ATP	$\kappa_a = 0.4 \times 10^3$	4°C, 0.1 M K-phosphate, pH 7.0	Equilibrium dialysis	22

TABLE III (*Continued*)

Synthetase for	Molar E/S ratio[a]	Substrate	κ^b	Conditions of measurement[a]	Method of ES complex detection	References
Methionine, *E. coli*, mol. wt. 96,000	1:2	Methionine	$\kappa_a = 2.2 \times 10^4$	0.1 M K-phosphate, pH 7.0, + 10 mM ME	Equilibrium dialysis	28
	1:2	ATP	$\kappa_a = 3 \times 10^3$	0.1 M K-phosphate + 10 mM ME + 10 mM, MgCl$_2$, pH 7.0	Equilibrium dialysis	28
	1:2	tRNA$_\mathrm{M}^{\mathrm{Met}}$	$\kappa_a = 2.5 \times 10^6$	0.01 M tris–phosphate, pH 8.0	Fluorimetry	28
	1:2	tRNA$_\mathrm{F}^{\mathrm{Met}}$	$\kappa_a = 2.0 \times 10^6$	0.01 M tris–phosphate, pH 8.0	Fluorimetry	28
Methionine, *E. coli*, mol. wt. 173,000	1:2	L-Methionine	$\kappa_a = (2.4-3.1) \times 10^4$	0.02 M K-phosphate, pH 7.6, + 0.01 M ME	Equilibrium dialysis	88
	1:2		$\kappa_a = 5 \times 10^4$	0.1 M K-phosphate, pH 7.6, + 5 mM MgCl$_2$ + 0.01 M ME	Equilibrium dialysis	88
	1:4	ATP	$\kappa_a = 0.18 \times 10^4$	0.02 M K-phosphate, pH 7.6,	Equilibrium dialysis	88

	Ratio	Association constant	Conditions	Method	Ref.
	1:4	$\kappa_a = 0.3 \times 10^4$	+ 0.1 M KCl + 0.5 mM MgCl$_2$ + 0.01 M ME	Equilibrium dialysis	88
L-methioninyladenylate	1:2	—	0.1 M K-phosphate, pH 7.6, 1 mM EDTA (Na)	Fluorimetry	88
			0.02 M K-phosphate, pH 7.6, + 0.01 M ME	Difference-absorbance spectroscopy	88
Serine (yeast) tRNA^{Ser}	1:2	$\kappa_s = 3 \times 10^{-7}$ to 3×10^{-8}	5°C, 0.03 M K-phosphate, pH 7.3, 5 mM MgCl$_2$, 0.5 mM EDTA, 1 mM GSH, 10% (v/v) glycerol	Fluorimetry with a reporter group	106
	1:1.2-1.5	$\kappa_a = (1-3) \times 10^7$	21°C, 0.03 M K-phosphate, pH 7.3, 5 mM MgCl$_2$, 0.5 mM EDTA, 1 mM GSH, 10% (v/v) glycerol	Fluorimetry with a reporter group	106
tRNA^{Phe}	1:2	$\kappa_a = 10^6$	21°C, 0.03 M K-phosphate, pH 7.3, 5 mM MgCl$_2$, 0.5 mM EDTA, 1 mM GSH, 10% (v/v) glycerol	Fluorimetry with a reporter group	106

TABLE III *(Continued)*

Synthetase for	Molar E/S ratio[a]	Substrate	κ[b]	Conditions of measurement[a]	Method of ES complex detection	References
Serine (yeast)	1:2	tRNA^{Ser}	$\kappa_a = 2.1 \times 10^6$	20°C, 0.03 u K-phosphate, pH 7.5, 0.01 u uqSOy, 0.07 M K$_2$ SOy, 1 mM DDT, 0.1 mu EDTA, 8% (v/v) glycerol	Fluorimetry	87
	1:1	tRNA^{Ser}	$\kappa_a = 2 \times 10^7$	20°C, 0.03 M K-phosphate, pH 7.2, 0.5 mM EDTA, 0.5 mM DTE, 5 mM MgCl$_2$	Fluorimetry	100e
			$\kappa_a = 4 \times 10^6$	20°, 0.03 M K-phosphate, pH 7.2, 0.5 mM EDTA, 0.5 mM DTE, 5 mM MgCl$_2$, 0.1 M KCl	Fluorimetry	100e
Serine (E. coli)	1:2	Serine	$\kappa_s = 2.3 \times 10^{-5}$	4°C, 0.05 M Tris-HCl, pH 7.4, 0.01 M MgAc, 0.02 M,	Equilibrium dialysis	98a

	1:2	ATP	$\kappa_s = 6.5 \times 10^{-6}$	4°C, 0.05 M Tris-HCl, pH 7.4, 0.01 M MgAc, 0.02 M ME, 10% (v/v) glycerol	Equilibrium dialysis	98a
Tyrosine (E. coli)	1:1	Tyrosine	$\kappa_a = 2.1 \times 10^5$	4°C, 0.1 M K-phosphate, pH 7.1, 5 mM ME, 10% (v/v) glycerol	Equilibrium dialysis	24c
	1:2	Tyr-tRNA	$\kappa_a = 3.1 \times 10^8$	20°C, 0.05 M cacodylate buffer, pH 6.0, 5 mM MgCl$_2$, 0.01 M ME	Adsorption on nitro-cellulose filters	24b
Valine (E. coli) mutant T918	—	tRNA^{Val}	$\kappa_a = 1.3 \times 10^7$	25°C, 0.04 M tris-HCl, pH 7.8, 0.07 M NH$_4$Cl, 6 × 10^{-3} M MgAc, 6 × 10^{-3} M ME, 0.2 mg/ml BSA	Thermal inactivation	31
Valine (E. coli)	—	L-valine	$\kappa = 5 \times 10^{-5}$	37°C (extrapolated), 0.1 M tris buffer, pH 8.0, 0.5 mg BSA	Thermal inactivation	186

TABLE III (*Continued*)

Synthetase for	Molar E/S ratio[a]	Substrate	κ[b]	Conditions of measurement[a]	Method of ES complex detection	References
			$\kappa = 6 \times 10^{-6}$	37°C (extrapolated), 0.1 M tris buffer, pH 8.0, 0.5 mg BSA, 10 mM MgCl$_2$, 4 mM ATP	Thermal inactivation	186
	—	D-valine	$\kappa = 1 \times 10^{-1}$	37°C (extrapolated), 0.1 M tris buffer, pH 8.0, 0.5 mg BSA	Thermal inactivation	186
			$\kappa = 6.4 \times 10^{-3}$	37°C (extrapolated), 0.1 M tris buffer, pH 8.0, 0.5 mg BSA, 10 mM MgCl$_2$, 4 mM ATP	Thermal inactivation	186
	—	L-α-Aminobutyrate	$\kappa = 10^{-1}$	37°C (extrapolated), 0.1 M tris buffer, pH 9.0, 0.5 mg BSA	Thermal inactivation	186
			$\kappa = 7 \times 10^{-3}$	37°C (extrapolated), 0.1 M tris buffer, pH 8.0, 0.5 mg	Thermal ainctivation	186

166

			BSA, 10 mM MgCl$_2$, 4 mM ATP		
1:1	Valine	$K_a = 2 \times 10^4$	0.03 M cacodylate, pH 6.5	Fluorimetry	38
		$K_a = 1.5 \times 10^4$	0.03 M cacodylate, pH 6.5, mM MgAc$_2$	Fluorimetry	38
1:1	ATP	$K_a = 1.2 \times 10^5$	0.03 M cacodylate, pH 6.5	Fluorimetry	38
		$K_a = 2 \times 10^5$	0.03 M cacodylate, pH 6.5, 6 mM MgAc$_2$	Fluorimetry	38
1:1	tRNA$_1^{Val}$	$K_a = 1 \times 10^6$	0.03 M cacodylate, pH 7	Fluorimetry	38
		$K_a = 3.5 \times 10^6$	0.03 M cacodylate, pH 6.5	Fluorimetry	38
		$K_a = 7 \times 10^6$	0.03 M phosphate, pH 6	Fluorimetry	38
		$K_a = 3.5 \times 10^7$	0.03 M phosphate, pH 5.5	Fluorimetry	38
		$K_a = 5 \times 10^6$	6 mM MgCl$_2$ + 0.03 M cacodylate, pH 7	Fluorimetry	38
		$K_a = 7 \times 10^6$	6 mM MgCl$_2$ + 0.03 M cacodylate, pH 6.5	Fluorimetry	38

TABLE III (*Continued*)

Synthetase for	Molar E/S ratio[a]	Substrate	K[b]	Conditions of measurement[a]	Method of ES complex detection	References
			$K_a = 2 \times 10^7$	6 mM MgCl$_2$ + 0.03 M phosphate, pH 6	Fluorimetry	38
			$K_a = 5 \times 10^7$	6 mM MgCl$_2$ + 0.03 M phosphate, pH 5.5	Fluorimetry	38
	1:1	tRNA$_{2A,B}^{Val}$	$K_a = 5 \times 10^7$	6 mM MgCl$_2$ + 0.03 M phosphate, pH 6.5	Fluorimetry	38
			$K_a > 10^8$	6 mM MgCl$_2$ + 0.03 M phosphate, pH 5.5	Fluorimetry	38
Valine, yeast	1:1	Valine	$K_a = 8.4 \times 10^3$	4°C, 0.1 M K-phosphate, pH 7.0	Equilibrium dialysis	22
	1:1	ATP	$K_a = 0.4 \times 10^3$	4°C, 0.1 M K-phosphate, pH 7.0	Equilibrium dialysis	22

[a] Extrapolated values except for heterologous E-tRNA pairs.

[b] The constants K listed in this table are as follows: K_a, equilibrium association (or binding, or affinity) constant (M^{-1}); K_s, dissociation constant (M); κ, protection constant (M).

[c] Abbreviations: Ac, acetate; BSA, bovine-serum albumin; EDTA, ethylenediaminetetraacetate; DTE, dithioerythritol; GSH, glutathione (reduced form), ME, 2-mercaptoethanol; MES, morpholinoethane sulphonic acid.

[d] Determinations with two chromatographically different enzyme fractions.

There are two types of such correlation. The number of substrate-binding centers can be equal to the number of subunits. Examples are the dimer form of methionyl-tRNA synthetase, which binds two molecules each of ATP, methionine, and tRNA^{Met} (28); the lysyl-tRNA synthetases of yeast and *E. coli*, which consist of two subunits and bind two molecules each of ATP and lysine (22); and the monomer valyl-, isoleucyl-, and leucyl-tRNA synthetases of *E. coli*, which bind one molecule each of the substrates (38,43,109). Alternatively the whole oligomer may contain a single substrate-binding site. For example, *E. coli* phenylalanyl-tRNA synthetase composed of four identical subunits binds a single molecule of phenylalanine and a single molecule of tRNA^{Phe} (104). The tetrameric form of methionyl-tRNA synthetase is an interesting exception. According to equilibrium-dialysis data (88), it contains four ATP-binding sites, but only two methionine-binding sites. The reason for this "functional asymmetry" is unknown.*

IV. Catalysis

A. THE ORDER OF SUBSTRATE BINDING

The study of the order of substrate binding with multisubstrate enzymes is not a trivial one; knowledge of the order of substrate addition may shed light on the mechanism of the enzymatic reaction. The problem can be subdivided into two parts: the order of binding in the absence of tRNA, measured by means of ATP–PP$_i$ isotope-exchange reaction, and the order of binding of all three substrates in the tRNA aminoacylation reaction.

1. ATP–PP$_i$ Exchange

Three somewhat different approaches have been applied to determine the order of substrate binding in the ATP–PP$_i$ reaction: (*a*) analysis of the dependence of reaction rate on the concentration of

---

*With the same enzymes isolated in various laboratories non-coincidence in determinations of the number of binding sites could be visualized even in the cases when similar techniques were applied. For instance, one (100*f*, 100*g*) or two (100*h*) binding sites were found for both phenylalanine and tRNA^{Phe} with yeast phenylalanyl-tRNA synthetase ($\alpha_2\beta_2$); one (100*e*) or two (87, 100*g*) binding sites were found for tRNA^{Ser} with yeast serine enzyme. This circumstance makes it premature to discuss possible relations between subunit composition and number of binding sites.

substrates from the viewpoint of a given hypothesis on the order of binding (130); (b) graphic treatment of the kinetic data obtained by the use of substrate analogs that are inhibitors of the reaction (see ref. 131); (c) use of the statistical method of "sequential experiment planning" (132) to treat the set of kinetic points in a manner similar to that outlined in approach b after representation of the complete reaction-kinetics scheme (133). A disadvantage of approaches b and c is the necessity of the postulate that the inhibitors form complexes with the same forms of the enzymes as those involved in the binding of substrates, although this assumption seems to be a reasonable one. Most often, anamorphoses are employed to treat the dependence of the reaction rate on substrate and inhibitor concentrations. However, the use of anamorphoses for the interpretation of the kinetics sometimes leads to equivocalities, which were noticed for the first time with some other enzymes (134,135). Cole and Schimmel (130), who studied the order of ATP and isoleucine binding to $E.$ $coli$ isoleucyl-tRNA synthetase, arrived at a similar conclusion. The main source of the ambiguity is the nonlinearity of the dependence of reciprocal reaction rate on reciprocal substrate concentration predicted by some of the substrate-binding schemes. To overcome the uncertainty due to the graphic treatment, method c was devised; according to this approach, similar experimental data on the dependence of the rate of ATP–PP$_i$ exchange in the presence and in the absence of substrate analogs are treated without application of the anamorphoses. The main idea of the approach is a comparison of the experimentally observed dependence of the initial rate of exchange on the concentrations of ATP, PP$_i$, amino acid, and inhibitor with that following from the kinetic equations describing the postulated mechanism.

This approach, which seems to be the most rigorous one, was applied (133,136) to the ATP–PP$_i$ exchange reaction catalyzed by beef-pancreas tryptophanyl-tRNA synthetase. Tryptamine was used as the inhibitor. Three possible mechanisms were considered: (1) ATP is added first, tryptophan second; (2) tryptophan is added first, ATP second; (3) the order of binding is random. The probabilities of the three hypotheses were at first assumed to be equal; subsequently, using a computer program, the effect of addition of new experimental points to the first 25 experimental points on the probabilities of all the hypotheses was investigated. After addition of seven new experimental points the probability of mechanism 1 increased to about 100%, whereas the probabilities of the two remaining ones became smaller than 10^{-20}. Further increase in the number of experimental points taken into calculation did not change the result. The same set of data may be taken for calculations of the dissociation constants for various complexes involved in the reaction mechanism (133,143). This approach, although crude enough, establishes the set of Kd in a complete reaction mixture, not in a simple enzyme–substrate complex.

More detailed analysis (143) of eight possible mechanisms of binding reveals an interesting feature—in addition to the mechanism where ATP binds first, another mechanism was found to be highly probable. In this second mechanism tryptophan may bind before ATP; however, this complex is unable to react with ATP followed by adenylate formation. In other words, enzyme-tryptophan "dead-end" complex may be formed. The existence of such a type of complex is an additional argument against attempts to study the order of substrate binding by means of complex formation with individual substrates instead of applying an adequate kinetic approach (see below). The kinetic equations and the computer program are independent of the nature of the synthetase, and therefore the method can be easily applied to other purified synthetases. It must be emphasized that the conclusion on the order of substrate binding in the ATP–PP$_i$ exchange reaction cannot be considered valid for the tRNA^{Trp} acylation reaction because the kinetic equations in this case should be based on a different and more complicated reaction scheme. To the same extent the conclusion must not be considered valid for other synthetases because the properties of different synthetases may be essentially different, as already emphasized in the literature (1,2).

Rouget and Chapeville (125) used approach b and concluded for the first time that ATP binds with enzyme first; this study was made with the leucyl-tRNA synthetase of *E. coli*. Later on application of the same approach to tyrosyl-tRNA (137) and phenylalanyl-tRNA (138) synthetases of *E. coli* suggested a random order of binding. Approach a applied to *E. coli* isoleucyl-tRNA synthetase failed to discriminate unequivocally between the possible orders of bindings (130,164). Explanations for the discrepancy between these results are not yet evident. This problem might be pursued fruitfully by studies of a wide variety of synthetases by the sequential planning method similar to that employed in work on tryptophanyl-tRNA synthetase.

Sometimes the ability of the enzyme to form complexes with the substrates is considered as an argument in favor of a given order of substrate binding. Studies of the complexes make it possible to obtain direct information concerning one of the intermediate stages of the whole reaction route; especially valuable is knowledge of the affinity constants. However, studies of the complexes alone cannot prove any reaction route because direct information to this end can be obtained only by kinetic methods. Moreover, it has been demonstrated for a number of synthetases that each substrate alone (in the absence of other substrates) is bound by the enzyme (28,38,85,88). At the same time the evidence obtained by investigation of the complexes does not contradict the results of kinetic studies: it was found by means of gel-filtration techniques that ATP forms complexes with leucyl-tRNA (125), threonyl-tRNA (124), and tryptophanyl-tRNA (117,118) synthetases.

2. The Acylation Reaction

Studies of the order of substrate binding and of the order of product release in the tRNA acylation reaction have been performed mainly by the method proposed by Cleland (139). This approach is based on the analysis of the initial rates of reactions run at a fixed concentration of a given substrate and at different concentrations of the other substrates, in the presence and in the absence of inhibitors (140). Further treatment is performed by graphic methods using double-reciprocal values and is not devoid of the disadvantages discussed above.

The following reaction scheme has been proposed by Allende et al. (124) for the threonine enzyme of rat liver:

A similar scheme was proposed by Papas and Mehler (141) for the proline enzyme of *E. coli*. In both cases unfractionated tRNA homologous with the enzymes studied was employed.

Rouget and Chapeville (111) studied the complexes of the substrates and inhibitors with *E. coli* leucyl-tRNA synthetase and outlined a scheme that was essentially the same as that proposed for the threonine and the proline enzymes; however, the kinetic analysis in this case has been made only for the ATP–PP$_i$ exchange, and not for the tRNA acylation reaction as a whole.

Hence the aminoacylation reaction is claimed to follow the bi-uni-bi Ping-Pong mechanism, according to Cleland's (139) nomenclature.

However, the Ping-Pong mechanism at present cannot be considered as finally proved. Henderson et al. (142) noticed that dependences graphically similar to those characteristic of the Ping-Pong mechanism can hold over a range of reaction conditions in the case of the sequential mechanism.

Myers, Blank, and Soll (112), who treated graphically their kinetic data obtained with leucyl-, valyl-, and seryl-tRNA synthetases of *E. coli*, arrived at first at an opposite conclusion: they claimed that the first substrate added was the amino acid rather than ATP. Later on, after treating the results according to Cleland's procedure, they revised the first notion and suggested a sequence in which ATP was bound first and the amino acid second. At the same time it was proposed that "if the reaction is concerted or both concerted and Ping-Pong (mixed mechanism), then either interpretation of the addition order of substrates is invalid" (112). With phenylalanyl-tRNA synthetase from *E. coli* it was concluded on the basis of similar evidence that the release of pyrophosphate is unnecessary for the addition of tRNA—that is, that the sequential mechanism may coexist with the Ping-Pong mechanism (138). It was believed that the binding of tRNA may occur at any moment before and after the addition of the amino acid and ATP. Thus random addition of substrates to phenylalanyl-tRNA synthetase is claimed. A similar interpretation was suggested for the arginine enzymes of *E. coli* (144) and *B. stearothermophilus* (156), which are known to need tRNA for the ATP–PP$_i$ exchange.

Evidently there is a discrepancy between the conclusions concerning the order of substrate binding. Different orders of addition are claimed for different enzymes. Obviously the explanation of the discrepancy may be the real difference between the mechanisms involved in the action of different synthetases, which are known to have widely varying properties. On the other hand, it is also possible that the graphic method of treatment of experimental data does not provide sufficient discriminating power to distinguish between alternative hypotheses.

B. MECHANISM

1. Nature of the Intermediates

Aminoacyladenylates have been proposed as the intermediates of tRNA acylation immediately after the discovery of the enzymatic reaction of ATP with amino acids (145,146) and of amino acid with tRNA (147,148). Aminoacyladenylates have been obtained with numerous synthetases from various sources after tryptophanyladenylate was isolated from reaction mixtures containing the beef-pancreas enzyme, tryptophan, ATP, and Mg^{2+} ions (149,150). It cannot be doubted now that the majority of synthetases do catalyze in the absence of tRNA the carboxyl activation of amino acids, resulting in the formation of aminoacyladenylates. An additional argument is the fact that aminoalkyladenylates are very strong competitive inhibitors of high specificity (125,151,152). However, the ability of synthetases to catalyze *in the absence of tRNA* and in the presence of divalent cations (Mg^{2+}, Mn^{2+}, Co^{2+}, etc.) the formation of aminoacyladenylates does not necessarily mean that the synthesis of aminoacyl-tRNA proceeds via this intermediate. From this viewpoint the transfer of the aminoacyl residue to tRNA from the enzyme–aminoacyladenylate complexes is also not a convincing argument, because this transfer is frequently not very efficient (in some cases half the aminoacyl residues are transferred to water instead of tRNA), and because the rate of this reaction, as calculated, is smaller in some cases than the overall molecular activity of the enzymes (cf. ref. 5).

To prove unequivocally the participation of aminoacyladenylates as intermediates it would be necessary at least to measure directly the rates of the intermediate stages and to compare them with that of the synthesis of aminoacyl-tRNA. Conventionally the rate of ATP–$^{32}PP_i$ exchange serves as the measure of the activation rate. However, Knorre and Malygin (153), who made a detailed kinetic analysis, found that the minimum apparent rate of ATP–PP_i exchange is not directly proportional to the rate of activation. For this reason, in order not to overcomplicate the kinetic treatment, it would be necessary to compare the reaction rates (*a*) of the synthesis of aminoacyladenylate in the presence of tRNA; (*b*) of the transfer of aminoacyl residue from the enzyme–aminoacyladenylate complex to tRNA; and (*c*) of the synthesis of aminoacyl-tRNA. As far as we know, such a comparative study has not yet been performed. It is not difficult to measure the rates of reactions *b* and *c*, but there is no kinetic method for measuring directly the formation of aminoacyladenylate in the presence of tRNA.

Eldred and Schimmel (154) compared the initial rate of the formation of isoleucyl-tRNA from isoleucine, ATP, and $tRNA^{Ile}$ with that of the formation of the same product from (Ile$\sim$AMP)·E complex. The initial rates appeared to be practically the same, which suggested that the rate-limiting stage was the same in both cases. This evidence is in line with (but does not finally prove) the role of aminoacyl-adenylate as the intermediate in tRNA acylation at least in the case of isoleucyl-tRNA synthetase.

Loftfield and Eigner (155) were the first to doubt the role of amino-acyladenylates as obligate intermediates in the tRNA aminoacylation reaction. Their argumentation rested on the results of the inhibition of the ATP–PP$_i$ exchange by some nucleophiles and of the inhibition of hydroxamate formation by pyrophosphate. However, a different interpretation of these data that does not contradict the role of amino-acyladenylates as obligate intermediates is probably possible. For example, the kinetic analysis of Loftfield and Eigner did not consider the possibility of the formation of an inhibitory complex between the enzyme and pyrophosphate, whereas Papas and Mehler (141) found for the prolyl-tRNA synthetase of E. coli that pyrophosphate is a strong inhibitor competitive with tRNA. It was also found with the phenylalanyl-tRNA synthetase of E. coli (138) that pyrophosphate has a greater affinity for the phenylalanyladenylate–enzyme complex than does the free enzyme.

Furthermore, the hydroxamate test, as applied by Loftfield and Eigner to measure the rate of amino acid activation, cannot be used for the purpose without detailed kinetic analysis, as demonstrated by Knorre (158) in a theoretical paper and confirmed experimentally (159) in a study of the stimulating effect of tRNA on the valylhydrox-amate synthesis catalyzed by the synthetase of E. coli. Hirsh and Lip-mann (160) and Iaccarino and Berg (161) found for several amino-acyl-tRNA synthetases of E. coli that the rate-limiting stage of isoleu-cylhydroxamate synthesis is the nonenzymatic hydroxylaminolysis of the enzyme–adenylate complex, rather than the enzyme-catalyzed synthesis of the adenylate.

The role of aminoacyladenylates as intermediates of tRNA acylation was also made doubtful by more recent studies. Examples are papers on the inhibition of the glutamyl-tRNA synthetase of rat liver with salicylates (162), studies of the kinetics of lysyl-tRNA synthesis cat-alyzed by the enzyme of E. coli, and investigations of the effect of different ions on the stages of this reaction (48,121). The observations discussed in the papers mentioned do not rule out the participation of aminoacyladenylates as intermediates in aminoacyl-tRNA synthesis be-cause their interpretation does not seem to be unequivocal. It appears that the interpretation of the experiments involving the inhibition or stimulation of the synthetases needs a more sophisticated kinetic considerations. The rate-limiting stage of the overall process should not be ignored either. However, a number of recent observations strongly suggest that the traditional concepts must be accepted with caution. It was found that polyamines like spermine, spermidine, and putrescine stimulate the formation of aminoacyl-tRNA in the reaction with the aminoacyl-tRNA synthetases of *E. coli* in the absence of Mg^{2+} ions or of other bivalent cations (165,166). It was found also that spermine binds to tRNA in the absence of synthetases to give a complex active in the acylation reaction in the absence of Mg^{2+} ions (167). These facts were not very surprising because it was already known that in the acylation reaction Mg^{2+} ions can be replaced by other ions, particularly polyvalent cations. However, the most interesting finding was the fact that no ATP–PP_i exchange occurs in the presence of spermine with any of the amino acids studied (168,169).

Many trivial explanations of the phenomenon can be proposed, such as desorption of the ^{32}P–ATP from charcoal by spermine, inhibition of the exchange with spermine by a reduction of the affinity of the amino acid for the enzyme, and a strict requirement for tRNA for exchange in the presence of spermine. All of these possible explanations and some other ones were ruled out by means of appropriate controls. The aminoacyl-tRNA synthesized in the presence of spermine does not differ from that obtained in the presence of Mg^{2+} ions in the ability to transfer the amino acid residues into the polypeptide chain on ribosomes; the acylation reaction needs ATP; the kinetics of the acylation reaction in in the presence of spermine is similar to that observed in the presence of Mg^{2+} ions; the amino acids are incorporated into the same tRNAs in the presence of any of the two ions, so that no mischarging occurs (170). All this evidence has been obtained with the well-studied enzyme *E. coli* isoleucyl-tRNA synthetase, for which the usual mechanism (154) is proposed based on detailed kinetic analysis of the reaction run in the presence of Mg^{2+} ions. These data have been completely confirmed for the same enzyme in another laboratory (171). In both studies it was found that the rate of the acylation is insignificantly reduced on substitution of spermine for Mg^{2+} ions, whereas the rate of ATP–PP_i exchange drops to 0.5% (or less).

More quantitative data have been obtained with the valyl-tRNA synthetase of *E. coli* (172). The rate of the acylation of tRNA with valine in the presence of spermine is 16 times lower than that of the reaction in the presence of Mg^{2+} ions; when tRNA was not dialyzed against EDTA prior to acylation, the rate of the reaction in the presence of spermine was the same as that observed in the presence of Mg^{2+} ions, as in the experiments of Igarashi, Matsuzaki, and Takeda (168). At the same time the rate of $ATP-PP_i$ exchange in the presence of spermine was 20 times lower than the rate of tRNA acylation under the same conditions and about 10,000 times smaller than the Mg^{2+}-catalyzed exchange. Hence the quantitative evidence is in line with the results of Igarashi et al. The phenomenon is not unique with respect to the polyamines. No exchange was observed on replacing Mg^{2+} by K^+, NH_4^+, Na^+ (173), or Nd^{3+} (171) ions, although the rate and the extent of tRNA acylation remained high. For example, appropriate choice of the concentration of NH_4^+ ions resulted in yields of leucyl-tRNA and valyl-tRNA that are the same as those observed at optimal Mg^{2+} concentration, whereas no $ATP-PP_i$ exchange could be measured (173). It is very important that no complex was detected, under the conditions employed, between E. coli isoleucyl-tRNA synthetase and isoleucyladenylate; the complex is readily detectable in the presence of Mg^{2+} ions. In the presence of spermine instead of Mg^{2+} no formation of amino acid hydroxamates takes place and ATP is not hydrolyzed (173a).

The ability of spermine to effectively support the aminoacyl-tRNA formation depends strongly on the nature of the enzyme. With valine and arginine enzymes from E. coli (173b) and with threonyl-tRNA synthetase from rat liver (173c) a very weak, if at all, aminoacylation reaction takes place in the presence of spermine. Contrary to this Chousterman and Chapeville (173d) found with E. coli tyrosyl-tRNA synthetase that the rate of tyrosyl-tRNA formation in the presence of spermidine was only 15 times lower than in the presence of Mg^{2+}.

Somewhat different data have been obtained with the tyrosyl-tRNA synthetase of E. coli (119,173d): tyrosyladenylate was formed in the absence of Mg^{2+} ions, and the tyrosine residue was subsequently transferred to $tRNA^{Tyr}$ in the presence of polyamines. The presence of firmly bound Mg^{2+} in the enzyme preparation was not ruled out in these experiments. No $^{32}PP_i-ATP$ exchange was observed in the absence of added Mg^{2+} ions. An explanation of this somewhat paradoxical result could be the inability of pyrophosphate in the non-magnesium form to produce a functionally active ES complex due to a dramatic decrease in the affinity of the active center or due to the inability of non-magnesium pyrophosphate to induce an appropriate conformational change in the enzyme. The result would be the apparent irreversibility of the reaction due to considerable shift in the equilibrium to the formation of AMP and pyrophosphate and the concomitant decrease in the rate of ATP–PP$_i$ exchange. Cole and Schimmel (174) demonstrated that monomagnesium pyrophosphate is the actual substrate in the reaction. This makes understandable the behavior of tyrosyladenylate (obtained in the absence of Mg^{2+} ions), which does not pyrophosphorylize but hydrolyze on addition of pyrophosphate to the aminoacyladenylate–enzyme complex (173d).

The pyrophosphorolysis reaction catalyzed by some synthetases has an interesting peculiarity noted by the first workers in the field: it is less specific with some synthetases than the direct reaction. For example, leucyl-tRNA synthetase catalyzes the pyrophosphorolysis of not only the specific L-leucyladenylate but also of D-leucyladenylate and even of alanyladenylate (175), although the direct reaction takes place only with the specific substrate L-leucine. Similar evidence has been obtained with tryptophanyl-tRNA synthetase (176,177). An earlier explanation (16) of the phenomenon seems to be unconvincing now. A more reasonable explanation proposed independently by Loftfield (4) and by Kisselev (9), is as follows: It is known that the K_m values for amino acids are 10^{-3}–10^{-5} M, whereas those for aminoacyladenylates are 10^{-7}–10^{-8} M. The noncognate amino acids are characterized by K_m values greater by a factor of 10^3–10^4 (i.e., $K_m \sim$ 0.1–1 M). Evidently the usual concentrations of nonsubstrate amino acids are not sufficient to stimulate the ATP–PP$_i$ exchange. Decrease in the K_m values of nonnatural aminoacyladenylates by the same factor of 10^3–10^4 would result in $K_m \sim 10^{-3}$–10^{-4}. This affinity would be enough to explain the observed rate of pyrophosphorolysis. Therefore the idea essentially is that the phenomenon of "lack of specificity" is mainly due to the retention of high affinity of the adenylate moiety of the noncognate aminoacyladenylate for the enzyme, although it decreases to a certain extent due to the change in the aminoacyl moiety.

Turning back to the problem of aminoacyladenylate as an intermediate in aminoacylation reaction, one should remember that arginyl-tRNA synthetase from *E. coli* (178,179), yeast (180), and *B. stearothermophilus* (156) as well as glutamyl- and glutaminyl-tRNA synthetases from *E. coli* (21,181,182) and rat liver (183) fail to catalyze the ATP–PP$_i$ exchange reaction at all, even in the presence of high concentrations of amino acid and Mg^{2+}. Attempts to reveal any arginyladenylate in complex with arginine enzyme by different methods were unsuccessful (156,184).

However, the criticism (5) of the role of aminoacyladenylates in the Mg^{2+}-dependent synthesis of aminoacyl-tRNAs catalyzed by "classical," "tRNA-independent" synthetases seems at present not enough convincing to reject the traditional viewpoint. At the same time we share Loftfield's (5) opinion that there is no direct and unequivocal proof of the role of aminoacyladenylates in the aminoacylation reaction.

An alternative to the adenylate pathway is the concerted mechanism proposed by Loftfield and Eigner (155), which does not involve as a separate stage the formation of the aminoacyladenylate intermediate in the presence of tRNA, but rather postulates a simultaneous (concerted) formation of aminoacyl-tRNA, AMP, and PP$_i$. A similar mechanism not involving the discrete formation of acyladenylate had been proposed earlier by Green and Goldberger (185) for the enzymatic synthesis of acetyl-CoA. According to the concerted mechanism (155) Me^{2+} ions directly participate in the reaction. However, isoleucyl-tRNA synthetase (*E. coli*) passed through Sephadex G–150 was completely free of Mg^{2+}, Zn^{2+}, Ca^{2+}, Mn^{2+}, Cr^{2+}, Co^{2+}, Fe^{2+}, Ni^{2+}, Cu^{2+}, and Cd^{2+}, but in spite of this aminoacyl-tRNA was formed in the reaction mixture containing polyvalent and/or monovalent cations (171).*

A description of the concerted mechanism and detailed argumentations against the traditional point of view were given recently by Loftfield (5). Although there is no experimental evidence in favor of a concerted mechanism in the case of "classical" synthetases, some type of concerted mechanism may be involved in catalysis with "tRNA-dependent" synthetases (144,156). It should be stressed that if the

---

*It was shown by Mehler et al (389) with *E. coli* valyl-tRNA synthetase that under nonequilibrium conditions an AMP-dependent exchange of tRNA $\leftrightarrow$ Val-tRNA is completely inhibited by low PP$_i$ concentration and with dAMP is only partially inhibited even by high PP$_i$ concentrations. These data indicate that the inhibition is caused by the reaction of PP$_i$ with an intermediate complex, not by participation in a concerted reaction. Although these and some other results of this group do not prove finally the role of adenylate in the aminoacylation reaction, they are in clear disagreement with the concerted mechanism.

180   LEV L. KISSELEV AND OL'GA O. FAVOROVA

nonparticipation of the aminoacyladenylate as an intermediate in the tRNA aminoacylation reaction were to be proved for some synthetases, it would not necessarily prove the correctness of the concerted mechanism. The shortcoming of the latter concept lies in the fact that it is based at present almost exclusively on the negative arguments concerning the role of adenylates. However, in principle one may imagine other transitory intermediates besides the aminoacyladenylates.

Irrespective of its validity, the appearance of the concerted mechanism in the literature has been highly stimulating to many workers in this field. The sharp increase in a number of publications during last year devoted to the mechanism of catalysis by the synthetases is at least partially due to Loftfield's provocative paper.

A chemical model of the enzymatic synthesis of aminoacyladenylates and of transaminoacylation reaction between aminoacyladenylate and tRNA has been proposed recently by Krayevsky, Kisselev, and Gottikh (187). The idea which forms the basis of this scheme consists in the postulate that the imidazole rings of the histidine residues of the enzyme protein participate both in aminoacyladenylate formation and in the transfer of the amino acid residue to tRNA in the transaminoacylation reaction. It is suggested that in both reactions the nucleophilic catalysis mechanism functions in the formation of intermediate covalent complexes of AMP and of aminoacyl residues with the imidazole group of the histidine residues of aminoacyl-tRNA synthetase. The proposed conversions allow the transmission of the high-energy bond derived from ATP through all the intermediate steps leading to the aminoacyl-tRNA.

Recently data on the transfer of acetylglycine from acetylglycyladenylate to imidazole in aqueous imidazole buffer have been communicated (191). In this work a suggestion has been also made concerning the participation of the imidazole ring of histidine in the transfer reaction between aminoacyladenylate and tRNA. In the next communication White, Lacey, and Weber (191a) reported the transfer of the N-acetylglycine from the adenylate anhydride to the 2'OH groups along the backbone of homopolyribonucleotides. This transfer involves an N-acetylglycylimidazole intermediate. These results are consistent with a model involving a histidine residue in the active site of aminoacyl-tRNA synthetases. In our view the shortcoming of this work from the viewpoint of the chemical model for synthetase action lies in taking advantage of N-protected amino acids. The reactivity of both the adenylates and imidazolyl derivatives of N-protected amino acids differs sharply from that of N-protonated amino acids (192,193).

The mechanism is attractive because the catalytic function of the synthetases is outlined in chemical terms; intermediate covalent ES complexes are postulated for the synthetases for the first time. Unlike the concerted mechanism, it does not fully depend on bivalent cations. The known properties of aminoacyl-tRNA synthetases do not contradict the hypothesis, although at present there is no direct proof of it. As emphasized by the authors themselves, the fact that the chemical model seems reasonable from the chemical viewpoint does not necessarily mean that the enzyme follows the same logic, but the detailed character of the mechanism provides the possibility of verifying or rejecting it experimentally and thus makes this working hypothesis useful as a tool for studying the catalytic function of the synthetases.

Very recently Kisselev and Kochkina (191b) isolated a complex of the beef-pancreas tryptophanyl-tRNA synthetase with AMP. The ability of the complex to form ATP in the presence of pyrophosphate proves the existence of the high-energy covalent bond between the enzyme and AMP. These data are in favor of the aforementioned concept although at the moment they do not prove the participation of the imidazole in the catalytic mechanism. Other indirect evidence may be found in a rapid kinetic investigation of $E.$ $coli$ isoleucyl-tRNA synthetase (164); the interaction of L-isoleucine and ATP, respectively, with the enzyme does not follow simple bimolecular kinetics but presumably a two-step mechanism where the second process is an isomerization of the first ES complex.

2. The Rate-Limiting Stage of the Reaction

Determination of the rate-limiting stage of multistage reactions facilitates the understanding of their mechanism and of the mode of action of different effectors on the overall rate. It was mentioned in Section III that the association constants of the synthetases with tRNAs are very high (10^7–10^9 M). Acylation of tRNA does not change the association constant. For example, K_A values of the leucyl-tRNA synthetase of $E.$ $coli$ with both tRNA^{Leu} and leucyl-tRNA^{Leu} are $1.2 \cdot 10^8$ M^{-1} (111). Thus the product compared with the substrate does not have a decreased affinity for the enzyme. The equal affinity of aminoacyl-tRNA and nonacylated tRNA for the enzyme probably reflects the absence of strong interactions between the aminoacyl moiety of aminoacyl-tRNA and of enzyme protein. Instructive to this end is the ready formation of complex with the enzyme of *E. coli* valyl-tRNA^{Val} phenoxyacetylated at the α-amino group of the aminoacyl residue (105).

It was noted for the isoleucyl-tRNA synthetase of *E. coli* that the liberation of isoleucyl-tRNA^{Ile} from the complex in the absence of other substrates takes about several minutes, which is much slower than the acylation reaction (109). These facts indicate that the most probable candidate for the rate-limiting stage is the release of the reaction product—that is, the liberation of aminoacyl-tRNA from the ES complex. Appropriate kinetic studies confirming this hypothesis proposed for the first time by Yarus and Berg (109) were made with the isoleucyl-tRNA (154) and leucyl-tRNA (111) synthetases of *E. coli*.

Another important finding of Yarus and Berg (109) was the effect of substrates (amino acid, amino acid + ATP and aminoacyladenylate) on the rate of liberation of aminoacyl-tRNA from the ES complex; the finding was confirmed for a number of enzymes by other workers (38,111,154). Isoleucine and isoleucyladenylate increased the rate of isoleucyl-tRNA liberation by a factor of 6–10. The ES complex of valyl-tRNA dissociates about 10 times faster in the presence of valine or of valinol-AMP. The rate of the dissociation at pH 5.5 and 20°C appears to be high enough to provide the observed overall rate of the acylation of tRNA with valine under the conditions. These data explain well the earlier observation (188) according to which the yield of the tRNA-synthetase complex decreases in the presence of ATP and valine. It was found for the leucine enzyme that the affinity constant of leucyl-tRNA^{Leu} does not change in the presence of leucine + ATP, but the rates of both association and dissociation increase by a factor of 5 in the combined presence of the two substrates.

Determination of the rate-limiting stage of the process makes it possible to revise numerous data on the stimulation of the rate of tRNA acylation by different compounds. For example, Yarus and Berg (109) point out that interpretation of the data (189) on the stimulating effects of imidazole, tris buffer, and other bases in the aminoacylation reaction should be revised. The ample evidence concerning the stimulating and inhibitory effects of different ions on the overall reaction must be also reconsidered in view of the already known rate-limiting stage.

The aforementioned data were obtained mainly by two methods: (1) by an adaptation of the nitrocellulose filter assay (109,111) and (2) by a careful study of the kinetics of transfer of isoleucine from an isoleucyl-tRNA synthetase—isoleucyladenylate complex to tRNA^{Ile} (154). Since both approaches have shortcomings, release of aminoacyl-tRNA from synthetase was studied by rapid molecular sieve chromatography (173e). The results confirmed previously published findings that the rate of release of newly synthesized Ile-tRNA^{Ile} from synthetase is very slow in the absence of isoleucine or isoleucyl-AMP, but that the release is greatly enhanced by these ligands.

Rate-limiting stage of the aminoacylation reaction was examined also at pH 7.2 in stopped-flow experiments with yeast seryl- (100e) and phenylalanyl- (100f) tRNA synthetases. In both cases it was found that release of the aminoacyl-tRNA is not limiting, at least at neutral pH. Since in these and the previous works other enzymes, experimental conditions, and techniques were used, a comparison of these two groups of data is rather useless.

V. Synthetase–tRNA Interactions

A. RECOGNITION AND MISRECOGNITION; MISACYLATION AND DEACYLATION REACTIONS

The level of errors in protein synthesis *in vivo* is very low (190), which suggests an extremely selective recognition in the synthesis of aminoacyl-tRNAs. For this reason the interaction of tRNA with synthetases was regarded until recently as an example of a highly specific interaction of proteins with nucleic acids. However, the selectivity of the synthetases in the synthetase–tRNA interactions *in vitro* appeared to be nonabsolute, and hence we shall now discuss the interaction of synthetases with both cognate and noncognate tRNAs. In recent literature emphasis is placed on noncognate interactions since this type of recognition was poorly understood, even less than the cognate type. Thus before passing to the evidently more important cognate recognition we shall discuss some investigations of noncognate interactions, which can be demonstrated (a) by physical methods; (b) by intraspecies and interspecies misacylation; and (c) by the deacylation of aminoacyl-tRNAs with synthetases.

1. Noncognate E–tRNA Complexes*

It is known that the intensity of the fluorescence of the tryptophan residues of synthetases decreases when they form complexes with the cognate tRNAs (38,83). The same effect is sometimes observed when noncognate tRNAs are added instead of the cognate ones—for example, when yeast tRNA^{Phe} is added to yeast seryl-tRNA synthetase or tRNA^{Ser} is added to phenylalanyl-tRNA synthetase (68,84), or when E. coli tRNA^{Leu} and tRNA^{Val} are added to E. coli glutamyl-tRNA synthetase (101). Yeast tRNA^{Asp} forms a complex with both aspartyl- and arginyl-tRNA synthetases (41). The number of such cases will certainly soon become much greater because formerly the complexes of cognate pairs were the principal objects of study.

2. The Intraspecies and Interspecies Misacylation Reactions

Synthetases are able to acylate in vitro the noncognate tRNAs of both the same (194,195) and other (see ref. 196) species under appropriate conditions (large excess of enzyme; the presence of organic solvents; optimum ATP-Mg^{2+} ratio, ionic strength, and pH). Interspecies misacylation reactions have been considered in detail recently by Jacobson (196,197) and by Ebel and colleagues (198–200), and for this reason the corresponding data will not be discussed here.

The possibility of misacylation and the fact that in spite of this not all the tRNAs are interchangeable in the reactions with synthetases even under special conditions suggest the occurrence of misrecognition along with correct recognition. Misrecognition is less selective than correct recognition, but it is different from the nonspecific interaction, such as, for example, the electrostatic interaction of tRNA with methylated serum albumin (201).

Under the "usual" conditions correct acylation by far prevails over misacylation, so that the latter cannot be detected. Increase in the affinity of tRNAs for the synthetases, presumably due to increased contribution of noncognate interactions, may result in a high level of misacylation. Yarus (202) found that addition of organic solvents, such as methanol or dioxane, or decrease in the concentration of bivalent cations led to increased affinity of E. coli tRNA^{Ile} for the isoleucyl-tRNA synthetase of the same organism. It is possible that under the

---

*Very recently Ebel et al. (191c) collected comprehensive data on noncognate E-tRNA complexes, including many new findings made in their laboratory.

conditions the three-dimensional structure of tRNA becomes less rigid and the internal mobility of the molecular segments becomes higher. Yarus believes that this effect facilitates the adaptation of the tRNA molecule to the molecule of the enzyme and increases the area of the contacting surfaces of the two macromolecules due to increased contribution of noncognate interactions. In line with this speculation is the fact that addition of methanol increases the affinity of *E. coli* isoleucyl-tRNA synthetase for tRNA^{Phe} and for tRNA$_f^{Met}$, and makes it possible to obtain isoleucyl-tRNA^{Phe} (203) and isoleucyl-tRNA$_f^{Met}$ (204), respectively. Thus one of the factors affecting the relative contributions of cognate and noncognate interactions may be the compactness (or the stability) of the three-dimensional structure of tRNA. A more detailed discussion of misrecognition can be found in the papers by Yarus (202–206).

The interest in misacylation reactions is stimulated also by reasons other than their role as a proof of misrecognition. First, these reactions make available "hybrids" of aminoacyl-tRNAs composed of noncognate tRNA and aminoacyl residues; these hybrids can be employed to confirm the adaptor hypothesis by an approach similar to that proposed in the classical studies of Chapeville et al. (207). Second, comparison of the primary structure of tRNAs specific to different amino acids but acylated by the same synthetase under appropriate "unusual" conditions may shed light on the location of "misrecognition sites" (208,209). The misacylation reactions make available new tRNA-synthetase pairs for studies of the forces involved in the formation of ES complexes, of the mechanisms of the selection of correct pairs, and of the role of noncognate interactions.

With the intraspecies misacylation reactions a problem also arises as to the possible extent of misacylation *in vivo*. It seems that misacylation does not take place *in vivo* to any considerable degree: the concentrations of the synthetase and of the cognate tRNAs in the cells are, presumably, about 10^{-5}–10^{-6} M, and their ratios are approximately equimolar (210,351,352). Hence it follows from the known affinity constants that both the tRNAs and the synthetases exist within the cells mainly in a form of cognate ES complexes, whereas the concentrations of the free synthetases and tRNAs are much lower than their total concentrations. For this reason the concentrations of noncognate tRNAs and synthetases *in the presence of cognate pairs* must be very small. Since the misacylation rate is a few orders of magnitude smaller than that of correct acylation (203,204), it appears that the probability of the formation of misacylated tRNA is negligible. However, this misacylated tRNA, if formed, would be immediately destroyed by the deacylation reaction discussed in the following section. Finally, it has been already mentioned that ionic strength as well as the intracellular concentrations of Mg^{2+} and of monovalent cations do not favor the misacylation reaction.

Instructive is the strong decrease in the rate of misacylation of *E. coli* tRNA[Phe] by *E. coli* isoleucine-tRNA synthetase caused by the addition of cognate tRNA[Ile] devoid of the acceptor activity due to periodate oxidation (206). Yarus (206) presents a table illustrating the decrease in the rates of misacylation reactions caused by the addition of cognate reagents.

It seems that any specific interaction of the tRNA with the synthetase is a superposition of the cognate and the noncognate interactions, whose contributions are determined by the composition of the reaction mixture and by the reaction conditions, by the presence of noncognate synthetases and tRNAs, and, probably, by other factors.

3. Enzymatic Deacylation of tRNA

A number of papers have described the enzymatic deacylation of aminoacyl-tRNA by synthetases in the absence of AMP and PP_i (i.e., by a pathway other than the reversed aminoacylation reaction). Berg et al. (163), Lagerkvist, Rymo, and Waldenström (188), Yaniv and Gros (105), and Befort et al. (114) pointed to the possibility of this reaction, but at that time it was not studied in detail.

A number of factors have stimulated interest in this reaction: its contribution could help to understand the incomplete acylation of tRNA in heterologous and homologous systems (211), and the incomplete transfer of the aminoacyl residue to tRNA from the enzyme–aminoacyladenylate complex (see ref. 5); more generally this would shed light on the mechanism of the acylation reaction.

Detailed studies of the deacylation reaction made by Schreier and Schimmel (212) with isoleucyl-tRNA synthetase led these authors to the following conclusions:

1. The deacylation reaction does in fact occur; it proceeds in the absence of AMP and pyrophosphate, and it is catalyzed by the synthetase itself rather than by interfering enzymes.

2. The affinity of the synthetase for isoleucyl-tRNA^{Ile} is high enough (K_m in the deacylation reaction is 10^{-7} M), and, in spite of this, the reaction is slow (turnover number 0.8 min^{-1} at pH 7.0 and 37°C).

3. The rate of deacylation does not change in the presence of either AMP or pyrophosphate alone, but deacylated tRNA^{Ile} is a competitive inhibitor.

4. The dependences of the rates of reverse aminoacylation and deacylation reactions on Mg^{2+} concentrations are essentially different.

Eldred and Schimmel (213) found that the deacylation reaction, which is slow for the homologous substrate, becomes very rapid with misacylated valyl-tRNA^{Ile}. This finding explains the long-known inability of isoleucyl-tRNA synthetase to catalyze the synthesis of valyl-tRNA^{Ile} in spite of the ready synthesis of valyladenylate by this enzyme. The same type of observation was made by Yarus (205), who found a rapid deacylation of isoleucyl-tRNA^{Phe} ($E.\ coli$) by phenylalanyl-tRNA synthetase.

It was supposed that the role of the deacylation reaction is to increase additionally the correctness of protein synthesis due to the "verification" (205) mechanism that removes misacylated tRNAs.

Bonnet, Giege, and Ebel (213a) synthesized a number of aminoacyl-tRNAs using highly purified yeast tRNA preparations specific to different amino acids. All of them, independently of the specificity of tRNA and of the specificity of the amino acid incorporated (in the case of misrecognition), are subject to enzymatic deacylation in the presence of valyl- and phenylalanyl-tRNA synthetases. The authors concluded that the deacylating activity of synthetases is rather nonspecific to either tRNA or the amino acids. It is interesting that valyl-tRNA^{Phe} is deacylated by phenylalanyl-tRNA synthetase 150 times faster than phenylalanyl-tRNA^{Phe}, but phenylalanyl-tRNA^{Val} is deacylated by valyl-tRNA synthetase 3.5 times slower than valyl-tRNA^{Val}. Hence misacylation is not always opposed by rapid deacylation.

The deacylation reaction, which is nonspecific for aminoacyl-tRNA synthetases, suggests that synthetases also recognize noncognate tRNAs.

4. The Specifying and the Binding Nucleotides

Existence of the cognate and the noncognate interactions discussed in the preceding section prompts one to look for the nucleotide residues

of the tRNA molecules that are responsible for the two functions. The
best candidates for the sites of noncognate interaction are the numerous
nucleotide residues present in different tRNAs at similar positions.
These "binding" nucleotide residues may be common either to all the
tRNAs or to groups of tRNAs (e.g., to tRNAs of the same organism or
to tRNAs specific for a given group of amino acids). It is now generally
believed that the number of nucleotides interacting with the synthetases ("recognizing nucleotides") is greater than the number of the
nucleotides that determine the ability of the synthetases to distinguish
one tRNA from another ("specifying nucleotides") because there are
also "binding nucleotides," which take part in the interaction but do
not determine the selection of the tRNA by the synthetase.

Presumably one of the binding nucleotides is the 3'-terminal adenosine residue present in all the active tRNAs. Removal of the terminal
adenosine residue of tRNA[Phe] affords an inhibitor of the acylation
with $K_i = 400$ pM, whereas the K_m value of native tRNA[Phe] in the
reaction with homologous phenylalanyl-tRNA synthetase is 40 pM.
Removal of the cytidine residue adjacent to the terminal adenosine
does not change the K_i value, but removal of the second cytidine residue results in an increase in the constant ($K_i = 900$ pM) (214).

B. THE POSSIBLE LOCATION OF THE SPECIFYING NUCLEOTIDES

The problem of the recognition sites (i.e., of the location along the
tRNA chain of the nucleotide residues that enable the synthetase to
distinguish the cognate tRNA from the noncognate species) has been
many times discussed, both from an optimistic viewpoint (11) and
also with a tinge of doubt, irony, and scepticism (5,215). To date three
hypotheses of the location of the specifying nucleotides in the primary
structure have been proposed: the anticodon hypothesis (216,217), the
amino acid stem hypothesis (11,218), and the dihydrouridine stem
hypothesis (219). In addition, two more hypotheses have been proposed, according to which the specifying nucleotides are located (a) in
the anticodon and in the amino acid stem (220), and (b) in the dihydrouridine stem and at position 4 counting from the acceptor end
(209).

1. The Anticodon

To study the role of the anticodon in the acceptor function is analogous to studying the possibility of the presence of the specifying nucleo-

tides in it because the anticodon moiety is necessarily specific for a tRNA of a given amino acid specificity. The majority of workers until 1971 definitely disagreed with the anticodon hypothesis and believed that the anticodon moiety neither contains the specifying nucleotides nor takes any part in the interaction of tRNAs with the synthetases (see refs. 1, 4, 5, 11, and 12). The minority constantly adhered to the opposite opinion (216,217,221–224), but the corresponding facts were ignored. To the end of 1972, sufficient evidence has been accumulated in favor of both viewpoints, and each of them may now be considered correct for certain synthetase-tRNA pairs. Hence it seems that the dramatic collision of opinions will soon come to an end.

Data in favor of anticodon participation in recognition have been obtained for tRNA^{Gly} (E. coli), for tRNA$_1^{Val}$ (yeast), and for tRNA^{Trp} (E. coli), as well as for some other tRNAs. Yeast valine tRNA was the first one for which the participation of the anticodon in the acceptor function was postulated on the basis of the experimental evidence obtained in the course of modification of the cytidine residues of this tRNA with O-methylhydroxylamine (216). Subsequently it was found that this tRNA contains four to five exposed cytidine residues (225), and the positions of these exposed cytidine residues were determined (226). Jilyaeva and Kisselev demonstrated that the modification of cytidine in the anticodon I–A–C of yeast tRNA$_1^{Val}$ leads to a loss of acceptor activity (227,228). Removal of the dinucleotide AC of the anticodon produces the same result (220).

According to the genetic code, both bases of the main part of the anticodon of tRNA^{Gly} are cytidine residues. In addition, cytidine is present in the Tψ-arm and in the miniloops of E. coli tRNA^{Gly} (234), but these must be buried in the three-dimensional structure, like many other base residues (cf. refs. 226 and 229). Thus reactive in the chemical modification must be first of all the cytosine residues of the anticodon and those of the ACC-terminus, which is also exposed (226). All tRNAs contain the same probably exposed ACC-terminus, and thus the difference between the functions describing the inactivation of tRNA^{Gly} and the inactivation of other tRNAs by cytidine-specific reagents may be attributed to the effect of anticodon modification. In fact, the glycine-acceptor activity decreased to the greatest extent compared with 12 other acceptor activities in the course of modification of the cytosine residues of E. coli tRNA with hydroxylamine (230). Special experiments made in the same studies showed that the modified cytosine residues were not involved in intramolecular hydrogen bonding (i.e., belonged to the loops of the secondary structure). A definite conclusion was made that the anticodon takes part in the acceptor function of tRNA^{Gly} (and of some other tRNAs).

Recently Carbon and coworkers (231–234) analyzed the suppressor mutations of E. coli at tRNA^{Gly} and clearly demonstrated that change of the base residue in the 5′-position of the anticodon (i.e., in the wobble position) does not considerably affect the kinetic parameters of the aminoacylation reaction. On the contrary, change of the base residues of the main part of the anticodon results in a decrease in the affinity of glycyl-tRNA synthetase for tRNA^{Gly} by a factor of 10^3–10^4. It is interesting that the primary structures of E. coli tRNA$_3^{Gly}$ (234) and tRNA$_{2B}^{Val}$ (235) differ in the second letter of the anticodon, the rest being very similar. The similarity of the structures is much more pronounced than that of the structures of tRNA$_1^{Val}$ and tRNA$_{2A,B}$ (235).

Very important results have been obtained by Yaniv, Söll, Folk, and Berg (236), who studied the tRNA of the E. coli SU7$^+$ mutation, which manifests itself by reading the terminating UAG codon by mutant tRNA as the codon for glutamine. It was believed at first that the mutation involved the anticodon of one of the glutamine tRNAs. However, combined genetic and structural studies led to the conclusion that the locus of the SU7$^+$ mutation is the structural gene coding for tRNA^{Trp} rather than that coding for tRNA^{Gln}. As a result of the point mutation C → U in the anticodon (the second letter), tRNA^{Trp} became capable of reading the codon U–A–G instead of the codon U–G–G and, furthermore, to incorporate glutamine instead of tryptophan, presumably due to the action of glutaminyl-tRNA synthetase. Although one of the authors (236) of this study refuses to regard these data as evidence in favor of the participation of the anticodon in the acceptor function, it seems very difficult to give them a different explanation.

Comparison of the primary structures of E. coli tRNA$_f^{Met}$ and tRNA^{Met}, and comparison of these structures with the other primary structures, suggested that the participation of the anticodons of these tRNAs in the acceptor function is possible (9). Chemical modification of E. coli tRNA$_f^{Met}$ with bisulfite, resulting in the C → U transformation, leads to a loss of the methionine-acceptor activity (237,237a). It was found that the inactivation is due to the transformation C → U in the acceptor end and/or in the main part of the anticodon. It is interesting that the modification of the wobble letter of the anticodon of this tRNA does not lead to a decrease in the rate or in the yield of the acylation (238).

Similar modification of the cytosine residues of the anticodon loop of tRNA^{Glu} (*E. coli*) also leads to a loss of acceptor activity (239). The participation of the anticodon of this tRNA in the acceptor function had been postulated earlier (230) on the basis of the high sensitivity of its acceptor function to NH$_2$OH treatment followed by modification of the exposed cytosine residues. There is also evidence pointing to the possibility of the participation of the anticodons in the acceptor functions of tRNA^{Pro} (224), tRNA^{Lys} (221,240,241), and tRNA^{Ile} (242) with cognate synthetases. Hence the anticodon hypothesis proposed for the first time in 1964 (216) and developed during the following years (see refs. 9 and 243) has now been proved in a number of cases by decisive experimental evidence based on selective chemical modification, by comparison of the primary structures of isoacceptor and of suppressor tRNAs, and by the "dissected-molecule" approach.

We believe that it is timely to abandon the formerly common viewpoint that the anticodon moiety *never* participates in the acceptor function. However, nonparticipation of the anticodon in the acceptor function is quite probable for some of the ES pairs. Complete removal of the anticodon triplets of yeast tRNA^{Ala} (244), tRNA^{Phe} (245), and tRNA^{Tyr} (246), though leading to a decrease of acceptor activity, does not result in complete inactivation.

Escherichia coli isoacceptor tRNAs^{Ser} and tRNAs^{Leu}, in spite of the profound differences in their anticodons, are capable of being charged by a single aminoacyl-tRNA synthetase with a similar K_m value (247, 248). More indirect evidence was presented for some other tRNAs (1,4,11,12).

2. The Amino Acid Stem

The participation of the amino acid stem of the cloverleaf secondary structure of tRNA in the acceptor function was postulated for the first time by Schulman and Chambers (218). Very recently it was found for the suppressor tRNA^{Tyr} (*E. coli*) that mutations at the bases occupying positions 2, 81, and 82 result in missuppressions, presumably due to the ability of tRNA^{Tyr} to be acylated by some other amino acid(s). These data (249,250) can be regarded as important evidence in favor of the presence of the specifying nucleotide residues in positions 2, 81, and 82 of the acceptor stem (237b).

These studies are in line with the earlier findings of Beltchev and Grunberg-Manago (251) on the competitive inhibition of the acceptor activity of tRNA$_2$^{Tyr} (E. coli) by a combination of oligoguanylate and a CCA-terminated 19-residue oligonucleotide of tRNA$_2$^{Tyr}. Mehler and Chakraburtty (252) found that relatively small 5'- and 3'-fragments of tRNA^{Arg} (E. coli) associate to give a complex that specifically inhibits the acylation of tRNA^{Arg} with arginyl-tRNA synthetase. However, the 3'-fragment contained 20 nucleotides, and the 5'-fragment 28 nucleotides (and no 5'-terminal residue). For this reason the location of the specifying nucleotides has not been established more precisely.

Two arguments were presented in favor of location of the recognition site of yeast tRNA^{Ala} in the acceptor stem: the acceptor activity of the complex of 3'- and 5'-terminal quarters of the molecule (244) and the behavior in photochemical inactivation (218). However, to date no other laboratory has yet found any acceptor activity in a complex of tRNA quarters, although the corresponding studies were made with fragments of different tRNAs. As for photochemical inactivation, it appeared to be due to lesions at the cytosine residues of the ACC-terminus rather than to changes of the specifying nucleotides. Comparison of the primary structures of the three valine tRNAs of E. coli (235) and of yeast tRNA$_1$^{Val} (253,254) reveals a difference of the base pairs in positions 5–7 counting from the acceptor terminus (postulated to be the specifying nucleotides by Schulman and Chambers), although these tRNAs are interchangeable substrates of both E. coli and yeast synthetases. The base pairs that must be the recognition sites according to Chambers are identical in the tRNA^{Gly} (234) and in the tRNA$_{2B}$^{Val} of E. coli (235).

Yeast tRNA^{Trp} (255) and E. coli tRNA^{Trp} (256) have considerably different stems, although they can be acylated with the synthetase of E. coli. The two isoacceptor leucine tRNAs of E. coli are acylated with almost the same K_m and V_{max} values, although their stems are considerably different (257). The serine tRNAs of yeast (258), rat liver (259), and E. coli (260) have very different acceptor stems, but the difference does not hinder the interaction with the cognate synthetases because the tRNA$_1$^{Ser} of E. coli is acylated by the synthetases of yeast and E. coli. These data as well as some less direct evidence strongly suggest that at least in some of the tRNAs the base pairs of the acceptor stem at positions 5–7 from the acceptor end do not contain the specifying nucleotides.

3. The Stem of the Dihydrouridine Loop

Dudock et al. (208,209,219) proposed a hypothesis according to which the specifying nucleotides are contained in the stem of the dihydrouridine loop. This hypothesis was based on the results of studies on heterologous aminoacylation. The same enzyme, yeast phenylalanyl-tRNA synthetase, is capable of acylating with phenylalanine under appropriate conditions the cognate $tRNA^{Phe}$ and the noncognate $tRNA^{Ala}_{1,2}$ and $tRNA^{Val}_{1,2}$ of E. coli. These tRNAs have very different primary structures, but similar bihelical regions adjacent to the dihydrouridine loops. In accord with the hypothesis is the fact that removal of the nucleosides of this region in yeast $tRNA^{Phe}$ results in a complete loss of acceptor activity, whereas removal of the loop nucleotides of the same arm and of the nucleotides of the anticodon arm does not result in complete inactivation (245).

However, detailed studies of the heterologous reactions catalyzed by the synthetases of yeast, E. coli, and Neurospora crassa, performed with unfractionated and with individual tRNAs, make doubtful the approach to the identification of the specifying nucleotides based on heterologous aminoacylation (197). For example, under appropriate conditions yeast phenylalanyl-tRNA synthetase acylates almost all the tRNAs of E. coli (199). It is scarcely possible to find the specifying nucleotides under these conditions, which lead to complete loss of specificity.

Second, the tRNAs, which are acylated to a high extent with the same synthetase, contain the same nucleotides not only in the bihelical region of the dihydrouridine arm but also in the regions of the extra arm, of the anticodon, of the dihydrouridine loops, and also in the 3′-terminal region (198–200). Hence the number of common nucleosides in tRNAs acylated by the same enzyme is greater than that considered by Dudock et al. Furthermore, the theoretical basis of the approach itself remains unclear.

Presumably the misacylation reactions are more valuable for identifying the binding, rather than the specifying, nucleotides, and thus the region traced by Dudock et al. (219) may be in fact the binding site (or a part of it). As for the yeast $tRNA^{Phe}$–yeast phenylalanyl-tRNA synthetase pair, there is as yet no direct evidence contradicting the

possibility that the specifying site is located in the stem of the dihydrouridine loop. However, this site is certainly not the specifying one in
some other tRNAs. For example, *E. coli* tRNA^{Met} and tRNA$_f^{Met}$ exhibit similar characteristics in the acylation with *E. coli* methionyl-
tRNA synthetase (28), although they have considerably different stems
of the dihydrouridine loops. One of the two isoacceptor tRNA^{Leu} of
E. coli contains three G-C pairs in the bihelical region of the D-arm,
whereas the other isoacceptor tRNA^{Leu} has two G-C and one A-U pair;
in spite of this, the two tRNAs have almost identical kinetic characteristics in the reaction with the corresponding synthetase (257).

4. The Recognition Problem

The available evidence leads to the conclusion that candidates for
the specifying sites are at present the root of the anticodon, the stem
of the D-loop, and the amino acid stem of the cloverleaf with the adjacent unpaired-base residue of the acceptor terminus. As for the other
regions, the Tψ-arm can hardly contain the specifying nucleotides, because its structure is very similar or the same in different tRNAs (see
refs. 13 and 261). The specifying nucleotides could be present also in
the stem of the anticodon arm or in the extra arm, which are different
in different tRNAs, but at present there is no evidence confirming or
contradicting this possibility.

Whatever the location of the specifying nucleotides and whatever
the structure of the recognition site, some general conclusions can be
made at present. First, the number of both the specifying and the binding nucleotides is small compared with the total number of nucleotides
in the tRNA molecule. This conclusion is based on the fact that removal of large regions of the molecule, modification of many bases,
and marked differences in the primary structures involving tens of
bases do not prevent tRNAs from interacting with the cognate synthetases. On the other hand, the ability of phenylalanyl-tRNA synthetase to protect completely the cognate tRNA^{Phe} from the action of
ribonucleases (128) points to the large area of the contact surfaces of
the two macromolecules. Second, intactness of the *whole* three-dimensional structure of tRNA is unnecessary for recognition. This follows
from the fact that profound changes in the primary structure, which
certainly result in at least local changes in the three-dimensional
structure, do not lead to a loss of acceptor activity.

However, a definite local three-dimensional structure is a necessary prerequisite because inactive, or denatured, tRNAs have been obtained (262,263). Determination of the amino-acid-acceptor activity of tRNA as a function of the temperature of the aminoacylation reaction showed a strong correlation between the loss of acceptor activity and the thermal denaturation profile of the tRNA (264). There is evidence that the loss in acceptor activity is most likely due to a change in tRNA structure as opposed to the denaturation of the enzyme. Besides the local three-dimensional structure, recognition needs also a definite (not too great and not too little) stability of conformation (see ref. 202). Third, it seems most probable that the topography of the specifying nucleotides (i.e., their positions along the chain and in the three-dimensional structure) is different in different tRNA species. This follows from the fact that the local structural similarities of tRNAs of different organisms specific to the same amino acid occur in topographically different sites of the molecule (an exception is the anticodon). Change in the structure of topographically similar regions of different tRNAs leads to different functional consequences (265).

Proof of the participation of one or another region of the tRNA molecule in the specific interaction with the enzyme does not mean that this region contains *all* the specifying nucleotides. For example, the fact that the root nucleotides of the anticodon participate in recognition does not rule out the existence of other specifying nucleotides in the same tRNA. It remains unknown whether the specifying nucleotides form a compact "recognition center" or are scattered over both the primary and the three-dimensional structures. As for the binding nucleotides, at least some of them (e.g., the acceptor terminus) occupy a position distant from the specifying ones.

Finally, the discovery of the intraspecies misacylation and deacylation reactions prompts that studies of the cognate individual pairs be complemented by studies of the noncognate combinations. It is clear at present that the specific interaction of tRNAs with the synthetases includes a contribution of nonspecific interactions common for several ES pairs rather than a single pair. The existence of intraspecies misacylation and deacylation strongly suggests that strictly quantitative characteristics of the interaction of tRNAs with the synthetases can be obtained only with individual tRNAs and individual enzymes. For this reason the great bulk of the information obtained with unfractionated and with partially purified ES pairs must be regarded at present as only semiquantitative or qualitative.

On the other hand, it is obvious that studies of only individual ES pairs may lead to erroneous conclusions concerning, for example, the fidelity of the *in vivo* acylation. The quantitative characteristics obtained by investigating individual tRNAs and synthetases, no matter how accurate, may be misleading in the understanding of the real situation within the living cell, where both the tRNAs and the synthetases are involved in complicated relations.

VI. Genetically Altered Synthetases

The isolation and studies of mutants with altered synthetases provide valuable information on the relationship between synthetase structure and function because the new functional properties of the enzymes can be connected with the mutational changes in their molecules. On the other hand, the new phenotype arising from mutation provides a means of studying the *in vivo* functions of synthetases, particularly their regulatory role within the cell.

A. ISOLATION OF MUTANTS WITH DEFECTIVE SYNTHETASES

1. General Remarks

The synthetase-catalyzed reaction being the only way by which the organism obtains the necessary aminoacyl-tRNA, it is impossible to isolate a mutant with a completely inactive synthetase. Almost all the mutants with modified synthetases are conditionally expressed mutants (266) that grow normally under appropriate permissive conditions, but react by complete or almost complete cessation of growth in response to changes in the external medium (restrictive conditions).

Identification of synthetase mutants is performed according to the following standard scheme (266): The mutants are usually obtained by the action of mutagens like ethylmethanesulfonate or N-methyl-N-nitrosoguanine, and the conditionally expressed mutants are isolated using penicillin selection or without any selection. Among these, the mutants with changed protein synthesis are detected. Normally, of several tens of mutants with protein-synthesizing-system lesions, two or three exhibit decreased activity in one of the synthetases.

Further, it is necessary to demonstrate, especially in the case of strong mutagens, that the phenotypic effect is in fact due to a change in synthetase activity. The corresponding test usually involves infecting the mutant with the transducing phage P1 previously grown on normal cells. The transductants are harvested capable of growth under restrictive conditions (reversion to the wild type), and the activity of the corresponding synthetase is estimated. Restoration of the activity is regarded as an indication of correct identification. However, one may assume also the possible presence of a synthetase inhibitor in the mutant strain. To eliminate this possibility, extracts of mutant and wild-type bacteria are mixed, and, if the activity of the combined extract is intermediate between those of the two separate extracts, it is concluded that no inhibitor is present. The isolation of a stable merodiploid of the mutant with the bacteria of the wild type having normal synthetase activity is also regarded as evidence of the absence of inhibitor in the mutant strain (267) and thus helps to identify the mutation as a synthetase lesion.

2. *Types of Synthetase Mutants*

The use of temperature as the restrictive factor in the external medium made it possible to isolate a large group of temperature-sensitive mutants whose growth ceases at increased temperature. In a number of cases this phenotype appeared to be due to a mutational change in synthetases that were active *in vivo* only at low temperatures. Temperature-sensitive mutants of *E. coli* have been isolated with damaged valyl- (268,269), phenylalanyl- (266,270), alanyl- (47,271,272), and glycyl- (273) tRNA synthetases. Temperature-sensitive mutants with altered valyl- and phenylalanyl-tRNA synthetases are easily isolated without special selection. Several (three to seven) mutations identified as lesions of valyl- and phenylalanyl-tRNA synthetases have been isolated for each of the *E. coli* wild strains; the frequency of these mutations is relatively high compared both with the frequency of all the types of temperature-sensitive mutations and with the frequency of temperature-sensitive mutations of other synthetases. Among the 400 temperature-sensitive mutants of yeast (274), two strains contained altered isoleucyl-tRNA synthetase (275) and one contained altered methionyl-tRNA synthetase (276).

Increasing the temperature to 35–40°C resulted in cessation of the growth of these temperature-sensitive mutants. A reverse situation occurred with *Micrococcus cryophilus*, in which mutation resulted in increased thermal resistance (58). The wild type of this organism—obligate psychrophile—has a defective protein-synthesizing system and for this reason cannot grow at temperatures above 25°C. It was found that the increased thermal sensitivity of *M. cryophilus* is due to the thermal lability of two synthetases, namely, of those specific for proline and glutamic acid (277). The two enzymes were thermally stable in a mutant of this organism selected according to its ability to grow at 30°C. In this case the mutant and the wild type changed their roles as it were, because the conditionally expressed strain is in fact the wild type.

A method based on temperature-conditional resistance to thymineless death has been designed to facilitate the isolation of synthetase mutants (278). Sixty mutants of the thymine-requiring D_2 *E. coli* strain were selected resistant to thymineless death at 42°C but susceptible at 30°C. The important feature is that these selected strains, even in the presence of thymidine, did not grow at 42°C. Of the 16 mutants selected at random for aminoacyl-tRNA synthetase assays, 10 had reduced valyl-tRNA synthetase activities, one had reduced alanyl-tRNA synthetase activity, and one had reduced phenylalanyl-tRNA synthetase activity. The authors (278) suggest that the resistance to thymineless death at 42°C appears to be due to the thermal lability of the mutationally altered synthetases.

Another factor employed to obtain conditionally expressed mutants is the presence in the medium of amino acids, of their metabolic precursors or analogs. By transferring bacteria from a rich medium to a poor one deficient in an amino acid, one can obtain mutants that are auxotrophs with respect to this amino acid. In many cases cessation of growth in a medium deficient in a given amino acid is due to a change in the activity of the corresponding synthetase. Thus of 70 mutants of *E. coli* that do not grow in the absence of glycine (Gly⁻ phenotype), 40 contained defective glycyl-tRNA synthetases (279). Strains with damaged synthetases were found among *E. coli* mutants—histidine (280), tryptophan (281–284), tyrosine (285), glycine (273,279), serine (286), and isoleucine (267) auxotrophs. The mutants of *Salmonella typhimurium* that do not grow in the absence of exogenous methionine (287) and isoleucine (288) contain altered aminoacyl-tRNA synthetases for these amino acids; the tryptophan auxotroph of eukaryote *N. crassa* contains a defective tryptophanyl-tRNA synthetase (289).

In some strains mutation imparts a resistance to an analog of an amino acid that is incorporated in the wild strain into protein (e.g., canavanine or fluorophenylalanine) or inhibits protein synthesis. These mutations are of the nonconditional type. It is noteworthy that a synthetase mutation has been traced for the first time with a strain resistant to an amino acid analog. This finding was made in Neidhardt's laboratory, where it was shown that a p-fluorophenylalanine-resistant mutant of E. coli contained an altered phenylalanyl-tRNA synthetase (290). More recent reports have described other mutants of E. coli that are resistant to amino acid analogs and contain modified synthetases specific for isoleucine (291), threonine (292), and serine (293), as well as a mutant of S. typhimurium with an altered leucyl-tRNA synthetase (294).

The sensitivity of some wild-type strains of S. typhimurium to histidine analogs is due to an effect on one of the biosynthesis enzymes rather than on the synthetase. Mutations characterized by a depressed level of histidine biosynthesis lead to a resistance to analogs. Some of them are the histidyl-tRNA synthetase mutations (295–297). This is possible because at least some of the synthetases participate in repressing the biosynthesis enzymes of the appropriate amino acids catalyzing their transformation into a repressor form; the matter will be discussed in Section VIII in more detail.

Mutants with depressed amino acid biosynthesis have been used also to select strains of E. coli with defective histidyl- (280) and arginyl-tRNA synthetases (298).

Strains with altered synthetases have been found also among conditionally expressed streptomycin-suppressible lethal mutants of E. coli that grow in the complete medium only in the presence of streptomycin (299).* The method of selecting for streptomycin-conditional mutants is a general one for all the indispensable enzymes, including the synthetases. Among the 41 mutants obtained in this work, three contained an altered glutamyl-tRNA synthetase.

Sometimes it is possible to obtain altered synthetases without mutagenesis and selection. It appears that there are natural genetic variations in E. coli glycyl-tRNA synthetase. The wild-type strain K10 contains a glycyl-tRNA synthetase whose K_m value for glycine is 100 times greater than those of the enzymes of other strains (300). The glycyl-tRNA synthetase of different strains of E. coli K12 is found in one of the

---

*The effect of phenotypic suppression (partial correction of mutation at the translation stage) is due to a change in the normal meaning of the genetic code determined by altered ribosomes. The supression resulted in the ability of the mutant with a defective gene for an indispensable enzyme to synthesize this enzyme in the active state in the presence of streptomycin.

two forms coded by the same *glyS* structural gene. The $glyS_H$ strains exhibit a high activity of the enzyme in extracts, and the $glyS_L$ strains, a low activity (about 15% of the activity of $glyS_H$ extracts). The K_m for glycine of the $glyS_L$ strains' synthetase is about 10 times greater; the enzyme is more sensitive to high and low temperatures (273,279).

An unusual valyl-tRNA-synthetase mutant, *E. coli* NP2907 (301), arose spontaneously from the NP29 strain with thermosensitive valyl-tRNA synthetase (268). Unlike the initial strain, which fails to grow at 40°C, the new mutant was characterized by a more rapid growth at 40°C typical for normal cells, but could not grow exponentially in any medium. The valyl-tRNA synthetase of this secondary mutant differed from that of the temperature-sensitive strain NP29 by retaining activity *in vitro* and from the wild-type strain by stability and by a different apparent K_m value for ATP.

A number of bacterial mutants are characterized by a complex phenotype. As a rule, altered temperature sensitivity is one of the new properties of the strain after mutation. A leucyl-tRNA-synthetase mutant of *S. typhimurium* is characterized by two new phenotypes: it is resistant to the leucine analog azaleucine and grows at 27°C rather than at 35°C (302). A 5-fluorotryptophan-resistant mutant of *Bacillus subtilis* grows well at 30°C, but not at 42°C (303). These mutants were obtained as secondary ones: initial mutants resistant to the analog were selected, and the thermolabile strain was found among the primary mutants. An unusual thiosine-resistant lysyl-tRNA-synthetase mutant of *E. coli* K12 has been isolated in Zamecnik's laboratory (304,305). The mutant is characterized by decreased lysyl-tRNA-synthetase activity and at the same time by increased thermostability as compared with the wild-type enzyme. In this respect it is similar to the above-mentioned *M. cryophilus* thermoresistant mutants (58,277). A complicated phenotype of *N. crassa* mutant—the thermolabile leucine auxotroph 45208 strain—is due to the synthesis of an altered cytoplasmic leucyl-tRNA synthetase combined with almost complete loss of activity of the corresponding mitochondrial enzyme (306,307).

Thus mutants with altered synthetases have been obtained to the date mainly from bacteria. A few genetically modified synthetases have been found in two eukaryotes: yeast (275,276) and *N. crassa* (289,306, 307). Genetic changes in the synthetases of multicellular organisms remain completely unexplored. The isolation of a purified phenylalanine-tRNA synthetase from *Drosophila* (66), the classical object of multicellular-organism genetics, makes it probable that genetic variations of the synthetase will be soon found in this species.

B. CHANGES IN THE STRUCTURE AND FUNCTION OF SYNTHETASES DUE TO MUTATION

Alterations in the activities of mutant enzymes can be due to changes in their amounts in the cells, or in their structures, although phenotypically the two effects are indistinguishable. For example, the specific activity of the threonyl-tRNA synthetase of crude extracts of two *E. coli* K12 mutants resistant to borrelidine (a specific inhibitor of the enzyme) is about five times greater than that obtained from the wild-type bacteria. However, the resistance to analog in one of the strains is due to a fivefold increase in the synthetase production, whereas that of the other strain is due to a change in the structure of the enzyme (292). However, the second case occurs much more frequently, independently of the nature of the agent causing the mutation and of the specificity of the synthetase.

1. Enzymological Characteristics of Mutant Synthetases in vitro

Structural alterations in mutant synthetases are frequently claimed on the basis of change in the kinetic constants of crude extracts or of partially purified preparations. Extensive purification of the mutant synthetases is difficult due to their lability and has been achieved only in a few cases. The differences in the apparent V_{max} values found for the crude preparations obtained from the mutant and the wild-type organism may be attributed not only to the structural changes in the enzyme molecule but also to some other reasons, such as different extractability or rate of transcription of mutant gene. The K_m values are mainly used as the criteria of substrate affinity because it is difficult to measure the K_s values. As a rule there is a direct dependence of the phenotypic expression of the mutation on the type of the alterations in the mutant synthetase. This correlation can be illustrated by several examples.

The majority of the auxotrophs with mutant synthetases are characterized by an increase in the K_m value for the amino acid. The K_m value of the mutant enzyme can sometimes be only a few times greater than that of the normal one [e.g., two to three times in the case of tryptophanyl-tRNA synthetases from *Neurospora* (289) and *E. coli* (282)], whereas in many cases it becomes several orders of magnitude greater [e.g., mutant methionyl- and isoleucyl-tRNA synthetases from *S. typhimurium* (287,288) and mutant isoleucyl-tRNA synthetase from *E. coli* (267)]. An increase in the K_m value means a decrease in the rate of the acylation of the corresponding tRNA by the amino acid at the normal endogenous level of the latter; increase in the amino acid endogenous level after the addition of the amino acid to the medium results in an increased rate of acylation and thus in normalization of the growth. Introduction of the mutant isoleucyl-tRNA-synthetase allele into the strain with overproduction of isoleucine (due to dere-pression of one of the enzymes of the biosynthesis pathway, namely, threonine deaminase) resulted in ability of hybrid to grow without added isoleucine (267).

As a rule mutants selected for resistance to amino acid analogs are characterized by a considerable increase in the K_m (or K_i) values as well as by a decrease in the V_{max} of the corresponding synthetases for the analogs. At the same time the affinity for the substrate amino acid also changes. For example, the phenylalanyl-tRNA synthetase from the *E. coli* PFP–10 *p*-fluorophenylalanine-resistant mutant exhibits a sixfold increase in the K_m value and a 25-fold decrease in the V_{max} for the analog. Its K_m for phenylalanine also increases, but the V_{max} value decreases only by a factor of 2. As a result of these changes phenyl-alanine has an advantage over its analog in aminoacylation reaction (308). An increase in the K_m value for the analog is sometimes ac-companied by a *decrease* in the K_m value for the natural amino acid. The borrelidine-resistant mutant of *E. coli* K12 has a threonyl-tRNA synthetase with $K_m = 1.7 \times 10^{-5}$ for threonine, compared with 8.5×10^{-5} of the wild-type enzyme (292). Hence with this group of mutants one may consider a change in substrate specificity that may sometimes be referred to amino acids or their analogs other than those used in the course of selection. For example, the activity of the phenyl-analyl-tRNA synthetase of the *E. coli* PFP–10 mutant mentioned in this paragraph is more inhibited by leucine than the enzyme ac-tivity of the wild type (309). However, the specificity sometimes re-mains unchanged, as in the case of the mutant isoleucyl-tRNA syn-thetase of the *E. coli* isoleucine auxotroph: a parallel increase in the K_m values both for isoleucine and for its analogs DL-valine, thioisoleucine, and O-methyl-DL-threonine was observed (310).

The synthetases of the thermolabile strains also exhibit remarkably altered kinetic constants *in vitro*. For example, the K_m value for valine in the acylation reaction at 20°C is greater for the thermolabile synthetase of *E. coli* than that for the wild type (311). The K_m value for glutamic acid of the altered glutamyl-tRNA synthetase of streptomycin-dependent mutants of *E. coli* is about 10 times greater than that of the enzyme from the parent strain (312).

Partially purified altered histidyl-tRNA synthetases have been obtained from five mutants of *S. typhimurium*, and their kinetic constants for three substrates have been compared with those of the wild-strain synthetase (297). Various types of lesions were anticipated because the mutants were selected for a change in a secondary function of the enzyme (i.e., a derepression of the histidine operon). In fact, three of the strains (S 1520, S 1210, and S 1595) exhibited a decreased affinity of the enzyme for the amino acid (K_m values from 905 to 82 μM, compared with 25 μM for the wild-type enzyme), which was insufficient for the normal aminoacylation of tRNA by the endogenous histidine pool [15 μM (313)]. Of the three mutants, the *his* S 1520 one had a synthetase with a considerably increased K_m for ATP, whereas its K_m for tRNA remained unchanged. As for the *his* 1210 and *his* 1595 mutants, their synthetases had approximately proportionally two to five times increased K_m values for all the three substrates. The fourth mutant (S 2280) had a synthetase with a normal affinity for histidine and for tRNA^{His}, but with an increased K_m for ATP. The fifth mutant (S 1587) had a synthetase with a decreased K_m for histidine and ATP, but with an increased K_m for tRNA^{His}.

Simultaneous determination of the K_m values for several substrates has been performed also by other workers. In some of the mutants a change in the K_m value for the amino acid is accompanied by a change in the K_m for ATP, although in many cases no proportionality can be seen (292,310,311,314). Yet the affinity of the enzyme for tRNA remains unchanged (310,311,314). On the other hand, mutant alanyl-tRNA synthetases from two temperature-sensitive mutants of *E. coli* have a changed affinity for tRNA (47,271,272) rather than for the amino acid. Presumably the enzymes, which retain *in vitro* the ability to activate valine (268), phenylalanine (266), and tryptophan (282, 289) but do not acylate tRNA, are characterized by a decreased affinity only for the latter substrate. The independent changes in the kinetic constants for amino acid and ATP on the one hand, and for tRNA on the other hand, suggest a relative structural independence of the activation and the acylation functions. As for the parallel change in the

K_m values for all three substrates, which is sometimes the case (297), this cannot be regarded as an argument against the viewpoint, because a single mutation may result in a change of the protein conformation that is important for both functions.

Different thermal sensitivity is another property that can be used as a guide in evaluating structural changes in mutant synthetases. Sensitivity to heat or cold are characteristic of not only the temperature-sensitive conditionally expressed mutants but also of mutants with other phenotypic expressions (267,273,279,282,284,294,310,312). Substrates often protect the mutant thermally labile enzymes from inactivation *in vitro* (272,282,284,301,311). The Trp⁻ phenotype of an *E. coli* mutant with an altered tryptophanyl-tRNA synthetase was due to the increased stability of the mutant enzyme in the presence of endogenous tryptophan whereas the affinity of the enzyme for the amino acid remained unchanged (284). In many cases the defective protein is characterized *in vitro* by a change in the K_m value and, at the same time, in thermal stability.

Hence changes in some properties of mutant synthetases *in vitro* suggest alterations of their structures.

Mutation sometimes makes the synthetase so unstable that its activity cannot be assayed *in vitro* at all (266,302,315). For this reason a special technique has been elaborated to detect enzymatically inactive proteins (316) involving antibodies against the corresponding synthetase. A given amount of the normal synthetase is incubated with an amount of the antiserum that inactivates the enzyme almost completely. In another experiment the mutant inactive protein is added to the incubation mixture. This protein being a cross-reacting material, some of the antibodies are spent in binding with it; therefore the synthetase activity is inhibited to a smaller extent and thus can serve as a measure of the amount of modified enzyme.

2. Studies of the Structural Alterations in Mutant Synthetases

It has been shown for a number of mutants, mainly with altered thermal stability, that changes in synthetase properties are due to changes in their quaternary structures. For example, it was demonstrated with *E. coli* phenylalanyl-tRNA synthetase that the structural basis of its thermal lability is a temperature-induced dissociation of the enzyme, which has a molecular weight of 180,000 (317), into a cross-reacting material with a molecular weight of about 90,000 (315). Since the wild-type enzyme consists of four presumably identical subunits of molecular weight 44,000 (59), the cross-reacting material is most probably a dimer. The cross-reacting material present in the extracts of independently isolated mutants NP37, NP313, NP314, and NP51 is capable of complementation *in vitro* with the subunits obtained by treating the wild-type enzyme with urea (270). The enzymatically active complementation products have the same size as the native enzyme and differ from the latter in their greater thermal lability. Attempts to obtain active complementation products by mixing the cross-reacting materials of different mutants were a failure, presumably due to the similarity of the structural changes caused by the mutations.

Glutamyl-tRNA synthetases of *M. cryophilus* of both the wild type (strain 68) and the stable mutant (TMP 9 strain) divide into two peaks of cross-reacting material and of the accompanying enzymatic activity in gel filtration as well as in sucrose-gradient centrifugation. The major and minor peaks correspond to molecular weights of 190,000 and 95,000, respectively. Heating the wild-type enzyme at 30°C before chromatography results in an enzymatically inactive peak for cross-reacting material of 53,000 molecular weight, and the peaks for molecular weights of 190,000 and 95,000 disappear. The mutant enzyme can be also dissociated into inactive subunits of the same size, but under rather drastic conditions (incubation with guanidinium chloride) (277).

Mutant *E. coli* lysyl-tRNA synthetase is distinguished from the wild-type enzyme both by higher thermostability and higher resistance to urea denaturation (305). In extracts from the wild-type strain two enzyme forms were found after gel chromatography on a Sephadex G-200 column; the major and minor forms have molecular weights of about 135,000 and 95,000, respectively. The latter size is in the range (80,000–110,000) reported by other workers (23,48,314,330). However, the fraction with a molecular weight of 135,000 is in sharp contrast with all these publications. The reason for this discrepancy is unknown. The profile of a mutant strain revealed a third form with a molecular weight of 195,000, in addition to the other two. The various forms of the enzyme could be accounted for by a basic subunit of 45,000–50,000 molecular weight (305). Hence in the wild-type strain one would postulate a dimer and a trimer, and in the mutant these plus a tetramer. A conclusion was made (305) that the increased thermal stability of the mutant enzymes was the result of improved subunit interaction. Hence all the studies support the conclusion that the reason for the decreased thermal stability of the synthetases is weakening of the interaction between the subunits, leading to their dissociation at slightly elevated temperatures.

Dissociation into subunits has been demonstrated also for alanyl-tRNA synthetases from the thermally labile strains of *E. coli* T 140 (271) and BM 113 (47,272). In both cases mutation results in a transition of the dimer-to-monomer type, although the sedimentation constants of the wild-type enzyme as claimed in the papers are considerably different. This transition leads to decreased affinity for tRNA.

Some mutants of glycyl-tRNA synthetase, like the natural strain $glyS_L$ and the auxotroph $glyS_{87}$, also contain this enzyme in a modified form that readily dissociates into subunits (279). For example, the glycyl-tRNA synthetase of $glyS_L$ at 20°C exhibits a sedimentation constant of 6.8S, while at 4°C (the temperature at which the enzyme becomes inactive) the constant is 5.3S. The discovery of nonidentical α- and β-subunits of the synthetase of the wild type (see Table I) and subsequent application of the complementation method made it possible to identify the subunit altered by the mutation (60). The α- and β-subunits of the wild type were separated and subjected to complementation with pCMB-treated crude extracts of mutant cells. It was expected that the activity of the enzyme with an altered α- (β-) subunit would be restored in the presence of the α- (β-) subunit of the wild type. Activity of the synthetases of two strains, $glyS_L$ and $glyS_{87}$, was recovered after the addition of the α-subunit. The third strain, as well as the glycine auxotroph $glyS_{17}$, contains a synthetase that is stimulated by addition of the β-subunit rather than the α-subunit. The fourth strain, the temperature-sensitive mutant BF 134 (279), also contains a defective β-subunit. Thus the mutants with damaged glycyl-tRNA synthetase belong to two groups, one with a defective α-subunit and the other with a defective β-subunit. The authors cited have also observed complementation on mixing two glycyl-tRNA synthetases, one having a defective α-subunit and the other a defective β-subunit. The enzymatic activity of the complementation products, their thermal stability, and the K_m values do not differ considerably from those of the native enzyme.

Hence changes in the properties of the synthetases of some of the mutants in several cases can be explained by oligomeric transitions. However, this reason is not at all a general one, because the mutations frequently involve synthetases consisting of a single polypeptide chain (see Table I). A typical example is a valyl-tRNA synthetase whose mutations in *E. coli* occur more frequently than those of the other synthetases (266). Nevertheless the complementation technique is an important achievement insofar as it makes it possible to localize more exactly the sites of mutation in strains having modified oligomer synthetases.

3. Acylation of tRNA by Mutant Synthetases

Cessation of the growth of mutants with changed synthetases under restrictive conditions must be due to a decrease in the level of the growth-limiting product of the enzymatic reaction, presumably of aminoacyl-tRNA.

Analysis of the kinetic constants of mutant synthetases *in vitro* (see above) suggests that the mutants must have a decreased level of aminoacyl-tRNA at normal concentration of the amino acid pool.

Determination of the intracellular pool of the $tRNA^{Val}$ acylated with valine in a strain with temperature-sensitive valine-tRNA synthetase was performed (318) by the periodate oxidation method. Under the restrictive conditions, the level of valyl-$tRNA^{Val}$ in mutant cells was considerably decreased. Similar results were obtained for temperature-sensitive mutants with damaged alanyl-tRNA (272) and isoleucyl-tRNA (275) synthetases, for the regulatory histidyl-tRNA-synthetase mutants (319), and also for Trp⁻ (284) and Gly⁻ mutants (273). Special techniques of rapid growth cessation were applied in some studies (272,273,319) because the pool of the charged tRNA is very small and is involved in a very rapid exchange. The strong decrease in aminoacyl-tRNA levels in these cases must have been due to decreased acylation of the major isoacceptor tRNA species; as for the extent of *in vivo* acylation of minor isoaccepting species, this must be studied by more special methods.

In vitro acylation of isoaccepting $tRNA^{Ala}$ species by a mutant alanyl-tRNA synthetase has been investigated by Buckel, Lubitz, and Bock (272). The isoaccepting species were separated by chromatography on benzoylated DEAE-cellulose. The mutant enzyme charged the same fractions of homologous and of heterologous tRNA as the native synthetase, and the K_m values for all of the three species increased approximately to the same extent (by a factor of 10). Thus the mutation resulted in a proportional decrease in the affinity for all the isoacceptor tRNA species.

To evaluate the extent of acylation of a minor tRNA^{Gly} fraction in a Gly$^-$ mutant *in vivo* (273), the efficiency of the *trpA36* locus suppression was determined. Part of the tRNA^{Gly} in the parent suppressor strain $Su_{36}{}^+$ recognizes the arginine codon and incorporates glycine instead of arginine into the protein chain. The mutants of this strain with altered glycyl-tRNA synthetase, even when grown in the presence of glycine, are characterized by a threefold to fourfold smaller efficiency of the suppression compared with the parent strain. Hence the acylation level of at least one of the minor tRNA^{Gly} fractions (suppressor tRNA^{Gly}) decreased in parallel with the decrease of the extent of acylation of major fraction(s).

In other cases the mutant synthetases may acylate the isoacceptor fractions to a different extent. No decrease in the level of phenylalanyl-tRNA *in vivo* under restrictive conditions occurs in temperature-sensitive mutants at phenylalanyl-tRNA synthetase (266). It was proposed that the mutant synthetase acylates to a sufficient extent the major fraction of tRNA^{Phe}, but loses the affinity for a minor fraction that escapes detection by the periodate method. Absence of this acylated minor tRNA fraction leads to a cessation of protein synthesis in spite of the sufficient amount of the major fraction(s) of phenylalanyl-tRNA^{Phe}. Presumably the growth-limiting leucyl-tRNA^{Leu} in an *S. typhimurium* mutant with altered leucyl-tRNA synthetase is also a minor isoacceptor fraction because under restrictive conditions about 75% of the tRNA^{Leu} is present in the acylated form (294). Three mutant glutamyl-tRNA synthetases charge the major and the minor species of tRNA^{Gly} at a smaller rate compared with the synthetase of the parent strain. All the mutant synthetases acylate the minor fraction at the same rate, but differ from each other in the rate of acylation of the major fraction (312). Preferred interaction of the synthetases with some of the isoacceptor tRNA species may be an important mechanism involved into cellular regulation.

VII. Structural Genes of the Synthetases

Use of the transducing phages and of the interrupted matings of different strains of *E. coli* makes it possible to find the positions of structural synthetase genes on the circular chromosome of *E. coli*.

Figure 1 shows the linkage map of *E. coli* according to Taylor and
Trotter (320); the positions are indicated for some of the synthetase
structural genes. It is seen that the structural genes coding for different
synthetases do not occupy neighboring positions. Generally, they are
also not adjacent to the genes for the corresponding amino-acid-bio-
synthesis enzymes. Exceptions are the genes of tyrosyl-tRNA synthetase
(*tyrS*) and tryptophanyl-tRNA synthetase (*trpS*). Gene *tyrS* is located
(285) near the gene *aroD* controlling one of the enzymes leading to
shikimic acid in the pathway common to the aromatic amino acids.
Gene *trpS* also maps very close to two genes governing earlier steps in
aromatic amino acid biosynthesis, not specific for tryptophan (*aroB*
and *pabA*) (282). The genes of the aromatic pathway, however, are
not clustered into one operon. The meaning of this neighborhood is
not clear; possibly the mapping of *tyrS* and *trpS* genes is fortuitous.
Gene *trpS* is located on the *B. subtilis* genetic map between *argC* and
metA, and therefore is unlinked with structural and regulatory genes
governing the tryptophan-biosynthesis enzymes (303).

As already mentioned, several mutant strains have been isolated,
frequently by different methods, with lesions in the synthetase of the
same specificity. However, each of these mutations occupied the same

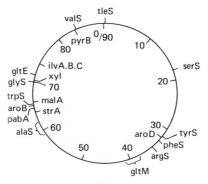

Fig. 1. Location of structural genes for some aminoacyl-tRNA synthetases in
E. coli. Modified from Neidhardt (321). The numbers on the map are from Taylor–
Trotter (320) determinations of marker entry times. References to synthetase mapping are as follows: isoleucine, *ileS* (267); serine, *serS* (286); tyrosine, *tyrS* (285);
phenylalanine, *pheS* (322); arginine, *argS* (320); glutamic acid, *gltE* and *gltM*
(299,312); alanine, *alaS* (269); tryptophan, *trpS* (282,284); glycine, *glyS* (279,300);
valine, *valS* (318,323).

position on the genetic map, suggesting that a single gene codes in these cases for the synthetase corresponding to a given amino acid. An exception is glutamyl-tRNA synthetase; two loci have been assigned to this enzyme on the genetic map of *E. coli* (312). Mutation of the EM 111 (and probably of the EM 102) strain is mapped to the *xyl* locus, whereas that of the EM 120 strain occupies the position between 38 and 41 min (see Fig. 1). The designation *gltS* had been used earlier for the structural gene of glutamate permease. For this reason the loci corresponding to the 111 and 120 mutations were called *gltE* and *gltM*, respectively. Attempts attribute the phenomenon to the presence of two genes coding for two different glutamyl-tRNA synthetases, for two different subunits, or, finally, for the synthetase and a specific modification factor have failed. One may recall here that both types of glycyl-tRNA-synthetase mutations, one leading to a defective α-subunit and the other to a defective β-subunit (60), are localized in the same position of the genetic map, close to the *xyl* locus (279). The cistrons coding for the two polypeptide chains are most probably immediately adjacent.

The positions of several synthetase genes have been determined on the genetic map of *S. typhimurium* (324), closely related to *E. coli*. These are the *hisS* gene of histidyl-tRNA synthetase (79 min, between *guaA* and *strB*) (296), the *leuS* gene of leucyl-tRNA synthetase (18 min, near the *gal* locus (31), and the *ilvS* gene of isoleucyl-tRNA synthetase (between *pyrA* and *thr*) (288). Mutations of *S. typhimurium* are also described at the *metG* gene (67 min), resulting in methionine auxotrophy (325); recent studies demonstrated that *metG* is the structural gene of methionyl-tRNA synthetase (287).

The position of the structural gene of yeast isoleucyl-tRNA synthetase has been determined (276). The *ILS* gene occupies a position about 22 units to the left of the P_9-marker of chromosome II on the yeast genetic map (327). As with prokaryotes, there is no coupling between the genes of different synthetases; the gene of methionyl-tRNA synthetase is not adjacent to the *ILS* gene.

Knowledge of the location of structural genes of enzymes on genetic maps is sometimes useful in the solution of some practical problems. For example, Cassio, Lawrence, and Lawrence (328) attempted to isolate a strain of *E. coli* providing a better yield of methionyl-tRNA synthetase compared with the usual haploid *Hfr H* strain. At that time the position of the structural gene on the map of *E. coli* was unknown, but by analogy with the *S. typhimurium* map it was proposed that the gene, which was homologous to the methionyl-tRNA synthetase *metG* gene of *S. typhimurium* (325), would be located on the F 32 episome of *E. coli*. For this reason the episome-bearing strains were investigated (328) in the hope of obtaining a two times greater yield of methionyl-tRNA synthetase due to the diploidy of the structural gene of the enzyme. Comparison of the activities of the synthetases in the haploid and in the merodiploid revealed in fact an increased level of only one synthetase specific for methionine. However, the increase in activity was fourfold instead of the expected twofold. To find the reason for this unpredicted, though certainly desired, increase, the parent strains of one of the merodiploids were investigated. It was found that the mother strain had the same level of the methionine-activating enzyme as the haploid *Hfr H* strain, whereas the father strain *Hfr AB* 311, which was the source of the F 32 episome, had a three times greater level of enzyme activity. Thus the merodiploid contained the sum of the activities of the enzyme characteristic of the parents; as for the reason of the increase of activity in *Hfr AB* 311, it remains unknown.

A duplication (or, probably, amplification) of one more mutant synthetase gene (*glyS*) (329) has also been discovered. The authors of this finding noticed that mutants with altered glycyl-tRNA synthetase that were auxotrophs to glycine (Gly$^-$) revert at a high frequency ($> 10^{-5}$) to the Gly$^+$ phenotype, that is, become independent of exogenous glycine supply. However, this reversion does not involve a change in the *glyS* mutant allele. Genetic analysis revealed that the reversions are due to duplication (or amplification) of the *glyS* allele which involves probably also other genes mapped within 4 min from the *glyS* gene. Increase in the number of structural genes for glycyl-tRNA synthetase leads to an increase in the intracellular concentration of the defective glycyl-tRNA synthetase, providing an almost normal rate of the acylation of tRNA^{Gly} at the same endogenous concentration of glycine. Amplification of the structural genes of threonyl-tRNA synthetase may be responsible for the fivefold increase in the production of the enzyme by a borrelidine-resistant strain compared with the wild type (292).

Hence it seems probable that strains with an increased number of structural genes for synthetases do occur in nature and will be soon found by genetic studies. The use of such strains for increasing the yield of bacterial synthetases would be very convenient.

VIII. Aspects of *in Vivo* Functioning of Aminoacyl-tRNA Synthetases

Although our knowledge of the enzymology of isolated aminoacyl-tRNA synthetases is incomplete, their functioning in living cells is far more obscure. Neidhardt was the first to consider the role of synthetases in cell physiology in the review (266) that remains a basic source of information in this field. Later on definite progress was achieved in the study of two problems that are the subject of this section: (*a*) the role of aminoacyl-tRNA synthetases in regulating amino acid biosynthesis in *Prokariota* and (*b*) the regulation of the synthesis and activity of the synthetases themselves. Presumably the synthesis of these enzymes is not of a constitutive type as was thought before but is subject to control by a repressionlike mechanism. The quantity of the synthetases in cells depends also on the rate of their inactivation, which in turn depends on various factors.

Other cellular events may be mentioned here in connection with the *in vivo* functioning of aminoacyl-tRNA synthetases. A correlation was revealed between synthetase activity and stable RNA synthesis. In a number of mutants having altered synthetases there is a sharp cessation of tRNA and ribosomal RNA synthesis under restrictive conditions (266,278,303) even at a large excess of free amino acids. These observations suggest (but do not prove) the participation of the products of synthetase action (presumably aminoacyl-tRNA) in the regulation of RNA biosynthesis. Dependence of the RNA accumulation on the synthetase activity is observed only in cells with stringent amino acid control, and not in relaxed strains lacking this control. The different behavior of RC^{str} and RC^{rel} strains both containing the same altered synthetase was shown for valine (268,318), glycine (279), and serine (293) enzymes. At the nonpermissive temperature *E. coli* RC^{str} strain containing temperature-sensitive valyl-tRNA synthetase is capable of producing (331) ppGpp, which is an inhibitor of stable RNA synthesis (332). De Boer, Raue, and Grunber (331) observed also, at variance with others (268,333), the resumption of RNA synthesis in a mutant strain at restrictive temperature after the addition of chloramphenicol or sparsomycine.

Aminoacyl-tRNA synthetases may play a role in regulating protein biosynthesis by controlling the relative amounts of charged and uncharged tRNA isoacceptors. This possibility was postulated among others by Strehler, Hendley, and Hirsch (334) and Anderson (335) in their theories of development. In fact, during the course of differentiation in plants and animals certain changes have been observed in the chromatographic and electrophoretic properties of some synthetases as well as in the range of their substrate specificity (337,338).* Those changes together with alterations in the set of isoacceptor species (see ref. 336) may lead to the cell's inability to read certain code words and therefore to synthesize some polypeptides.

Although observations of this type continue to accumulate in the literature, there is no crucial evidence in favor of this hypothesis at least in regard to the role of synthetases.

Probably in the same line are certain events taking place in bacterial cells after phage infection. It has been shown in Neidhardt's laboratory that a new modified valyl-tRNA synthetase develops in *E. coli* during infection by T-even phages. The enzyme is characterized by improved stability and increased molecular weight (339–342), other synthetases remaining unchanged (343). After the modified enzyme was isolated in a purified state, it was discovered to contain a phage-directed component that can be dissociated from the core resembling normal host synthetase by treatment with denaturing agents (342). This component, called the τ-factor, seems to be a polypeptide with a molecular weight of 10,000. It is reasonable to suppose that phage-induced changes in the translating apparatus of the host cell are in some way related to the biology of the infecting agent. However, no differences were found between the enzymological properties of the host and phage-induced valyl-tRNA synthetases, including the relative extent of acylation of *E. coli* tRNA^{Val} isoacceptors (342). By means of phage mutants that are unable to convert the valine enzyme into the modified form it was shown that the modified enzyme is not essential for normal phage development and production (340). Thus the biological significance of the appearance of the modified valyl-tRNA synthetase after phage infection remains mysterious.

A. POSSIBLE ROLE OF AMINOACYL-tRNA SYNTHETASES IN THE REGULATION OF AMINO ACID BIOSYNTHESIS

Participation of aminoacyl-tRNA synthetases in the repression of the synthesis of the corresponding amino acids has been found in bac-

*For additional references see Kanabus and Cherry (336).

teria for valine (268,344,345), isoleucine (288,291), leucine (294), histidine (295,297,319,346), and tryptophan (284). The phenomenon has been demonstrated by both indirect and direct methods. The former approach makes use of the fact that specific inhibitors of synthetases, but not of the biosynthesis enzymes, derepress the pathway of biosynthesis of corresponding amino acids. The latter approach is based on the findings of mutant strains where in all probability the same mutation resulted in both alteration of synthetase and derepression of biosynthesis enzymes. In accord with the concept of multivalent repression of the isoleucine–valine pathway (368), the strains with mutant valine-tRNA synthetase or isoleucyl-tRNA synthetase are characterized by derepressed levels of the enzymes participating in valine and isoleucine synthesis, whereas the strains with mutant leucyl-tRNA synthetase have increased levels of enzymes involved in leucine, valine, and isoleucine synthesis. Derepressed levels of biosynthesis enzymes have been also found in similar mutants of eukaryotes, such as a yeast mutant with modified isoleucyl-tRNA synthetase (275) or a Trp⁻ auxotroph of *Neurospora* (289).

Summarizing these data it can be concluded that amino acids with branched chains, as well as histidine and tryptophan, cannot act as repressors themselves, but interact somehow with the synthetases. Two types of such interaction can be distinguished: (*a*) involvement of the synthetase in a process whereby the amino acid is converted to an active repressor and (*b*) a repressor function of the synthetase together with the amino acid or its derivative.

As for the first possibility, it may well be that the role of valyl-tRNA synthetase in the repression of the valine–isoleucine biosynthesis pathway consists of the synthesis of valyl-tRNA, which acts as the corepressor. Although there is no direct evidence to this end, the known facts are in line with this proposal. The products of the first stage of the reaction seem to be not involved in the repression because derepression is observed in mutant strains that retain practically all the valine-activating ability (268). Furthermore, the valine analog DL-α-aminobutyric acid, which is activated by the synthetase but not transferred to tRNA, does not affect the synthesis of the valine–isoleucine biosynthesis enzymes. On the other hand, aminochlorobutyrate, which is able to charge the tRNA, blocks the synthesis of these enzymes (345). There is also a correlation between the intracellular level of valyl-tRNA and the repression of the biosynthesis enzymes (318,345).

Another candidate for the corepressor function of the *ilv* operon is leucyl-tRNA (348). This compound binds reversibly and specifically with the immature form of L-threonine deaminase of *S. typhimurium*, an enzyme specified by the A-cistron of the *ilv* operon. The native (mature) catalytically active L-threonine deaminase formed in the presence of maturation-inducing ligands does not bind tRNA. For this reason the proposed scheme (348) of the *ilv*-operon regulation involves the immature form of threonine deaminase as the aporepressor and leucyl-tRNA as the corepressor; the participation of valyl- or isoleucyl-tRNA was not discussed.

The role of histidyl-tRNA^{His} in the regulation of the *S. typhimurium* operon has been studied thoroughly by Ames et al. (295,319,349). Working with six classes of regulatory mutants of the histidine-bio-synthesis operon, one of them (*hisS*) with an altered histidyl-tRNA-synthetase gene (295), the authors found a correlation between the amount of charged tRNA^{His} and the repression of the histidine operon (319). Decrease in the level of histidyl-tRNA^{His} *in vivo* occurs also during physiological derepression caused by a partial defect in histidine biosynthesis. The repression depends on the absolute content of his-tidyl-tRNA^{His}, rather than on the charged/uncharged tRNA^{His} ratio or on the content of the uncharged tRNA^{His} (319).* The histidine analog triazolalanine, which acylates tRNA^{His}, also represses the histi-dine operon. Decreased levels of histidyl-tRNA *in vivo* were also ob-served in derepressed *B. subtilis* cells (350); the ratio of the two acylated isoacceptor tRNA^{His} species remained unchanged. These data suggest that histidyl-tRNA synthetase may participate in the regulation of the histidine operon via the synthesis of histidyl-tRNA^{His}; the histidine operon may be regulated also by the products of other genes, that is, the structural gene for tRNA^{His} *hisR*, and the *hisU* and *hisW* genes, which are presumably involved in tRNA maturation.

Possibly the regulatory role of histidyl-tRNA synthetase is not lim-ited to the production of histidyl-tRNA. Brenner et al. (351,352) meas-

---

*Phosphoribosyltransferase, the enzyme that catalyzes the first step in the path-way for histidine biosynthesis, has a high affinity for histidyl-tRNA *in vitro* in prefer-ence to deacylated tRNA^{His} and to other species of aminoacylated tRNA (347). It had been shown earlier that phosphoribosyltransferase plays a role in the regula-tion of the histidine operon (370); existence of the specific complex of this enzyme with histidyl-tRNA is consistent with the possibility that the regulatory role of phosphoribosyltransferase is carried out in a complex with histidyl-tRNA.

ured the intracellular concentrations of histidine tRNA and histidyl-tRNA synthetase, and the K_m value for the tRNA. They came to the conclusion that both the enzyme and tRNA[His] exist mainly in a form of an ES complex within the cell and proposed that the synthetase itself plays the role of a repressor in the complex with histidyl-tRNA[His]. Even if synthetase is not itself the aporepressor, its ability to form the complex with the major tRNA[His] fraction must be considered in discussions of the repression mechanism (352).

The situation was, however, complicated by the isolation of another group of *his T* mutants that are defective in the proper regulation of the histidine operon in spite of the normal level of histidyl-tRNA[His] (319). The tRNA[His] of this mutant differs from the wild-type tRNA in that it contains two uridylic acid, instead of pseudouridylic acid, residues in the anticodon loop (353), while retaining affinity for the synthetase. Allaudeen, Yang, and Söll (355) have established the nucleotide sequence of another tRNA species of *S. typhimurium*—the major fraction of the tRNA[Leu]. A comparison of the nucleotide sequence of the wild-type tRNA[Leu] with that from the *his T* mutant reveals that the mutant tRNA[Leu], as in the case of mutant tRNA[His], lacks two pseudouridines in the anticodon region and contains instead two uridines. The leucine-biosynthesis enzymes in the *his T* mutant are also derepressed (355). This work proves further the conclusion of Ames and others (353) that *his T* mutants are defective in enzyme(s) converting uridine to pseudouridine in the tRNA molecule. In this case the tRNA-modifying enzyme(s) is responsible for the regulation of the biosynthesis pathways for at least two amino acids. These data illustrate also that not only the absolute level of the histidyl-tRNA is essential in regulation but also the chemical nature of the tRNA[His] irrespective of the extent of its aminoacylation.

The participation of tryptophanyl-tRNA synthetase in the repression of the tryptophan operon and of the synthesis of tryptophan messenger RNA has been demonstrated using *E. coli* mutants with the altered-synthetase structural gene *trpS* (283,284). However, acylation of the major tRNA[Trp] fraction seems to be unnecessary for the repression because 5-methyltryptophan, which induces repression, is not incorporated into tRNA either *in vitro* (282) or *in vivo* (284). Both groups believe that tryptophanyl-tRNA synthetase participates in the repression due to other functions, presumably due to acylation of minor tRNAs. The data on the participation of *E. coli* tryptophanyl-tRNA synthetase in the regulation are in line with the results obtained with Neurospora (289).

However, it was also found that the synthetases specific for a number of other amino acids are not involved in the conversion of amino acids into compounds with repressor activity. These results were obtained by means of bacterial mutants with modified phenylalanyl- (321), tyrosyl- (285), arginyl- (298), glycyl- (279), and methionyl- (287) tRNA synthetases. In these cases no correlation was found between the level of tRNA acylation and repression. The possible regulatory role of minor tRNA fractions has been studied with tRNA^{Arg} (326). It seems that none of the five isoacceptor fractions takes part in the regulation, because their ratio in *E. coli* was the same in the wild type under repression and derepression conditions as well as in the mutant with damaged arginyl-tRNA synthetase. It is noteworthy that the negative results (i.e., failure to demonstrate the participation of some of the synthetases in repression) cannot serve as a basis for any final conclusions. Faanes and Rogers (390) and Williams (391) recently reported results implicating arginyl-tRNA synthetase in the repression by arginine of its own biosynthesis enzymes. In contrast to (326) it was found (392) that a mutant possessing an arginyl-tRNA synthetase with an altered K_m for tRNA^{Arg} has a reduced level of *in vivo* aminoacylation of two of the five isoaccepting species of tRNA^{Arg} and complete absence of aminoacylation of one of the isoaccepting species. Phenylalanyl-tRNA-synthetase mutants are characterized by normal repression of phenylalanine biosynthesis (321); in spite of this, it was found that the allosteric (first) enzyme of the common pathway of aromatic amino acid biosynthesis—phenylalanine-dependent 3-deoxy-D-*arabino*-heptulosonate-7-phosphate synthetase—forms a specific complex with phenylalanyl-tRNA (354).

Hence at present it can be claimed only that some of the synthetases are involved in the repression of biosynthesis enzymes. The concept of indirect participation via aminoacyl-tRNA formation has been confirmed experimentally for the isoleucine–valine and for the histidine pathways. As for the direct action of synthetases as repressors, this possibility remains unexplored.

B. CONTROL OF AMINOACYL-tRNA SYNTHETASES IN BACTERIAL CELLS

It is important to determine whether the synthesis of aminoacyl-tRNA synthetases is constitutive or whether there are mechanisms that regulate the intracellular levels of these enzymes. Early results published before 1967 suggested that the level of these enzymes did not vary under conditions leading to significant changes in the quantities of biosynthesis enzymes for the same amino acids (295,296,356–358). As for the modern, though still incomplete, knowledge of the regulation of the rate of synthesis of the synthetases in bacterial cells we are obliged first of all to the studies of F. C. Neidhardt and associates. Success in this very difficult field of *in vivo* regulation of the synthesis of indispensable enzymes was connected primarily with two approaches: (*a*) use of strains auxotrophic and braditrophic to a given amino acid which are very sensitive to a decrease in the intracellular concentration of the amino acid; (*b*) use of the fractionation method, which makes it possible to separate according to density preexisting enzyme molecules from newly synthesized ones.

The method proposed by Hu, Bock, and Halverson (359) for pure proteins is applicable also to crude extracts. Bacteria are grown in D_2O, and subsequently the culture is transferred to a different H_2O-containing medium. The preexisting "heavy" proteins and the newly synthesized "light" proteins are separable by buoyant density centrifugation in a CsCl gradient. Williams and Neidhardt (360) employed the method for some aminoacyl-tRNA synthetases and found that two processes take place after the strain is transferred to a medium restricted in the appropriate amino acid: the preexisting synthetases are very rapidly inactivated, and the rate of their *de novo* synthesis is considerably increased, resembling the derepression of biosynthesis enzymes. The superposition of the two processes must have been the reason for the former conclusion that the differential rate of the synthesis of some of the synthetases remains constant under amino-acid-restriction conditions.

Further investigation of *in vivo* synthetase inactivation was made with an unusual *E. coli* mutant at valyl-tRNA synthetase (301) (see Section VI.A.2), which appeared to be conditionally thermolabile *in vivo* (361). At 40°C the enzyme remains stable only so long as the endogenous levels of L-valine and ATP remain high. Furthermore, a correlation was found between the rate of growth and the rate of mutant-valyl-tRNA-synthetase inactivation *in vivo* at different temperatures. In trying to explain these results the authors speculated on the possible involvement of the enzyme–aminoacyladenylate complex in the phenomenon observed.

1. Amino Acid Regulation of the Synthesis of Aminoacyl-tRNA Synthetases

Nass and Neidhardt (362) were able to show a twofold to threefold increase in the activity of isoleucyl- and phenylalanyl-tRNA synthetases by growing *E. coli* braditrophs with limiting levels of the appropriate amino acid. They observed also an increase in antibody-binding activity tested by rabbit antiserum to a purified phenylalanyl-tRNA synthetase. Addition of excess amino acid resulted in a backward decrease in activity (362). The deuterium-labeling method revealed that the increase in synthetase activities under amino acid starvation was due to an increase in the *de novo* synthesis of the enzymes (360). Determination of the rate of inactivation of the "heavy" preexisting synthetase molecules and allowance for the inactivation of the newly made "light" molecules strongly suggested that the true differential rate of the synthesis of aminoacyl-tRNA synthetase becomes up to 50 times greater in the absence of the amino acid (360,363).

Increase in the differential rate of synthesis under growth-limiting conditions has been observed at least with 10 bacterial aminoacyl-tRNA synthetases: phenylalanine- (316,362), isoleucine- (362,364), leucine- and valine- (364), arginine- (360), histidine- (360,365), tyrosine- (366), methionine-, proline-, and threonine- (363) activating enzymes. In all cases restoration of the amino acid to the derepressed culture caused repression of the synthetase synthesis. It is thus possible to regard as general the regulation of aminoacyl-tRNA synthetase synthesis by means of amino-acid-mediated repression.

However, acceleration of synthetase formation at growth conditions that restrict the supply of the amino acid is not always observed, even for the synthetases just listed. The manner in which restriction is imposed and the nature of the bacterial strain used seem to have a pronounced effect on the response (cf. refs. 316, 362, and 366). This variability, combined with some aspects of synthetase formation during the recovery of cells from amino acid restriction, led Parker and Neidhardt (367) to believe that the response of individual synthetases to the supply of their cognate amino acids does not by itself define a clear and unambiguous control system for regulating synthetase formation.

Generally the type of regulation of aminoacyl-tRNA-synthetase synthesis in response to intracellular concentration of amino acids is similar to the regulation of amino acid biosynthesis. The similarity is strengthened by the fact that specific tRNAs seem to be involved in the regulation in both systems (see Section VIII.A above). Derepression of the rate of histidyl-tRNA-synthetase formation during growth under histidine restriction was observed with both wild-type and mutant strains of *S. typhimurium* (365). This mutation localized in the *hisR* gene led to a decrease in the tRNA^{His} amount to 52% as compared with the wild type. Addition of histidine to the wild-type derepressed culture resulted in repression of the synthesis of histidyl-tRNA synthetase, whereas the mutant continued to synthesize the enzyme at the same derepressed rate. These results suggested that histidine must be bound with tRNA^{His} to repress the synthesis of histidyl-tRNA synthetase. The idea was supported by the data obtained with *hisS* mutant of *S. typhimurium* (393). These results indicated an indispensable role of histidyl-tRNA synthetase (in addition to histidine and tRNA^{His}) in the control of synthetase formation and, as already mentioned, suggests the co-repressor (or repressor) function for histidyl-tRNA in histidyl-tRNA-synthetase formation. Possible involvement of the methionyl-tRNA^{Met} in the control of its cognate synthetase was postulated for *E. coli* and *S. typhimurium* methionyl-tRNA-synthetase mutants (394).

One more common feature of the regulation of synthetase and biosynthesis-enzyme synthesis is the multivalent manner of the repression. It has been shown (364) that the synthesis of valyl-tRNA synthetase is regulated by both valine and isoleucine. However, unlike the multivalent repression of biosynthesis enzymes (see ref. 368), this repression is due to valine and isoleucine alone. As for the other aminoacyl-tRNA synthetases specific to amino acids with branched side chains (e.g., isoleucine and leucine), the rate of their formation depends only on substrate amino acids (364).

Some other features distinguish the repression of synthetases from that of the biosynthesis enzymes. Partial derepression of the latter occurs when wild-type cells are grown on a minimal medium, whereas that of the synthetases takes place no sooner than the concentration of the amino acid becomes growth limiting (360,364). Kinetic differences have been discovered between the regulation of arginyl-tRNA synthetase and the regulation of arginine-biosynthesis enzymes: the synthetase remains derepressed for a short time after the addition of arginine while ornithine transcarbamylase is fully repressed (360). Histidyl-tRNA synthetase remains non-derepressed in three regulatory mutants of *S. typhimurium* derepressed at the histidine operon (296). Similarly *B. subtilis* strains derepressed in the synthesis of the biosynthesis enzymes of the tyrosine and histidine pathways because of a regulatory gene mutation have repressed levels of tyrosyl-tRNA synthetase whereas braditrophs, auxotrophs, and even wild-type strains are derepressed (366) under these conditions.

These data suggest that the components involved in synthetase regulation are not identical with those involved in the regulation of biosynthesis enzymes. It was mentioned in Section VII that the structural genes of the synthetases are situated far from those of the biosynthesis enzymes, and it is thus not surprising that their regulation is not coordinated. Probably the aporepressors of the two regulatory systems are different, whereas corepressors like aminoacyl-tRNAs (365,393, 394) may be common.

An important contribution was made by Clarke, Low, and Konigsberg by investigating the regulatory mechanisms of aminoacyl-tRNA-synthetase biosynthesis. They reported recently (395) the isolation and characterization of an E. coli mutant in which the level of seryl-tRNA synthetase was five times that of the parent strain and was not affected by the level of serine in contrast to the response in the parent strain. This mutation was found to map very closely to the structural gene for seryl-tRNA synthetase and to be cis dominant in a stable merodiploid. These results strongly suggest that this locus denoted SerO is an operator site involved in the control of the SerS gene. Although the enzyme levels for the biosynthetic pathways of serine are not known to be regulated, the close proximity of SerC to SerO and SerS was found.

2. Metabolic Regulation of Aminoacyl-tRNA-Synthetase Formation

Parker and Neidhardt (367) showed that the cellular levels of several aminoacyl-tRNA synthetases vary with the rate of growth of prototropic strains of S. typhimurium NT1 and E. coli NC1 in different media. Over a sevenfold range in growth rate the levels of arginyl- and of valyl-tRNA synthetases changed by a factor of approximately 2.5; that of leucyl-tRNA synthetase varied less. The apparent differential rate of formation of arginyl-tRNA synthetase changes to its new, rapid, steady-state rate immediately after cells are shifted from a poor to a rich medium. This new aspect of synthetase control is provisionally called metabolic regulation (367). Variations consistent with these, but over a lesser growth-rate range, may be seen in the data of Hirshfield and Zamecnik (304) for the glutamate, arginine, and lysine enzymes of E. coli K12.

Hirshfield and Zamecnik (304) also described thiosine-resistant mutants that had growth-medium-dependent lysyl-tRNA-synthetase activity (see Section VI). The activity was significantly higher (i.e., for some strains by a factor of 15–20) in an enriched medium (AC broth) than in a minimal medium. The variation in enzyme activity is due in all probability to changes in the amount of enzyme (304), and the properties of the lysyl-tRNA synthetase from mutant strains do not vary with the growth medium (305). Although the genetic characterization of the mutants is necessary for final conclusions, the phenomenon described may be attributed in some way to a metabolic regulation.

IX. The State of Aminoacyl-tRNA Synthetases within Aminal Cells

There is a little doubt that aminoacyl-tRNA synthetases in *Prokaryota* cells are present as individual molecules or as enzyme–substrate complexes with the corresponding tRNAs. This is suggested by the preparation of individual enzymes from the soluble fraction without any special extraction procedures. Furthermore, gel chromatography of the cell sap of *E. coli* reveals that the molecular weights of the synthetases present are the same as those of the enzymes in the purified state (369). Hence, except sometimes for tRNA and ribosomal factors, synthetases do not associate with any macromolecular components of *Prokaryota* cells.

As for multicellular organisms, their cells do contain synthetases associated with other macromolecules. It was reported long ago that the cytoplasmic membrane fraction contains activating enzymes (371, 372); it has also been noted that microsomes incorporate amino acids into proteins in the absence of the "pH5-enzyme" fraction.

Repeated washing of microsomes did not remove the phenylalanine- and tyrosine-activating enzymes. Treatment of the microsomal fraction with sodium deoxycholate resulted in solubilization of the lysine-, arginine-, and threonine-activating enzymes (373). Other workers mentioned the presence of tRNA in the microsomal or membrane fraction (374) and the association of tRNA with proteins (375). A high activity of aminoacyl-tRNA synthetases is characteristic of the rat-liver postmicrosomal fraction (376).

Chromatography on calcium phosphate gel and on DEAE-cellulose failed to remove the valyl-, leucyl-, and isoleucyl-tRNA-synthetase activities from methionyl-tRNA synthetase when it was attempted to obtain a highly purified preparation of the enzyme. The enzyme preparation did not penetrate into polyacrylamide gel in electrophoresis; its molecular weight was about 340,000 or 170,000, depending on the isolation conditions (65). It is quite possible in view of the more recent findings that the failure was due to the formation of a complex of synthetases. Association of different synthetases, as well as association between synthetases and ribosomes, has been reported for the same system (377).

More recent studies (378) demonstrated that about 90% of the total phenylalanyl-tRNA-synthetase activity is associated with ribosomes; presumably this synthetase forms a stable complex with ribosomes *in vivo*. Lysyl- and arginyl-tRNA synthetases are present in two fractions: in ribosomes and in particles with a sedimentation constant of about 14S. Washing with 0.5 M KCl removes the synthetases from ribosomes, but a decrease in ionic strength results in reassociation, leading, *interalia*, to an increase in the rate of ribosome sedimentation. The synthetases are bound mainly by the large ribosomal subunit. Attempts to purify rat-liver glutamyl-tRNA synthetase afforded a high-molecular-weight aggregate that contained other synthetases and RNA (379). The results of these and other studies made it doubtful that the synthetases of animal cells are present in the soluble fraction of animal-cell cytoplasm.

Recently the state of the synthetases in animal cells has been subjected to detailed investigation (380). It was found that the fraction immediately above the microsomes pellet after 90-min centrifugation at 105,000 g is strongly enriched in aminoacyl-tRNA-synthetase activity. In gel chromatography on Sephadex G–200 all the synthetase activities (18 amino acids, except asparagine and cysteine) were eluted in the excluded volume of the column. The same fractions contained considerable amounts of tRNA. Rechromatography of the peak on Sepharose 6B resulted in a similar pattern: all the 18 synthetases were eluted as a single symmetrical peak whose position corresponded to a molecular weight greater than 10^6. The complex was destroyed when the hand-driven Dounce homogenizer was replaced by a motor-driven Teflon blender. Alternatively it could be destroyed by freezing and thawing. These data make it highly improbable that the complex is an artifact of isolation, but rather point to its possible physiological importance *in vivo*.

Essentially similar results have been obtained in another laboratory (381,382). Glutaminyl-, isoleucyl-, leucyl-, lysyl-, and methionyl-tRNA-synthetase activities considerably greater than those in the supernatant fraction have been found in the postmicrosomal fraction that sedimented on 15-hr centrifugation at 105,000 *g* [this fraction was obtained for the first time by Hoagland (383) and named fraction X]. The enzymes were purified by gel chromatography on Sephadex G–200 followed by chromatography on DEAE-Sephadex A–50 and on hydroxyapatite columns. In all the fractionation procedures, as well as in isoelectric focusing and in sucrose-gradient centrifugation, the peaks of all of the above-mentioned activities coincided. The sedimentation constant of the complex was equal to 18S; that is, it was much greater than the sedimentation constant of any of the constituent individual synthetases. The isolated complex has been studied also by electron microscopy.

Phenylalanyl-tRNA synthetase of Ehrlich ascite cells is present in two states: in the form of an associate with other macromolecules with a sedimentation constant of about 25S and in the form of an associate with the 60S ribosome (384). Beef-liver valyl-tRNA synthetase was obtained, depending on the homogenization and fractionation conditions, either as a complex with a sedimentation constant of about 14S or in a solubilized form that was homogeneous in polyacrylamide-gel electrophoresis and had a molecular weight of about 100,000 (385). Electron microscopy revealed particles somewhat smaller than those described by Venegoor and Bloemendal (382). The particles contained little if any RNA. They were remarkably stable and survived repeated precipitation with ammonium sulfate and column chromatography on hydroxylapatite and on Sephadex G–200. Presumably the particles are similar to those present in reticulocytes (65); they may represent "subunits" of the higher structures found in rat liver (380).

In spite of the controversy between various workers (e.g., concerning the association of the complex with ribosomes, on the presence of tRNA within it, on whether all or only some of the synthetases are involved), it is clear at present that the synthetases may exist in animal cells in the form of intermolecular aggregates. Detailed investigation of this new class of particles is now in its very early stages. It is possible that these particles like informosomes or informofers are a characteristic feature that distinguishes animal cells from the cells of *Prokaryota*.

As all subcellular particles are now known by specific terms (chromosomes, ribosomes, etc.), the name "codosomes" can be proposed for the synthetase complexes. Aminoacyl-tRNA synthetases code for the amino acids by adding them to the cognate tRNAs, and V. Engelhardt (222) proposed the term "codases" for aminoacyl-tRNA synthetases; the suffix "some" points to the particulate nature of the complex. The idea that the synthetases are components of higher structures rather than "soluble" constituents of the cytoplasm was long ago suggested by Hendler (386) and Hoagland (387). Hradec and Dusek (388) found that the activity of the aminoacyl-tRNA synthetases isolated from rat liver is stimulated by cholesteryl-14-methyl-hexadecanoate. These data are not surprising in view of the existence of "codosomes" because lipids may well be involved in the complexes, bearing in mind their stability during chromatography and their lability to freezing and thawing and to detergents, as well as the greater activity of "codosomes" in comparison with the isolated synthetases.

Acknowledgments

It is a pleasure to express our thanks to Professor W. A. Engelhardt, who stimulated us to write this review, and to Dr. M. A. Gratchev for his great contribution in preparing the English version of the manuscript. We are indebted to Professors P. Berg, G. Dirheimer, B. Dudock, J.-P. Ebel, J. R. Fresco, C. Hélene, A. Kelmers, R. B. Loftfield, A. H. Mehler, D. Söll, S. Takemura, M. Yarus, and H. G. Zachau for making available to us their manuscripts before publication. We also thank all those who supplied us with reprints.

References

1. Mehler, A. H., *Prog. Nucleic Acid Res. Mol. Biol.*, *10*, 1 (1970).
2. Kisselev, L. L., and Favorova, O. O., *Usp. Biochim.* (*Advances in Biochemistry*), *11*, 39 (1970).
3. Mehler, A. H., and Chakraburtty, K., *Adv. Enzymol.*, *35*, 443 (1971).
4. Loftfield, R. B., in *Protein Synthesis*, McConkey, Ed., Dekker, New York, 1971, chapter 1, p. 1.
5. Loftfield, R. B., *Prog. Nucleic Acid Res. Mol. Biol.*, *12*, 87 (1972).
6. Favorova, O. O., Parin, A. V., and Lavrik, O. I., *Biophysica* (*Ann. Rev. Biophys.*), *2*, 6 (1972).
7. Lenguel, P., and Söll, D., *Bacteriol. Rev.*, *33*, 264 (1969).
8. von Ehrenstein, G., in *Aspects of Protein Biosynthesis*, C. B. Anfinsen, Jr., Ed., Academic Press, New York and London, 1970, p. 139.

9. Kisselev, L. L., in *Molecular Bases for Protein Biosynthesis*, "Nauka," Moscow, 1971, p. 121.
10. Yarus, M., *Ann. Rev. Biochem.*, *38*, 841 (1969).
11. Chambers, R. W., *Prog. Nucleic Acid Res. Mol. Biol.*, *11*, 489 (1971).
12. Gauss, D. H., Harr, F. von der, Maelicke, A., and Cramer, F., *Ann. Rev. Biochem.*, *40*, 1045 (1971).
13. Dirheimer, G., Ebel, J.-P., Bonnet, J., Gangloff, J., Keith, G., Krebs, B., Kuntzel, B., Roy, A., Weissenbach, J., and Werner, C. , *Biochimie*, *54*, 127 (1972).
14. Zachau, H. G., in *The Mechanism of Protein Synthesis and Its Regulation*, L. Bosch, Ed., North-Holland, Amsterdam (in press).
15. Venkstern, T. V., *Primary Structure of Transfer Ribonucloic Acids*, Plenum Press, New York, 1973.
16. Novelli, G., *Ann. Rev. Biochem.*, *36*, 449 (1967).
17. Zamecnik, P. C., *Cold Spring Harbor Symp. Quant. Biol.*, *34*, 1 (1969).
18. Muench, H., in *Protein Biosynthesis in Bacterial Systems*, New York, 1971, p. 213.
18a. Lea, P. J., and Norris, R. D., *Phytochemistry*, *11*, 2897 (1972).
19. Joseph, D. R., and Muench, K. H., *J. Biol. Chem.*, *246*, 7602 (1971).
20. Folk, W. R., *Biochemistry*, *10*, 728 (1971).
21. Lapointe, J., and Söll, D., *J. Biol. Chem.*, *247*, 4966 (1972).
22. Rymo, L., Lundvik, L., and Lagerkvist, U., *J. Biol. Chem.*, *247*, 388 (1972).
23. Stern, R., and Peterkofsky, A., *Biochemistry*, *8*, 4346 (1969).
24. Chlumecka, V., von Tigerström, M. D'Obrenan, P., and Smith, C. J., *J. Biol. Chem.*, *244*, 5481 (1969).
24a. Calendar, R., and Berg, P., *Biochemistry*, *5*, 1681 (1966).
24b. Chousterman, S., and Chapeville, F., *Eur. J. Biochem.*, *35*, 51 (1973).
24c. Krajewska-Grynkiewicz, K., Buonocore, V., and Schlesinger, S., *Biochim. Biophys. Acta*, *312*, 518 (1973).
24d. Heinrikson, R. L., and Hartley, B. S., *Biochem. J.*, *105*, 17 (1967).
25. Lemaire, G., Rapenbusch, V. R., Gros, C., and Labouesse, B., *Eur. J. Biochem.*, *10*, 366 (1969).
26. Favorova, O. O., Kochkina, L. L., Sajgo, M., Parin, A. V., Khilko, S. N., Prasolov, V. S., and Kisselev, L. L., Molek. Biol., (in press).
27. Seifert, W., Nass, G., and Zillig, W., *J. Mol. Biol.*, *33*, 507 (1968).
28. Bruton, C. J., and Hartley, B. S., *J. Mol. Biol.*, *52*, 165 (1970).
29. Frolova, L., Yu., Krayevsky, A. A., and Kisselev, L. L., unpublished data.
30. Frolova, L. Yu., Kovalyeva, G. K., Agalarova, M. B., and Kisselev, L. L., *FEBS Letters*, *34*, 213 (1973).
31. Ohta, T., Shimada, I., and Imahori, K., *J. Mol. Biol.*, *26*, 519 (1967).
32. Bolotina, I. A., Kochkina, L. L., Favorova, O. O., and Kisselev, L. L., unpublished data.
33. Santi, D. V., Marchant, W., and Yarus, M., *Biochem. Biophys. Res. Commun.*, *51*, 370 (1973).
33a. Epely, S., Lemaire, G., and Gros, C., *Ninth Inter. Congr. Biochem.*, Stockholm, Abstract Book, 3q15, p. 199, 1973.
33b. Kisselev, L. L., *tRNA: Structure and functions*. EMBO workshop, Aspenas-garden, Sweden, June 1973, abstracts.

33c. Durekovic, A., Flossdorf, J., and Kula, M.-L., *Eur. J. Biochem.*, *36*, 528 (1973).
33d. Tscherne, J. S., Lanks, K. W., Salim, P. D., Grunberger, D., Cantor, C. R., and Weinstein, I. B., *J. Biol. Chem.*, *248*, 4052 (1973).
33e. Prasolov, V. S., Favorova, O. O., and Kisselev, L. L., in preparation.
34. Chirikjian, J. G., Kanagalingam, K., Lau, E., Haga, J., and Fresco, J. R., *tRNA Structure and Functions*, Princeton Conference, May 1972.
35. Cassio, D., and Waller, J.-P., *Eur. J. Biochem.*, *20*, 283 (1971).
36. Waller, J.-P., Risler, J.-L., Monteilhet, C., and Zelwer, C., *FEBS Letters*, *16*, 186 (1971).
37. Dimitrijevic, L., *FEBS Letters*, *25*, 170 (1972).
38. Hélene, C., Brun, F., and Yaniv, M., *J. Mol. Biol.*, *58*, 349 (1971).
39. Hélene, C., *FEBS Letters*, *17*, 73 (1972).
40. Hélene, C., *Nature New Biol.*, *234*, 120 (1971).
41. Gangloff, J., and Dirheimer, G., *Biochim. Biophys. Acta*, *294*, 263 (1973).
42. Arndt, D. J., and Berg, O., *J. Biol. Chem.*, *245*, 665 (1970).
43. Berthelot, F., and Yaniv, M., *Eur. J. Biochem.*, *16*, 123 (1970).
44. Hayashi, H., Knowles, J. R., Katze, J. R., Lapointe, J., and Söll, D., *J. Biol. Chem.*, *245*, 1401 (1970).
45. Rouget, P., and Chapeville, F., *Eur. J. Biochem.*, *14*, 498 (1970).
46. Beikirch, H., Haar, F. von der, and Cramer, F., *Eur. J. Biochem.*, *26*, 182 (1972).
47. Böck, A., *Arch. Microbiol.*, *68*, 165 (1969).
48. Marshall, R. D., and Zamecnik, P. C., *Biochim. Biophys. Acta*, *181*, 454 (1969).
49. Österberg, R., Sjöberg, B., Rymo, L., and Lagerkvist, U., *J. Mol. Biol*, *77*, 153 (1973).
50. Bruton, C. J., *tRNA: Structure and functions*. EMBO workshop, Aspenasgarden, Sweden, June 1973, abstracts.
50a. Chirikjian, J. G., Kanagalingam, K., Lau, E., and Fresco, J. R., *J. Biol. Chem.*, *248*, 1074 (1973).
50b. LeMeur, M. A., Gerlinger, P., Clavert, J., and Ebel, J. P., *Biochimie*, *54*, 1391 (1972).
50c. DeLorenzo, F., DiNatale, P., Guarini, L., and Schechter, A., *Ninth Int. Congr. Biochem.*, Stockholm, Abstract Book, 3q12, p. 198, 1973.
50d. Hartley, B. S., *Ninth Int. Congr. Biochem.*, Stockholm, Abstract Book, P. 1, p. 7, 1973.
51. Lee, M. L., and Muench, K. H., *J. Biol. Chem.*, *244*, 233 (1970).
52. Katze, J. R., and Konigsberg, W., *J. Biol. Chem.*, *245*, 923 (1970).
53. Heider, H., Gottschalk, E., and Cramer, F., *Eur. J. Biochem.*, *20*, 144 (1971).
54. Joseph, D. R., and Muench, K. H., *J. Biol. Chem.*, *246*, 7610 (1971).
55. Favorova, O. O., Stelmastchuk, V., Parin, A. V., Kchilko, S., Kiselev, N. A., and Kisselev, L. L., *Abstr. Commun. 7th Meeting Eur. Biochem. Soc.*, p. 149, 1971.
56. Lemaire, G., Gross, C., Rapenbusch, R. van, and Labouesse, B., *J. Biol. Chem.*, *247*, 2931 (1972).
57. Lapointe, J., and Söll, D., *J. Biol. Chem.*, *247*, 4982 (1972).
58. Malcolm, N. L., *Biochim. Biophys. Acta*, *190*, 337, 347 (1969).
59. Kosakowski, M. H. J. E., and Böck, A., *Eur. J. Biochem.*, *12*, 67 (1970).
59a. Kucan, Z., and Chambers, R. W., *J. Biochem.* (*Tokyo*), *73*, 811 (1973).

59b. Muench, K. H., Lipscomb, M., Kuehl, G. V., and Penneys, N. S., *Ninth Int. Congr. Biochem.*, Stockholm, Abstract Book, 3h12, p. 160, 1973.

60. Ostrem, D. L., and Berg, P., *Proc. Natl. Acad. Sci. U.S.*, *67*, 1967 (1970).

61. Fasiolo, F., Béfort, N., Boulanger, Y., and Ebel, J.-P., *Biochim. Biophys. Acta*, *217*, 305 (1970).

62. Schmidt, J., Wang, R., Stanfield, S., and Reid, B., *Biochemistry*, *10*, 3264 (1971).

63. Lemoine, F., Waller, J. P., and Rapenbusch, R., *Eur. J. Biochem.*, *4*, 213 (1968).

64. Surgutchev, A. P., Surgutcheva, I. G., and Saitseva, G. N., *Dokl. Akad. Nauk SSSR*, *192*, 923 (1970).

65. Neale, S., *Chem. Biol. Interactions*, *2*, 349 (1970).

66. Christopher, C. W., Jones, M. B., and Stafford, D. W., *Biochim. Biophys. Acta*, *228*, 682 (1971).

67. Rouget, P., and Chapeville, F., *Eur. J. Biochem.*, *23*, 452, 459 (1971).

68. Pachmann, U., Cronvall, E., Rigler, R., Hirsch, R., Wintermeyer, W., and Zachau, H. G., *Eur. J. Biochem.*, *39*, 265 (1973).

69. Preddie, E. C., *J. Biol. Chem.*, *244*, 3958 (1969).

70. Burgess, R. B., *Ann. Rev. Biochem.*, *40*, 711 (1971).

71. Davie, E. W., Koningsberger, V. V., and Lipmann, F., *Arch. Biochem. Biophys.*, *65*, 21, (1956).

72. Makman, M., and Cantoni, G., *Biochemistry*, *4*, 1434 (1965).

73. Rymo, L. ,and Lagerkvist, U., *Nature*, *226*, 77 (1970).

74. Rymo, L., Lagerkvist, U., and Wynacott, A., *J. Biol. Chem.*, *245*, 4308 (1970).

75. Lagerkvist, U., Rymo, L., Lindqvist, O., and Andersson, E., *J. Biol. Chem.*, *247*, 397 (1972).

75a. Risler, J. L., Monteilhet, C., Zelwer, C., and Waller, J. P., *tRNA Structure and functions*. EMBO workshop, Aspenasgarden, Sweden, June 1973, abstracts.

75b. Reid, B. R., Koch, G. L. E., Boulanger, Y., Hartley, B. S., and Blow, D. M., *J. Mol. Biol.*, *80*, 199 (1973).

76. McNeil, M. R., and Schimmel, P. R., *Arch. Biochem. Biophys.*, *152*, 175 (1972).

77. Chirikjian, J., Wright, H., and Fresco, J. R., *Proc. Natl. Acad. U.S.*, *69*, 1638 1972.

78. Kisselev, L. L., Favorova, O. O., Parin, A. V., Stel'mashchuk, V., and Kiselev, N. A., *Nature New Biol.*, *233*, 231 (1971).

79. Favorova, O. O., Gretchko, V. V. Kisselev, L. L., and Sacharova, N. K., *Dokl. Akad. Nauk SSSR*, *171*, 742 (1966).

80. Chuang, H., Atherly, A., and Bell, F., *Biochem. Biophys. Res. Commun.*, *28*, 1013 (1967).

81. Lemaire, G., Dovizzi, M., and Labouesse, B., *Biochim. Biophys. Acta*, *132*, 155 (1967).

82. Kukhanova, M. K., Favorova, O. O., and Kisselev, L. L., *Biokhimiya*, *31*, 71 (1968).

83. Hélene, O., Braun, E., and Yaniv, M., *Biochem. Biophys. Res. Commun.*, *37*, 393 (1969).

84. Rigler, R., Cronvall, E., Hirsch, R., Pachmann, U., and Zachau, H. G., *FEBS Letters*, *11*, 320 (1970).

85. Penzer, G. R., Bennett, E. L., and Calvin, M., *Eur. J. Biochem.*, *20*, 1 (1971).
86. Farrelly, J. G., Longworth, J. W., and Stulberg, M. P., *J. Biol. Chem.*, *246*, 1266 (1971).
87. Engel, G., Heider, H., Maelicke, A., Haar, F. von der, and Cramer, F., *Eur. J. Biochem.*, *29*, 257 (1972).
88. Blanquet, S., Fayat, G., Waller, J.-P., and Iwatsubo, M., *Eur. J. Biochem.*, *24*, 461 (1972).
89. Kisselev, L. L., and Gottikh, B. P., *Abstracts VIIth Int. Congr. Biochem.*, *Tokyo*, p. 815, 1967.
90. Kisselev, L. L., *Abstracts Int. Symp. Protein Biosynthesis, Olstyn, Poland*, p. 23, 1968.
91. Nelidova, O. D., and Kisselev, L. L., *Molek. Biol.*, *2*, 60 (1968).
91a. Engelhardt, W. A., Kisselev, L. L., and Nezlin, R. S., *Monatsh. Chem.*, *101*, 1510 (1970).
92. Bartkowiak, S., and Pawelkiewicz, J., *Biochim. Biophys. Acta*, *272*, 137 (1972).
93. Remy, P., Birmele, C., and Ebel, J.-P., *FEBS Letters*, *27*, 134 (1972).
94. Avdonina, T. A., Baturina, I. D., and Kisselev, L. L., *Abstracts 11th Allunion Biochem. Congr.*, *Tashkent*, Section 4, p. 31, 1969.
95. Baturina, I. D., *Urk. Zh. Biokhim.*, *44*, 338 (1972).
96. Letendre, C., Humphreys, J., and Grunberg-Manago M., *Biochim. Biophys. Acta*, *186*, 46 (1969).
97. Berry, S. A., and Grunberg-Manago, M., *Biochim. Biophys. Acta*, *217*, 83 (1970).
98. Denburg, J., and DeLuca, M., *Proc. Natl. Acad. Sci. U.S.*, *67*, 1057 (1970).
98a. Waterson, R. M., Clarke, S. J., Kalousek, F., and Konigsberg, W. H., *J. Biol. Chem.*, *248*, 4181 (1973).
98b. Jacubowski, H., and Pawelkiewicz, J., *FEBS Letters*, *34*, 150 (1973).
99. Mehler, A. H., *J. Cell. Physiol.*, *74*, Suppl. 1, 117 (1969).
100. Mitra, S. K., Chakraburtty, K., and Mehler, A. H., *J. Mol. Biol.*, *49*, 139 (1970).
100a. Parfait, R., *FEBS Letters*, *29*, 323 (1973).
100b. Robert-Gero, M., and Waller, J.-P., *Eur. J. Biochem.*, *31*, 315 (1973).
100c. Forrester, P. I., and Hancock, R. L., *Can. J. Biochem.*, *51*, 231 (1973).
100d. Haar, F. von der, *Eur. J. Biochem.*; *34*, 84 (1973).
100e. Pigoud, A., Riesner, D., Boehme, D., and Maass, G., *FEBS Letters*, *30*, 1 (1973).
100f. Krauss, G., Römer, R., Riesner, D., and Maass, G., *FEBS Letters*, *30*, 6 (1973).
100g. Hörz, H. S., and Zachau, H. G., *Eur. J. Biochem.*, *37*, 203 (1973).
100h. Fasiolo, F., Remy, P., and Ebel, J. P., *Ninth Int. Congr. Biochem.*, Stockholm, Abstract Book, 3q16, p. 199, 1973.
101. Lapointe, J., and Söll, D., *J. Biol. Chem.*, *247*, 4975 (1972).
102. Kedzierski, W., and Pawelkiewicz, J., *Acta Biochim. Polon.*, *17*, 41 (1970).
103. Charlier, J., and Grosjean, H., *Eur. J. Biochem.*, *25*, 163 (1972).
104. Kosakowski, H., and Böck, A., *Eur. J. Biochem.*, *24*, 190 (1971).
105. Yaniv, M., and Gros, F., *J. Mol. Biol.*, *44*, 17 (1969).
106. Rigler, R., Cronvall, E., Ehrenberg, M., Pachmann, U., Hirsch, R., and Zachau, H. F., *FEBS Letters*, *18*, 193 (1971).

107. Cittanova, N., Vincent, M., Lepretre, M., Alfsen, A., and Petrissant, G., *Proc. 4th Int. Biophys. Congr.*, 1972.
108. Yarus, M., and Berg, P., *J. Mol. Biol.*, *28*, 479 (1967).
109. Yarus, M., and Berg, P., *J. Mol. Biol.*, *42*, 171 (1969).
110. Yarus, M., and Berg, P., *Anal. Biochem.*, *35*, 450 (1970).
111. Rouget, P., and Chapeville, F., *Eur. J. Biochem.*, *23*, 443 (1971).
112. Myers, G., Blank, H. U., and Söll, D., *J. Biol. Chem.*, *246*, 4955 (1971) and Addendum, ibid., *247*, 6011 (1971).
113. Knowles, J. R., Katze, J. R., Konigsberg, W., and Söll, D., *J. Biol. Chem.*, *245*, 1407 (1970).
114. Befort, N., Fasiolo, F., Bollack, C., Ebel, J.-P., *Biochim. Biophys. Acta*, 217, 319 (1970).
115. Lagerkvist, U., and Rymo, L., *J. Biol. Chem.*, *244*, 2476 (1969).
116. Lagerkvist, U., and Rymo, L., *J. Biol. Chem.*, *245*, 435 (1970).
117. Parin, A. V., Savelyev, E. P., and Kisselev, L. L., *FEBS Letters*, *9*, 163 (1970).
118. Parin, A. V., Savelyev, E. P., Favorova, O. O., and Kisselev, L. L., *Studia biophys. 24/25*, 391 (1970).
119. Chousterman, S., and Chapeville, F., *FEBS Letters*, *17*, 153 (1971).
120. Dorizzi, M., Labouesse, B., and Labouesse, J., *Eur. J. Biochem.*, *19*, 563 (1971).
121. Hele, P., and Barber, H., *Biochim. Biophys. Acta*, *258*, 319 (1972).
122. Okamoto, J., and Kawade, Y., *Biochim. Biophys. Acta*, *145*, 613 (1967).
123. Allende, J. E., and Allende, C. C., in *Methods in Ezymology*, Vol. XX, Part C, p. 212, 1971.
124. Allende, C. C., Chaimovich, H., Gatica, M., and Allende, J. E., *J. Biol. Chem.*, *245*, 93 (1970).
125. Rouget, P., and Chapeville, F., *Eur. J. Biochem.*, *4*, 305, 310 (1968).
126. Norris, A., and Berg, P., *Proc. Natl. Acad. Sci. U.S.*, *52*, 330 (1964).
127. Hirsh, D., *J. Biol. Chem.*, *243*, 5731 (1968).
128. Hörz, W., and Zachau, H. G., *Eur. J. Biochem.*, *32*, 1 (1973).
129. Yarus, M., *J. Biol. Chem.*, *247*, 2738 (1972).
130. Cole, F. X., and Schimmel, P. R., *Biochemistry*, *9*, 480 (1970).
131. Cleland, W. W., in *The Enzymes*, 3rd ed., Vol. II, P. D. Boyer, Ed., Academic Press, New York, 1970, p. 1.
132. Fedorov, V. V., in *New Ideas in Planning Experiments*, Moscow, 1969, p. 209.
133. Zinoviev, V. V., Kisselev, L. L., Knorre, D. G., Kochkina, L. L., Malygin, E. G., and Favorova, O. O., *Biokhimiya*, *37*, 443 (1972).
134. Noat, G., Ricard, J., Borel, M., and Got, C., *Eur. J. Biochem.*, *II*, 106 (1969).
135. Rudolph, F. B., and Fromm, H. J., *J. Biol. Chem.*, *245*, 4047 (1970).
136. Knorre, D. G., Malygin, E. G., Zinoviev, V. V., Slinko, M. G., Timoshenko, V. I., Kisselev, L. L., Kochkina, L. L., and Favorova, O. O., *Biochimie*, in press.
137. Santi, D. V., and Pena, A. van, *FEBS Letters*, *13*, 157 (1971).
138. Santi, D. V., Danenberg, P. V., and Satterly, P., *Biochemistry*, *10*, 4804 (1971).
139. Cleland, W. W., *Biochim. Biophys. Acta*, *67*, 104, 173, 188 (1963).
140. Fromm, H. J., *Biochim. Biophys. Acta*, *139*, 221 (1967).
141. Papas, T., and Mehler, A. H., *J. Biol. Chem.*, *246*, 5924 (1971).

142. Henderson, J. F., Brox, L. W., Kelley, W. N., Rosenbloom, F. M., and Seegmiller, J. E., *J. Biol. Chem.*, 243, 2514 (1968).
143. Zinoviev, V. V., Kisselev, L. L., Knorre, D. G., Kochkina, L. L., Malygin, E. G., Slinko, M. G., Timoshenko, V. I., and Favorova, O. O., *Molek. Biol.*, (in press).
144. Papas, T. S., and Peterkofsky, A., *Biochemistry*, 11, 4602 (1972).
145. Hoagland, M. B., *Biochim. Biophys. Acta*, 16, 288 (1955).
146. Berg, P., *J. Biol. Chem.*, 222, 1025 (1956).
147. Hoagland, M. B., Zamecnik, P. C., and Stephenson, M. L., *Biochim. Biophys. Acta*, 24, 215 (1957).
148. Ogata, K., Nohara, H., and Morita, T., *Biochim. Biophys. Acta*, 26, 656 (1957).
149. Karasek, M., Castelfranco, P., Krishnaswamy, P. P., and Meister, A., *J. Am. Chem. Soc.*, 80, 2335 (1958).
150. Kingdon, H. S., Webster, L. T., and Davie, E. W., *Proc. Natl. Acad. Sci. U.S.*, 44, 757 (1958).
151. Cassio, D., Lemoine, F., Waller, J.-P., Sandrin, E., and Boissonnas, R. A., *Biochemistry*, 6, 827 (1967).
152. Chousteman, S., Sonino, F., Stone, M., and Chapeville, F., *Eur. J. Biochem.*, 6, 8 (1968).
153. Knorre, D. F., and Malygin, E. G., *Molek. Biol.*, 5, 364 (1971).
154. Eldred, E. W., and Schimmel, P. R., *Biochemistry*, 11, 17 (1972).
155. Loftfield, R. B., and Eigner, E. A., *J. Biol. Chem.*, 244, 1746 (1969).
156. Parfait, R., and Grosjean, H., *Eur. J. Biochem.*, 30, 242 (1972).
157. Holler, E., Bennett, E. L., and Calvin, M., *Biochem. Biophys. Res. Commun.*, 45, 409 (1971).
158. Knorre, D. G., *Molek. Biol.*, 2, 715 (1968).
159. Zinoviev, V. V., Knorre, D. G., and Lavrik, O. I., *Molek. Biol.*, 4, 673 (1970).
160. Hirsh, D., and Lipmann, F., *J. Biol. Chem.*, 243, 5724 (1968).
161. Iaccarino, A., and Berg, P., *J. Mol. Biol.*, 42, 151 (1969).
162. Burleigh, M., and Smith, M. J. H., *Biochem. J.*, 119, P69 (1970).
163. Berg, P., Bergmann, F. H., Ofengand, E. J., and Dieckmann, M., *J. Biol. Chem.*, 241, 1726 (1961).
164. Holler, E., and Calvin, M., *Biochemistry*, 11, 3741 (1972).
165. Takeda, Y., and Igarashi, K., *Biochem. Biophys. Res. Commun.*, 37, 917 (1969).
166. Doctor, B. P., Fournier, M. J., and Thornsvard, C., *Ann. N.Y. Acad. Sci.*, 171, 863 (1970).
167. Igarashi, K., and Takeda, Y., *Biochim. Biophys. Acta*, 213, 240 (1970).
168. Igarashi, K., Matsuzaki, K., and Takeda, Y., *Biochim. Biophys. Acta*, 254, 91 (1971).
169. Igarashi, K., Matsuzaki, K., and Takeda, Y., *Biochim. Biophys. Acta*, 262, 476 (1972).
170. Takeda, Y., Matsuzaki, K., and Igarashi, K., *J. Bacteriol.*, 111, 1, (1972).
171. Kayne, M. S., and Cohn, M., *Biochem. Biophys. Res. Commun.*, 46, 1285 (1972).
172. Pastuszyn, A., and Loftfield, R. B., *Biochem. Biophys. Res. Commun.*, 47, 775 (1972).

173. Igarashi, K., Yoh, M., and Takeda, Y., *Biochim. Biophys. Acta*, *238*, 314 (1971).
173a. Matsuzaki, K., and Takeda, Y., *Biochim. Biophys. Acta*, *308*, 339 (1973).
173b. Mehler, A. H., Midelfort, C. F., Chakraburtty, K., and Steinschneider, A., *Ninth Int. Congr. Biochem.*, Stockholm, Abstract Book, 3h7, p. 159, 1973.
173c. Aoyama, H., and Chaimovich, H., *Biochim. Biophys. Acta*, *309*, 502 (1973).
173d. Chousterman, S., and Chapeville, F., *Eur. J. Biochem.*, *35*, 46 (1973).
173e. Eldred, E. W., and Schimmel, P. R., *Anal. Biochem.*, *51*, 229 (1973).
174. Cole, F. X., and Schimmel, P. R., *Biochemistry*, *9*, 3143 (1970).
175. DeMoss, J., Genuth, S., and Novelli, D., *Proc. Natl. Acad. Sci. U.S.*, *42*, 325 (1956).
176. Sharon, N., and Lipmann, F., *Arch. Biochem. Biophys.*, *69*, 219 (1957).
177. Krishnaswamy, P. R., and Meister, A., *J. Biol. Chem.*, *235*, 408 (1960).
178. Mitra, S. K., and Mehler, A. H., *J. Biol. Chem.*, *241*, 5161 (1966).
179. Mitra, S. K., and Mehler, A. H., *J. Biol. Chem.*, *242*, 5490 (1967).
180. Mitra, S. K., and Smith, C. J., *Biochim. Biophys. Acta*, *190*, 222 (1969).
181. Revel, J. M., Wang, S. F., Heinemeyer, C., and Shive, W., *J. Biol. Chem.*, *240*, 432 (1965).
182. Lee, L. W., Ravel, J. M., and Shive, W., *Arch. Biochem. Biophys.*, *124*, 614 (1967).
183. Deutscher, M., *J. Biol. Chem.*, *242*, 1132 (1967).
184. Mehler, A. H., and Mitra, S. K., *J. Biol. Chem.*, *242*, 5495 (1967).
185. Green, D. E., and Goldberger, R. F., *Molecular Insights into the Living Process*, Academic Press, New York, 1967, p. 170.
186. Chuang, H. Y. K., and Bell, F. E., *Arch. Biochem. Biophys.*, *152*, 502 (1972).
187. Krayevsky, A. A., Kisselev, L. L., and Gottikh, B. P., *Molek. Biol.*, 7, 769 (1973).
188. Lagerkvist, U., Rymo, L., and Waldenström, J., *J. Biol. Chem.*, *251*, 5391 (1966).
189. Loftfield, R. B., and Eigner, E. A., *Biochemistry*, 7, 1100 (1968).
190. Loftfield, R. B., and Vanderjagt, D., *Biochem. J.*, *128*, 1353 (1972).
191. Lacey, J. C., Jr., and White, W. E., Jr., *Biochem. Biophys. Res. Commun.*, 47, (1972).
191a. White, W. E., Jr., Lacey, J. C., and Weber, A. L., *Biochem. Biophys. Res. Commun.*, *51*, 283 (1973).
191b. Kisselev, L. L., and Kochkina, L. L., *Dokl. Akad. Nauk SSSR*, (in press).
191c. Ebel, J. P., Giege, R., Bonnet, J., Kern, D., Befort, N., Bollack, C., Fasiolo, F., Gangloff, J., and Dirheimer, G., *Biochimie*, *55*, 547 (1973).
192. Moldave, K., Castelfranco, P., and Meister, A., *J. Biol. Chem.*, *234*, 841 (1959).
193. Krayevsky, A., Degterev, E. V., Gottikh, B. P., and Nicolenko, L. N., *Izv. Akad. Nauk SSSR, Ser. Khim.*, 1730 (1971).
194. Arca, M., Frontali, L., Sapora, O., and Tesse, G., *Biochim. Biophys. Acta*, *145*, 284 (1967).
195. Arca, M., Frontali, L., and Tecce, G., *J. Bot. Ital.*, *102*, 303 (1968).
196. Jacobson, K. B., *Prog. Nucleic Acid Res. Mol. Biol.*, *11*, 461 (1971).
197. Strickland, J. E., and Jacobson, K. B., *Biochim. Biophys. Acta*, *269*, 247 (1972).
198. Giege, R., Kern, D., Ebel, J.-P., and Taglang, R., *FEBS Letters*, *15*, 281 (1971).

199. Kern, D., Giege, R., and Ebel, J.-P., *Eur. J. Biochem.*, *31*, 148 (1972).
200. Giege, R., Kern, D., and Ebel, J.-P., *Biochimie*, *54*, 1245 (1972).
201. Goldin, H., and Kaiser, I. I., *Biochem. Biophys. Res. Commun.*, *36*, 2364 (1969).
202. Yarus, M., *Biochemistry*, *11*, 2050 (1972).
203. Yarus, M., *Biochemistry*, *11*, 2352 (1972).
204. Mertes, M., Peters, M. A., Mahoney, W., and Yarus, M., *J. Mol. Biol.*, *71*, 671 (1972).
205. Yarus, M., *Proc. Natl. Acad. Sci. U.S.*, *69*, 1915 (1972).
206. Yarus, M., *Nature New Biol.*, *239*, 106 (1972).
207. Chapeville, F., Lipmann, F., Ehrenstein, G., Weisblum, B., Ray, W., and Benzer, S., *Proc. Natl. Acad. Sci. U.S.*, *48*, 1086 (1962).
208. Roe, B., Sirover, M., Williams, R., and Dudock, B., *Arch. Biochem. Biophys.*, *147*, 176 (1971).
209. Roe, N., and Dudock, B., *Biochem. Biophys. Res. Commun.*, *49*, 399 (1972).
210. Jacobson, K. B., *J. Cell. Comp. Physiol.*, *74*, Suppl. 1, 99 (1969).
211. Bonnet, J., and Ebel, J.-P., *Eur. J. Biochem.*, *31*, 335 (1972).
212. Schreier, A. A., and Schimmel, P. R., *Biochemistry*, *11*, 1582 (1972).
213. Eldred, E., and Schimmel, P. R., *J. Biol. Chem.*, *247*, 2961 (1972).
213a. Bonnet, J., Giege, R., and Ebel, J.-P., *FEBS Letters*, *27*, 139 (1972).
214. Tal, J., Deutscher, M., and Littauer, U., *Eur. J. Biochem.*, *28*, 478 (1972).
215. Zachau, H. G., *FEBS Symposium*, *23*, 93 (1972).
216. Kisselev, L. L., and Frolova, L. Yu., *Biokhimiya*, *29*, 1177 (1964).
217. Frolova, L. Yu., and Kisselev, L. L., *Dokl. Akad. Nauk SSSR*, *157*, 1466 (1964).
218. Schulman, L. H., and Chambers, R., *Proc. Natl. Acad. Sci. U.S.*, *61*, 308 (1968).
219. Dudock, B., di Peri, C., Scileppi, K., and Reszelbach, R., *Proc. Natl. Acad. Sci. U.S.*, *68*, 681 (1971).
220. Mirzabekov, A. D., Lastity, D., Levina, E. S., and Bayev, A. A., *Nature, New Biol.*, *229*, 21 (1971).
221. Frolova, L. Yu., Kisselev, L. L., and Engelhardt, V. A., *Dokl. Akad. Nauk SSR*, *164*, 212 (1965).
222. Engelhardt, V. A., and Kisselev, L. L., in *Current Aspects of Biochemical Energetics*, N. O. Kaplan and E. P., Kennedy, Eds., Academic Press, New York and London, 1966, p. 213.
223. Hayashi, H., and Miura, K., *Nature*, *209*, 376 (1966).
224. Kuwano, M., Hayashi, Y., Hayashi, H., and Miura, K., *J. Mol. Biol.*, *32*, 659 (1968).
225. Jilyaeva, T. I., Tatarskaya, R. I., and Kisselev, L. L., *Molek. Biol.*, *5*, 139 (1971).
226. Jilyaeva, T. I., and Kisselev, L. L., *FEBS Letters*, *10*, 229 (1970).
227. Kisselev, L. L., and Venkstern, T., *FEBS Letters*, *11*, 73 (1970).
228. Jilayeva, T. I., and Kisselev, L. L., *tRNA: Structure and Functions*, IUPAC Pre-Symposium, Riga, 1970, Abstracts.
229. Cashmore, A. R., Brown, D. M., and Smith, J. D., *J. Mol. Biol.*, *59*, 359 (1971).
230. Kisselev, L. L., Frolova, L. Yu., and Alexandrova, M. M., *Molek. Biol.*, *1*, 123 (1967).

234   LEV L. KISSELEV AND OL'GA O. FAVOROVA

231. Carbon, J., and Curry, J. B., *J. Mol. Biol.*, *38*, 204 (1968).
232. Carbon, J., Squires, C., and Hill, C. W., *Cold Spring Harbor Symp. Quant. Biol.*, *34*, 505 (1969–1970).
233. Carbon, J., and Squires, C., *Cancer Res.*, *31*, 663 (1971).
234. Squires, O., and Carbon, J., *Nature New Biol.*, *233*, 274 (1971).
235. Yaniv, M., and Barrell, B. G., *Nature New Biol.*, *233*, 113 (1971).
236. Berg, P., *Harvey Lectures*, *Ser. 67*, 247 (1973).
237. Goddard, J. P., and Schulman, L. H., *J. Biol. Chem.*, *247*, 3864 (1972).
237a. Schulman, L. H., and Goddard, J. P., *J. Biol. Chem.*, *248*, 1341 (1973).
237b. Celis, J., Hooper, U., and Smith, J., *Nature New Biol.*, *244*, 261 (1973).
238. Schulman, L. H., *Proc. Natl. Acad. Sci. U.S.*, *66*, 507 (1970).
239. Singhal, R. P., *tRNA: Structure and Functions*, Princeton Conference, May 1972, Abstracts.
240. Gottschling, H., and Zachau, H. G., *Biochim. Biophys. Acta*, *103*, 418 (1965).
241. Harriman, P. D., and Zachau, H. G., *J. Mol. Biol.*, *16*, 387 (1966).
242. Schimmel, P. R., Uhlenbeck, O. C., Lewis, J. B., Dickson, L. A., Eldred, E. W., and Schreier, A. A., *Biochemistry*, *11*, 642 (1972).
243. Miura, K., *Prog. Nucleic Acid Res. Mol. Biol.*, *6*, 39 (1967).
244. Imura, N., Weiss, G. B., and Chambers, R. W., *Nature*, *222*, 1147 (1969).
245. Thiebe, R., Harbers, K., and Zachau, H. G., *Eur. J. Biochem.*, *26*, 144 (1972).
246. Hashimoto, S., Kawata, M., and Takemura, S., *J. Biochem.*, *72*, 1339 (1972).
247. Roy, K., and Söll, D., *J. Biol. Chem.*, *245*, 1394 (1970).
248. Kano, J., and Sueoka, N., *J. Biol. Chem.*, *246*, 2207 (1971).
249. Shimura, Y., Aono, H., Ozeki, H., Sarabhai, A., Lamfrom, H., and Abelson, J., *FEBS Letters*, *22*, 144 (1972).
250. Hooper, M. L., Russell, R. L., and Smith, J. D., *FEBS Letters*, *22*, 149 (1972).
251. Beltchev, B., and Grunberg-Manago, M., *FEBS Letters*, *12*, 27 (1970).
252. Mehler, A. H., *FEBS Symposium*, *23*, 103 (1972).
253. Bayev, A. A., Venkstern, T. V., Mirsabekov, A. D., Krutilina, A. I., Li, L., and Axelrod, V. D., *Molek. Biol.*, *1*, 754 (1967).
254. Bonnet, J., Ebel, J.-P., and Dirheimer, G., *FEBS Letters*, *15*, 286 (1971).
255. Keith, G., Roy, A., Ebel, J.-P., and Dirheimer, G., *FEBS Letters*, *17*, 306 (1971).
256. Hirsh, D., *Nature*, *228*, 57 (1970).
257. Blank, H. U., and Söll, D., *Biochem. Biophys. Res. Commun.*, *43*, 1192 (1971).
258. Zachau, H., Dutting, D., and Feldmann, H., *Hoppe-Seylers Z. Physiol. Chem.*, *347*, 212 (1966).
259. Staehelin, M., Rogg, H., Baguley, B. C., Ginsberg, T., and Wehrli, W., *Nature*, *219*, 1363 (1968).
260. Ishikura, H., Yamada, Y., and Nishimura, S., *FEBS Letters*, *16*, 68 (1971).
261. Staehelin, M., *Experientia*, *27*, 1 (1971).
262. Gartland, W., and Sueoka, W., *Proc. Natl. Acad. Sci. U.S.*, *55*, 948 (1966).
263. Fresco, J. R., Adams, A., Ascione, R., Henley, D., and Lindahl, T., *Cold Spring Harbor Symp.*, *31*, 527 (1966).
264. Johnson, L., and Söll, D., *Biopolymers*, *10*, 2209 (1971).
265. Kumar, S. A., Krauskopf, M., and Ofengand, J., *tRNA: Structure and Function*, Princeton Conference, May 1972.

266. Neidhardt, F. C., *Bacteriol. Rev.*, *30*, 701 (1966).
267. Iaccarino, M., and Berg, P., *J. Bacteriol.*, *105*, 527 (1971).
268. Eidlic, L., and Neidhardt, F. C., *J. Bacteriol.*, *89*, 706 (1965).
269. Yaniv, M., Jacob, F., and Gros, F., *Bull. Soc. Chim. Biol.*, *47*, 1609 (1965).
270. Yasakowski, M. H. J. E., Neidhardt, F. C., and Böck, A., *Eur. J. Biochem.*, *12*, 74 (1970).
271. Larzar, M., Yaniv, M., and Gros, F., *C. R. Acad. Sci. Paris*, *266*, 531 (1968).
272. Buckel, P., Lubitz, W., Böck, A., *J. Bacteriol.*, *108*, 1008 (1971).
273. Folk, W. R., and Berg, P., *J. Bacteriol.*, *102*, 204 (1970).
274. Hartwell, L. H., *J. Bacteriol.*, *93*, 1662 (1967).
275. McLaughlin, C. S., Magee, P. T., and Hartwell, L. H., *J. Bacteriol.*, *100*, 579 (1969).
276. McLaughlin, C. S., and Hartwell, L. H., *Genetics*, *61*, 557 (1969).
277. Malcolm, N. L., *Nature*, *221*, 1931 (1969).
278. Kaplan, S., and Anderson, D., *J. Bacterial.*, *95*, 991 (1968).
279. Folk, W. R., and Berg, P., *J. Bacteriol.*, *102*, 193 (1970).
280. Nass, G., and Neidhardt, F. C., *Bacteriol. Proc.*, *87* (1966).
281. Hiraga, S., Ito, K., Hamada, K., and Yura, T., *Biochem. Biophys. Res. Commun.* *26*, 522 (1967).
282. Doolittle, W. F., and Janofsky, C., *J. Bacteriol.*, *95*, 1289 (1968).
283. Kano, J., Matsushiro, A., and Shimura, J., *Mol. Gen. Genetics*, *102*, 15 (1968).
284. Ito, K., Hiraga, S., and Yura, T., *Genetics*, *61*, 521 (1969).
285. Schlesinger, S., and Nester, E. W., *J. Bacteriol.*, *100*, 167 (1969).
286. Hoffman, E. P., Wilhelm, R. C., Konigsberg, W., and Katze, J. R., *J. Mol. Biol.*, *47*, 619 (1970).
287. Gross, T. S., and Rowbury, R. J., *Biochim. Biophys. Acta*, *184*, 233 (1969).
288. Blatt, J. M., and Umbarger, H. E., *Biochem. Genetics*, *6*, 99 (1972).
289. Nazario, M., Kinsey, J. A., and Ahmad, M., *J. Bacteriol.*, *105*, 121 (1971).
290. Fangmann, W. L., and Neidhardt, F. C., *J. Biol. Chem.*, *239*, 1839 (1964).
291. Szentirmai, A., Szentirmai, M., and Umbarger, H. E., *J. Bacteriol.*, *95*, 1672 (1968).
292. Paetz, W., and Nass, G., *Abstracts VIIth FEBS Meeting, Varna*, p. 320 (1971).
293. Tosa, T., and Pizer, L. I., *J. Bacteriol.*, *106*, 972 (1971).
294. Alexander, R. R., Calvo, J. M., and Freundlich, M., *J. Bacteriol.*, *106*, 213 (1971).
295. Roth, J. R., and Ames, B. N., *J. Mol. Biol.*, *22*, 325 (1966).
296. Roth, J. R., Auton, D. N., and Hartman, P. E., *J. Mol. Biol.*, *22*, 305 (1966).
297. Lorenzo, F., Straus, D. S., and Ames, B. W., *J. Biol. Chem.*, *247*, 2302 (1972).
298. Hirshfield, I. N., DeDeken, R., Horn, P. C., Hopwood, D. A., and Maas, W. K., *J. Mol. Biol.*, *35*, 83 (1968).
299. Murgola, E. J., and Adelberg, E. A., *J. Bacteriol.*, *103*, 20 (1970).
300. Böck, A., and Neidhardt, F. C., *F. C.*, *Z. Verebungslehre.*, *98*, 187 (1966).
301. Anderson, J. J., and Neidhardt, F. C., *J. Bacteriol.*, *109*, 307 (1972).
302. Mikulka, T. W., Stielglitz, B. I., and Calvo, J. K., *J. Bacteriol.*, *109*, 584 (1972).
303. Steinberg, W., and Anagnostopoulos, C., *J. Bacteriol.*, *105*, 6 (1971).
304. Hirshfield, I. N., and Zamecnik, P. C., *Biochim. Biophys. Acta*, *259*, 330 (1972).

305. Hirshfield, I. N., Tomford, J. W., and Zamecnik, P. C., *Biochim. Biophys. Acta*, *259*, 344 (1972).
306. Printz, D. B., and Gross, S. R., *Genetics*, *55*, 451 (1967).
307. Weeks, C. O., and Gross, S. R., *Biochem. Genetics*, *5*, 505 (1971).
308. Fangman, W. L., Nass, G., and Neidhardt, F. C., *J. Mol. Biol.*, *13*, 202 (1965).
309. Kondo, M., *Naturwissenschaften*, *58*, 567 (1971).
310. Treiber, G., and Iaccarino, M., *J. Bacteriol.*, *107*, 828 (1971).
311. Yaniv, M., and Gros, F., *J. Mol. Biol.*, *44*, 31 (1969).
312. Murgola, E. J., and Adelberg, E. A., *J. Bacteriol.*, *103*, 178 (1970).
313. Ames, G. F., *Arch. Biochem. Biophys.*, *104*, 1 (1964).
314. Hirshfield, I. N., and Bloemers, P. J., *J. Biol. Chem.*, *244*, 2911 (1969).
315. Böck, A., *Eur. J. Biochem.*, *2*, 165 (1967).
316. Nass, G., in *Molecular Genetics*, H. G., Wittman and H. Shuster Eds., Springer, Berlin, 1968, p. 147.
317. Stulberg, M., *J. Biol. Chem.*, *242*, 1060 (1967).
318. Böck, A., Faiman, L. E., and Neidhardt, F. C., *J. Bacteriol.*, *92*, 1076 (1966).
319. Lewis, J. A., and Ames, B. N., *J. Mol. Biol.*, *66*, 131 (1972).
320. Taylor, A. L., and Trotter, C. G., *Bacteriol. Rev.*, *31*, 332 (1967).
321. Neidhardt, F. C., in *Molecular Genetics*, H. G., Wittman and H. Shuster, Eds., Springer, Berlin, 1968, p. 133.
322. Böck, A., and Neidhardt, F. C., *Science*, *157*, 78 (1967).
323. Tingle, M. A., and Neidhardt, F. C., *J. Bacteriol.*, *98*, 837 (1969).
324. Sanderson, K. E., *Bacteriol. Rev.*, *31*, 354 (1967).
325. Smith, D. A., and Childs, J. D., *Heredity*, *21*, 265 (1966).
326. Celis, T. F. R., and Maas, W. K., *J. Mol. Biol.*, *62*, *179* (1971).
327. Mortimer, R. K., and Hawthrone, D. C., *Genetics*, *53*, 165 (1966).
328. Cassio, D., Lawrence, F., and Lawrence, D. A., *Eur. J.*, *Biochem.*, *15*, 331 (1970).
329. Folk, W. R., and Berg, P., *J. Mol. Biol.*, *58*, 595 (1971).
330. Kisselev, L. L., and Baturina, I. D., *FEBS Letters*, *22*, 231 (1972).
331. De Boer, H. A., Raue, H. A., Ab, G., and Grunber, M., *Biochim. Biophys. Acta*, *246*, 157 (1971).
332. Travers, A., Kamen, R., and Cashel, M., *Cold Spring Harbor Symp. Quant. Biol.*, *35*, 415 (1970).
333. Ezekiel, D. H., and Elkins, B. N., *Biochim. Biophys. Acta*, *166*, 466 (1968).
334. Strehler, B. L., Hendley, D. D., and Hirsch, G. P., *Proc. Natl. Acad. Sci. U.S.*, *57*, 1751 (1967).
335. Anderson, W. F., *Proc. Natl. Acad. Sci. U.S.*, *62*, 566 (1969).
336. Kanabus, J., and Cherry, J. H., *Proc. Natl. Acad. Sci. U.S.*, *68*, 873 (1971).
337. Ceccarini, C. Maggio, R., and Barbata, G., *Proc. Natl. Acad. Sci. U.S.*, *58*, 2235 (1967).
338. Bick, M. D., and Strehler, B. L., *Proc. Natl. Acad. Sci. U.S.*, *68*, 224 (1971).
339. Neidhardt, F. C., and Earhart, C. F., *Cold Spring Harbor Symp. Quant. Biol.*, *31*, 557 (1966).
340. Neidhardt, F. C., Marchin, G. L., McClain, W. H., Boyd, R. F., and Earhart, C. F., *J. Cell. Physiol.*, *74*, 87 (1969).

341. Chrispeels, M. J., Boyd, R. F., Williams, L. S., and Neidhardt, F. C., *J. Mol. Biol.*, *31*, 463 (1968).
342. Marchin, C. L., Comer, M., and Neidhardt, F. C., *J. Biol. Chem.*, *247*, 5132 (1972).
343. Earhart, C. F., and Neidhardt, F. C., *Virology*, *35*, 694 (1967).
344. Yaniv, M., and Gros, F., in *Genetic Elements*, D. Shugar, Ed., Academic Press, New York, 1967, p. 157.
345. Williams, L., and Freundlich, M., *Biochim. Biophys. Acta*, 186, 305 (1969).
346. Schlesinger, S., and Magasanik, B., *J. Mol. Biol.*, *9*, 670 (1964).
347. Vogel, T., Meyers, M., Kovach, J. S., and Goldberger, R. F., *J. Bacteriol.*, *112*, 126 (1972).
348. Hatfield, G. W., and Burns, R. O., *Proc. Natl. Accad. Sci. U.S.*, *66*, 1027 (1970).
349. Silbert, D. F., Fink, G., and Ames, B. N., *J. Mol. Biol.*, *22*, 335 (1966).
350. Stulberg, M. P., Isham, K. R., and Stevens, A., *Biochim. Biophys. Acta*, *186*, 297 (1969).
351. Brenner, M., De Lorenzo, F., and Ames, B. N., *Fed. Proc.*, *29*, 1248 (1970).
352. De Lorenzo, F., and Ames, B. N., *J. Biol. Chem.*, *245*, 1710 (1970).
353. Singer, C. E., Smith, G. R., Cortese, R., and Ames, B. N., *Nature New Biol.*, *238*, 72 (1972).
354. Duda, E., Staub, M., Venetianer, P., and Denes, G., *Biochem. Biophys. Res. Commun.*, *32*, 992 (1968).
355. Allaudeen, H. S., Yang, S. K., and Söll, D., *FEBS Letters*, *28*, 205 (1972).
356. Boman, H. G., Boman, I. A., and Maas, W. K., in *Biological Structure and Function*, T. W. Goodwin and O. Lindberg, Eds., Vol. 1, Academic Press, New York, 1961, p. 297.
357. Ames, B. N., and Hartman, P. E., in *Molecular Basis of Neoplasia*, University of Texas Press, Austin, 1962, p. 322.
358. Coles, R. S., and Rogers, P., *Bacteriol. Proc.*, 87 (1964).
359. Hu, A. S. L., Bock, R. M., and Halverson, H. O., *Anal. Biochem.*, *4*, 489 (1962).
360. Williams, L. S., and Neidhardt, F. C., *J. Mol. Biol.*, *43*, 529 (1969).
361. Anderson, J. J., and Neidhardt, F. C., *J. Bacteriol.*, *109*, 315 (1972).
362. Nass, G., and Neidhardt, F. C., *Biochim, Biophys. Acta*, *134*, 347 (1967).
363. Archibold, E. P., and Williams, L. S., *J. Bacteriol.*, *109*, 1020 (1972).
364. McGinnis, E., and Williams, L. S., *J. Bacteriol.*, *108*, 254 (1971).
365. McGinnis, E., and Williams, L. S., *J. Bacteriol.*, *109*, 505 (1972).
366. Dale, B. A., and Nester, E. W., *J. Bacteriol.*, *108*, 586 (1971).
367. Parker, J., and Neidhardt, F. C., *Biochem. Biophys. Res. Commun.*, *49*, 495 (1972).
368. Umbarger, H. E., *Ann. Rev. Biochem.*, *38*, 323 (1969).
369. Nass, G., and Stöffler, G., *Mol. Gen. Genetics*, *100*, 378 (1967).
370. Kovach, J. S., Phang, Y. M., Ference, M., and Goldberger, R. F., *Proc. Natl., Acad. Sci. U.S.*, *63*, 481 (1969).
371. Nisman, B., Hirsch, M. L., and Bernard, A. M., *Ann. Inst. Pasteur*, 95, 615 (1958).
372. Hunter, G. D., Brookes, P., Crathorn, A. R., and Butler, J. A. V., *Biochem. J.*, *73*, 369 (1959).

373. Norton, S. J., Key, M. D., and Scholes, S. W., *Arch. Biochem. Biophys.*, *109*, 7 (1965).
374. Wilson, S. H., and Quincey, R. V., *J. Biol. Chem.*, *244*, 1092 (1969).
375. Hess, E. L., Herranen, A. M., and Lagg, S. E., *J. Biol. Chem.*, *236*, 3020 (1961).
376. Hird, H. J., McLean, E. J. T., and Munro, H. N., *Biochim. Biophys. Acta*, *87*, 219 (1964).
377. Irvin, J., Cimadevilla, J., and Hardesty, B., *Fed. Proc.*, *30*, 1166 (1971).
378. Irvin, J. A., and Hardesty, B., *Biochemistry*, *11*, 1915 (1972).
379. Deutscher, M. P., *J. Biol. Chem.*, *242*, 1123(1967).
380. Bandyopadhyay, A. K., and Deutscher, M. P., *J. Mol. Biol.*, *60*, *113* (1971).
381. Vennegoor, C. J. G. M., Stols, A. L. H., and Bloemendal, H. *J. Mol. Biol.*, *65*, 375 (1972).
382. Vennegoor, C., an Bloemendal, H., *Eur. J. Biochem.*, *26*, 462 (1972).
283. Hoagland, M. B., *Cold Spring Harbor Symp. Quant. Biol.*, 26, 153 (1961).
384. Roberts, W. K., and Coleman, W. H., *Biochem. Biophys. Res. Commun.*, *46*, 206 (1972).
385. Frolova, L. Yu., and Kisselev, L. L., *Biokhimiya*, in press.
386. Hendler, R. W., *Nature*, *193*, 821 (1962).
387. Hoagland, M. B., in *The Role of Nucleotides for the Function and Conformation of Enzymes*, H. M. Kalkar et al., Eds., Academic Press, London, 1969, p. 319.
388. Hradec, J., and Dusek, Z., *Biochem. J.*, *110*, 1 (1968).
389. Midelfort, C. F., Chakzaburrty, K., and Mehler, A. H., *Fed. Proc., Abstr.*, *32*, 460 (1973).
390. Faanes, R., and Rogers, P., *J. Bacteriol.*, *112*, 102 (1972).
391. Williams, L. S., *J. Bacteriol.*, *113*, *1419* (1973).
392. Williams, A. L., and Williams, L. S., *J. Bacteriol.*, *113*, 1433 (1973).
393. McGinnis, E., and Williams, L. S., *J. Bacteriol.*, *111*, 739 (1972).
394. Archibold, E. R., and Williams, L. S., *J. Bacteriol.*, *114*, 1007 (1973).
395. Clark, S. J., Low, B., and Konigsberg, W., *J. Bacteriol.*, *113*, 1096 (1973).

SOME ASPECTS OF THE STRUCTURE, IMMUNOCHEMISTRY, AND GENETIC CONTROL OF YEAST MANNANS

By CLINTON E. BALLOU, *Berkeley, California*

CONTENTS

I. Introduction

When first invited to write a chapter for this volume on our work on yeast cell-wall mannans, my immediate reaction was to question whether this was the time for such an article because it was my feeling that there is not too much of a definitive nature that one can say about the enzymology of mannan synthesis or degradation. On reflection, however, I realized that some of the new information obtained during the last few years on the structure, immunochemistry, and genetic control of yeast mannans might warrant review as providing a base from which to mount a renewed attack on the complex problem of mannan biosynthesis and on which to devise experiments that would tell something about mannan function. With this purpose in mind I have approached the following discussion. Although no attempt has been made to provide a detailed review of the work of others, in some instances I have sketched in some of the background in order to place our own studies in perspective. The reader wishing more detail and a broader appreciation should consult the excellent review by Phaff (1) and the references cited therein.

II. The Structure of Yeast Mannan

As anyone who works with yeast must soon realize, "yeasts" are many different things. Since I am not a trained microbiologist, my own interest in yeasts has been influenced strongly by others working in the field, so that the particular organisms treated in this chapter have been selected with a controlled bias. Most are genera belonging to the ascomycetous yeasts, and a few are asporogenous. Only time will tell how representative are the specific mannan structures we have investigated.

Chemists have generally studied bakers' yeast (*Saccharomyces cerevisiae*) mannan because it can be obtained in quantity from commercial sources. However, bakers' yeast is usually a mixture of strains, which may differ from one place to another, so one might expect some disagreement in the literature concerning yeast-mannan structure. Actually it was just such apparent disagreement that first directed my attention to species differences in mannan structure when we observed (2) that the acetolysis products of bakers' yeast mannan differed from those reported previously for the mannan from *Saccharomyces rouxii* (3). Subsequently we detailed several species-specific differences between yeast mannans (4–9) and even found differences between strains within the species *S. cerevisiae* (2,7). The following treatment applies generally to *Saccharomyces* and *Kluyveromyces* species; to *Candida albicans*, *Candida tropicalis*, *Candida parapsilosis*, and *Kloeckera brevis*; and, in a limited way, to *Hansenula angusta*, *Hansenula wingei*, and *Pichia bispora* (9).

A. THE MACROMOLECULAR STRUCTURE OF YEAST MANNAN

Yeast mannan is a covalently linked polysaccharide–protein complex (1), illustrated somewhat fancifully in Figure 1A, in which the protein contributes 5–50% of the weight, depending on the type of mannan. Coating the cell and interspersed with the rigid glucan layer that gives the cell its shape is a mannan-protein complex that contains

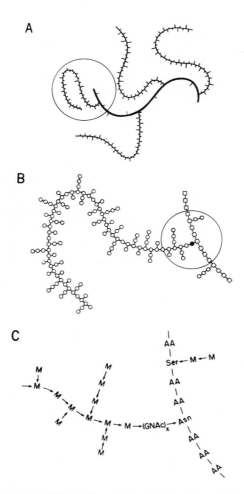

Fig. 1. Diagrammatic illustration of the proposed structure of one kind of a mannan-protein molecule. (A) A "tetrameric" form with four polysaccharide chains held together by the polypeptide chain, which also carries short oligosaccharides. The circled portion is enlarged in (B), which shows a "subunit," of the type produced by pronase digestion, with a polysaccharide chain of 150 mannoses attached to a peptide of 15 amino acids that is also substituted by a disaccharide and trisaccharide unit. The circled portion of this structure is enlarged in (C) to emphasize that the polymannose chain is attached via *N*-acetyl-D-glucosamine (GNAc) to asparagine (Asn), whereas the oligosaccharide chains are attached directly to serine (Ser).

241

on the average 5–10% protein and appears to have a structural role. Therefore I have chosen to call this "structural" mannan. In the peri-plasmic space, and perhaps intermixed with the structural mannan–glucan matrix, are mannan-protein enzymes (invertase, acid phospha-tase, α-glucosidase) (10), which contain 30–50% protein and whose hydrolytic activities presumably are important in cellular nutrition. The sexual agglutination factors of *H. wingei* (11) are mannan-protein complexes (10% protein) (11,12), and their action almost certainly requires that they be located in part on the cell surface. From these examples it is a reasonable conclusion that the mannan-protein com-ponent of the yeast cell is a mixture of many structurally and func-tionally differentiated macromolecules that come in a variety of sizes and shapes, the unifying feature being a high D-mannose content. In this sense they are unique as a class of natural products.

The form in which mannan is obtained from the yeast cell depends on the method of extraction. An original isolation procedure (13) was based on the easy solubility of mannan in alkali and the relative in-solubility of the cell-wall glucan. From the alkali extract the mannan can be precipitated by copper-ion complexation. Although this pro-cedure has been useful for studies on the major polysaccharide com-ponent, it obviously must lead to extensive degradation of alkali-labile structures. Since phosphodiester and glycosyl–serine linkages are known to exist in the mannan-protein complex, these would be broken in whole or in part by such a procedure. A milder isolation utilizes a hot-citrate-buffer extraction of the cell wall (14); although the mannan-protein enzymes are denatured, the phosphodiester and glycosyl–serine linkages do not seem to be affected seriously. A comparably mild pro-cedure uses extraction with ethylenediamine at 37°C (15). However, the isolation of biologically active mannan-protein fractions depends on solubilization by mechanical or enzymatic treatment. Thus man-nan invertase can be obtained from cells broken in a French press (16) or by grinding them with sand, whereas the *H. wingei* sexual ag-glutination 5-factor is released by treating whole cells with subtilisin (11). Solubilized mannan can be fractionated by ion exchange (6) and by gel filtration (17).

The size of mannan-protein complexes varies greatly. The "pure" mannan invertase has a molecular weight of 270,000 (18), whereas that of the *H. wingei* 5-factor is about 900,000 (19), a value we have confirmed (12). The total cell-wall mannan from *Kloe. brevis*, isolated by citrate extraction, can be resolved by gel filtration into three major fractions with average molecular weights of 25,000, 100,000, and 500,000 (17). The alkali used in Fehling's precipitation of mannan and even the conditions for the milder β-elimination reaction reduce the size of this material substantially. Pronase digestion reduced the larger two mannan components from *Kloe. brevis* to about 25,000, whereas the smallest component was not affected (17). The latter, which may represent a subunit of the two larger forms, is composed of about 150 mannose units and 15 amino acids (Fig. 1B). There is some evidence of mannan–protein–glucan complexes in the yeast cell wall (20), and it seems reasonable that there should be a close association between these components. However, it is not established that they form a covalently linked unit.

B. THE STRUCTURE OF A MANNAN-PROTEIN "SUBUNIT"

One simplified concept of a mannan-protein subunit, from which larger components might be constructed, is illustrated in Figure 1B. The basic idea is that some of the mannose is attached to the poly-peptide chain as short oligosaccharides, glycosidically linked to serine and threonine, but the majority is attached as polysaccharide chains with perhaps 150 or more mannose units linked via N-acetyl-D-gluco-samine to asparagine (15). Not too much can be said about the details of the latter linkage because the attachment of mannose to glucosamine and the number of glucosamine units at the linkage point are still uncertain. All mannan-protein complexes release a part of the carbo-hydrate as small oligosaccharides under the mildly alkaline conditions that promote β-elimination (15), and there is a concomitant loss of serine and threonine. In the case of bakers' yeast mannan the oligo-saccharide fraction is a mixture of mannose, mannobiose, mannotriose, and mannotetraose (6,15).

Early chemical studies on the polysaccharide component of bakers' yeast mannan hinted that it was composed of an $\alpha1\rightarrow6$-linked back-bone of D-mannose units to which were attached short side chains by $\alpha1\rightarrow2$- and $\alpha1\rightarrow3$-linkages (21). Recently we have developed two general procedures for the structural analysis of such polysaccharides. One utilizes the acetolysis reaction, which cleaves preferentially at $1\rightarrow6$-linkages and produces good yields of the mannan side chains (2,3,5). The other makes use of an *exo-α-mannanase* obtained from a soil organism that was isolated by enrichment culture on bakers' yeast mannan (22). This enzyme, which is secreted into the medium and can be isolated with ease, acts preferentially to remove the short side chains from the mannan and allows recovery of the intact backbone (23,24). Direct chemical analyses on the resistant core have established that bakers' yeast mannan, and many others, does have the $\alpha1\rightarrow6$ backbone proposed by Peat, Turvey, and Doyle (21). In several different mannans, the average chain length of the backbone is about 50 mannose units. Since 65% of the mannose in bakers' yeast mannan is in the side chains, the original mannan subunit must contain about 150 anhydromannose units. Figure 1 is an attempt to illustrate a mannan-protein molecule at the different levels of detail described here. Figure 2 shows a space-filling model, corresponding to a portion

Fig. 2.   Space-filling model of a portion of the "subunit" structure in Figure 1B, illustrating the brushlike shape of the polysaccharide chain. The first 10 side-chain units of the polysaccharide are shown attached via N-acetylglucosamine to asparagine, which is part of a pentapeptide (on the right side of the figure) with a serine unit substituted by mannobiose.

of the structure in Figure 1B, which emphasizes the brushlike shape
of the molecule.

C. POLYMORPHISM OF MANNAN STRUCTURE

Our attention (4,5) was first drawn to variability in mannan structure by the differences noted in the acetolysis patterns of several yeasts
(Fig. 3). While these patterns indicate gross differences in the amounts
and number of side chains, detailed chemical analysis of each fragment
revealed more subtle differences in the linkages and even in sugar
composition (2,8,25). From the analysis of these fragments and of the
exo-α-mannanase-resistant cores of the different mannans the various
structures shown in Figure 4 were derived. I have called these "mannan chemotypes." It is probable that all of these mannans have structures similar to that in Figure 1A, but that they differ slightly in the side
chains as indicated. As discussed below, these differences are reflected
in the serology of the whole yeast cell, since the carbohydrate part of

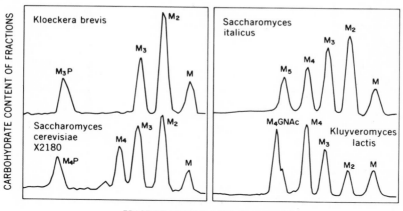

FRACTION FROM BIO-GEL P-2 COLUMN

Fig. 3. Acetolysis "fingerprints" of several yeast mannans. The fragments,
which were separated by gel filtration on Bio-Gel P-2, correspond to mannose
(M), mannobiose (M_2), mannotriose (M_3), mannotriose phosphate (M_3P), mannotetraose (M_4), mannopentaose (M_5), and the N-acetylglucosamine derivative of
the mannotetraose (M_4GNAc). The selective acetolysis of 1→6-linkages was carried out in a mixture of acetic anhydride–acetic acid–sulfuric acid (10:10:1) at
40°C for 12 hr.

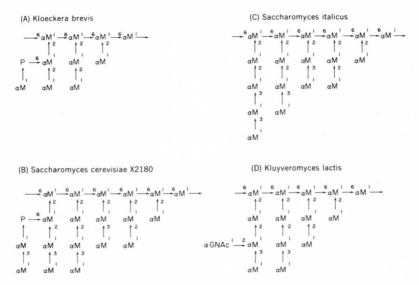

Fig. 4. Structures that illustrate the chemotypes corresponding to the mannans whose acetolysis patterns are shown in Figure 3. All mannose units are in α-anomeric linkage (αM). The relative amounts of the different side chains in the polysaccharides are not indicated, and their order, if any, is not known. *Saccharomyces cerevisiae* 4484–24D has chemotype A. The importance of the mannobiosylphosphate unit in chemotype B previously has been underestimated, although the presence of this unit had been suggested earlier (6). The postulated conversion of mannosylphosphate groups in chemotype A to mannobiosylphosphate groups in chemotype B is consistent with the immunochemical studies on the mannan mutants (Section IV).

the mannan-protein complex is the immunodominant component of the cell surface.

Certain structural features of these polysaccharides deserve mention. First, all have in common the α1→6-backbone, and the first and second D-mannose units in the side chains are in most cases attached by α1→2-linkages. The variability comes by modification of this common structure, either by attachment of one or two D-mannose units in α1→3-linkage, by substitution with an α-D-mannosylphosphate group, or by addition of the *N*-acetyl-D-glucosamine unit. Thus one would predict certain common biosynthetic pathways and, perhaps, a taxonomic relatedness. Second, reference to the acetolysis patterns shows that the relative amounts of the same side chains also differ between mannans. On the other hand, we have found that in the same strain grown under different conditions the ratios of the side chains are much more constant (26), although some variability has been noted when yeasts enter the stationary phase (8). Thus there must be some regulatory mechanism that determines the amount of each side chain that is made. This constancy of side-chain ratios raises another question, and that is whether there is a specific order of the side chains along the backbone. We are presently attempting to sequence a mannan chain, so this question may be answered.

One problem with interpreting much of the literature on mannan structure is that never yet has a complete sequential analysis of a homogeneous mannan been carried out in a manner that would allow reconstruction of the complete macromolecule. Based on what is now known about mannan structure, I have proposed a potential degradation scheme (Fig. 5) that would give much of the information one would like to have in order to define the polysaccharide component. Moreover, anyone investigating the biosynthesis of mannan, by following the incorporation of labeled precursors into the polymer, might profitably apply this kind of detailed analysis to the reaction product in order to determine which of the many possible enzymatic processes was occurring.

D. STRUCTURAL ANALYSIS MADE EASY

The kind of study we have carried out, in which detailed analysis of complex polysaccharides is used as a routine tool for comparing different yeasts and for following changes due to mutation or genetic recombination, is practical only because of the advances in methodology during the last decade. Although yeast mannans have structural features that make them a fortunate choice for this kind of investigation, the analyses that one can carry out in a week or two now might have taken the same number of years not too long ago.

I have already referred to the use of the *exo*-α-mannanase (22) that facilitates determination of the backbone structure, and the acetolysis procedure (3) that, in conjunction with gel filtration as a method for fractionation (2), allows isolation of the side chains in good yield. More important, however, have been the development of an improved method of methylation (27) and the application of gas chromatography coupled to mass spectrometry for the characterization of methylated fragments from polysaccharides (28). Finally, from the nuclear-mag-

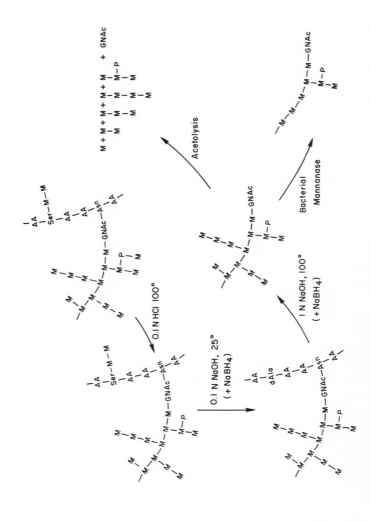

Fig. 5. Outline of a proposed scheme for selective degradation of yeast mannan which allows a distinction between sugars attached by phosphodiester linkage, sugars linked glycosidically to serine and threonine, and sugars in the polysaccharide chains linked to asparagine via N-acetylglucosamine. The use of the bacterial mannanase to isolate the backbone structure and the use of the acetolysis reaction to prepare the side chains are also illustrated.

248

netic-resonance spectrum one can usually obtain direct evidence for assignment of anomeric configurations (2,29,30).

The acetolysis reaction (31) deserves some comment. The seemingly "enzymatic" specificity of this chemical reaction, which cleaves the $\alpha1\rightarrow6$-linkages in the mannan at a much greater rate than the other linkages, is impressive. However, with time we have gained a better feeling for the limitations of the reaction and have concluded, from a study of the rates of degradation of a number of disaccharides (32), that yeast mannan is a very special case. Whereas the relative rates of acetolysis of mannose disaccharides with $\alpha1\rightarrow2$-, $\alpha1\rightarrow3$-, and $\alpha1\rightarrow6$-linkages are 1, 13, and 280, those for the corresponding glucosides are 1, 5, and 90. In the β-linked glucose series the relative rates for the same three linkages are 1, 0.8, and 25. Anomalously high rates are found with galactosides, possibly owing to participation by the axial 4-acetoxy group. Along with acetolysis, anomerization occurs to an appreciable extent with some glycosides (33), and gentiobiose is converted to isomaltose more rapidly than it is acetolyzed. Although these observations indicate that acetolysis cannot be applied indiscriminately, its use for the selective degradation of yeast mannans remains valid. Since the α-linked mannans are already in a conformationally stable form, anomerization is minimized. In some situations the acetolysis of other linkages may be enhanced, as with the Man$_4$GNAc sidechain of *Kluyv. lactis* mannan, which is converted to Man$_3$GNAc at an appreciable rate by acetolysis of the terminal $\alpha1\rightarrow3$-linked mannose unit (8). Finally, it will be remembered that a side chain such as the mannosylphosphorylmannotriose unit in *Kloe. brevis* is converted to mannose and mannotriose phosphate (6), whereas the $1\rightarrow2$-linkage in galactomannans may be split at an enhanced rate.

E. NATURE OF THE PROTEIN COMPONENT

There does not appear to be a single example in which an isolated mannan-protein complex has been demonstrated to contain a homogeneous protein component. The most promising case should be one of the mannan enzymes since the material can be purified by following enzymatic activity. However, the high carbohydrate content of the purified enzymes makes it difficult to establish protein homogeneity by standard criteria. Thus, while the external mannan invertase and the internal carbohydrate-free invertase of *S. cerevisiae* were found to cross-react immunologically and to be genetically related, the amino acid compositions were quite different (18). This may mean, as suggested, that both enzymes are aggregates of different subunits, with one or more in common; but it could also mean that the isolated mannan invertase was a mixture of macromolecules with similar properties. Until the carbohydrate can be removed completely from the mannan enzymes so that the protein part can be characterized in the usual way, this uncertainty will remain.

The amino acid compositions of the mannan enzymes do not seem to be unusual. On the other hand, the glycopeptides from some mannan preparations are characterized by unusually high contents of serine and threonine, 30–45% being commonly reported (15,17). In this respect this mannan-protein is reminiscent of blood-group substances which contain about 80% carbohydrate, with 40–45% of the amino acids as serine and threonine. Recent studies on the structure of the *H. wingei* sexual agglutination 5-factor reveal a still more extreme composition. Our preparations of this mannan-protein complex contain about 85% carbohydrate, with serine and threonine making up 70% of the amino acids in the protein part (12). In this case most of the carbohydrate is attached to the serine and threonine as short oligosaccharides, which are released by mild alkaline treatment. In the process 87% of the serine and 65% of the threonine are destroyed. Thus 50% of the amino acids in the protein must be substituted with carbohydrate. Such a mannan-protein complex would look quite different from that pictured in Figure 1A.

III. Immunochemistry of Yeast Mannans

While the acetolysis patterns gave a clear-cut demonstration of polymorphism in yeast mannans (5), it had been noted much earlier that there were serological differences between yeasts that presumably reflected differences in surface structure (34). The recognition that the mannan component was the immunodominant structure on the yeast cell came about 10 years ago (35), and during the intervening period considerable progress has been made in elucidating the chemical basis for the serological differences. This progress has been enhanced by two facts, the first being that the mannan is the principal antigen of these yeasts under study, and second that the acetolysis of these mannans gives large quantities of potential haptens for use as inhibitors of the precipitin reaction.

Rabbit antiserum raised against whole yeast agglutinates the cells, and this agglutination can be inhibited by mannan acetolysis fragments or by isolated mannan, which in turn will precipitate with the antiserum. The immunodominant structures in each mannan were identified by investigating the inhibition of precipitin reactions with acetolysis fragments, and in each of the examples cited in Figure 4 it was possible to obtain complete inhibition with carbohydrate fragments known to be present in the original mannan (8,36,37). From these studies we have identified three important haptenic groups: the mannotetraose unit, the *N*-acetyl-D-glucosaminylmannotetraose unit, and the α-D-mannosylphosphorylmannotriose unit (Fig. 6). It is noteworthy, and was recognized first by Suzuki, Sunayama, and Saito (38), that the terminal α1→3-linked unit is more immunogenic than a terminal α1→2-linked unit.

Some detailed results obtained for *Kluyv. lactis* mannan are of interest (8,9). The acetolysis pattern of this mannan before and after digestion with the *exo-α-mannanase* are shown in Figure 7. The intact mannan gave five major peaks corresponding to mannose, mannobiose, mannotriose, mannotetraose, and the *N*-acetylglucosamine derivative of mannotetraose, Man$_4$GNAc. After enzymatic digestion, only mannose and the latter fragment were obtained on acetolysis since the enzyme removes all other side chains. The precipitin curves of these two mannan preparations with the antiserum prepared against whole *Kluyv.*

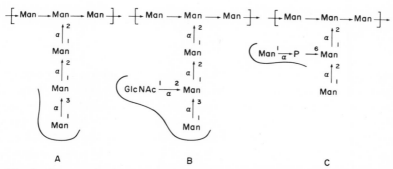

A   B   C

Fig. 6. Immunodominant side chains in (A) *Saccharomyces cerevisiae* X2180, (B) *Kluyveromyces lactis,* and (C) *Kloeckera brevis* and *Saccharomyces cerevisiae* 4484–24D. The curved lines indicate the presumed binding sites. From Raschke and Ballou (8).

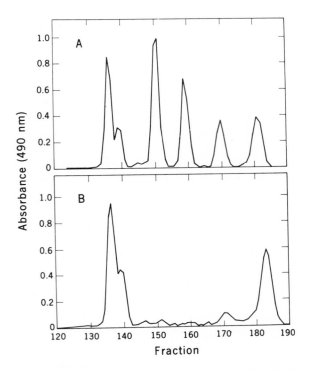

Fig. 7.   Acetolysis fingerprints of (A) intact and (B) mannanase-digested *Kluyv.*
lactis mannan. The peaks in A, right to left, are mannose, mannobiose, mannotriose,
mannotetraose, and *N*-acetylglucosaminylmannotetraose with a small amount of
N-acetylglucosaminylmannotriose formed by partial degradation of Man₄GNAc.
From Raschke and Ballou (8).

lactis cells (Fig. 8) show that the mannanase digestion decreased the
antigen–antibody reaction. The inhibition curves (Fig. 9) elucidate
the basis of these reactions. Thus the precipitin reaction with the digested mannan is completely inhibited by the Man₄GNAc fragment,
evidence that this was the only group remaining in the digested mannan that reacts with antibody. On the other hand, the intact mannan–
antimannan reaction was only partially inhibited by this pentasaccharide fragment and partially by the mannotetraose fragment, but it was
completely inhibited by a mixture of the two. Thus the antiserum made
against the whole cells contains two antibody specificities, one directed

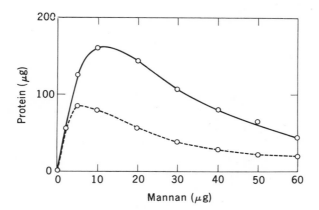

Fig. 8. Precipitin curves obtained with anti-*Kluyv. lactis* serum on reaction with intact (solid line) and mannanase-digested (dashed line) *Kluyv. lactis* mannan. From Raschke and Ballou (8).

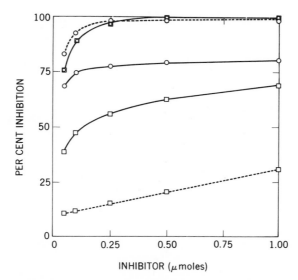

Fig. 9. Inhibition of the two precipitin reactions in Figure 8 by *Kluyv. lactis* M_4 (squares), M_4GNAc (circles), and a mixture of the two (circle in a square). The solid line is for the intact mannan, and the dashed line for mannanase-digested mannan. The smaller oligosaccharides obtained by acetolysis were much poorer inhibitors. Data from Raschke and Ballou (8).

against the tetrasaccharide side chain and the other against the pentasaccharide side chain.

The role of the mannosylphosphate group as an important yeast antigen has been confirmed in an interesting way (37). This phosphodiester structure is readily hydrolyzed by heating whole yeast cells in dilute acid, and the resulting cells possess a "transformed" cell-surface antigen (Fig. 10). As already mentioned, rabbit antiserum made with the untreated cells has a specificity for the mannosylphosphate group. On the other hand, antiserum made against the acid-treated cells is specific for the monoesterified mannotriose phosphate group. In the former case α-D-mannose 1-phosphate is a good inhibitor of the precipitin reaction, and in the latter case mannotriose phosphate is a good inhibitor and α-D-mannose 1-phosphate has no activity.

Some other yeasts we have studied (*H. angusta*, *H. wingei*) (9) give results suggesting that some unidentified acid-labile determinant is important in the mannan–antimannan reaction. Such mannans also appear to have what has been called a "block-type" structure (39) in

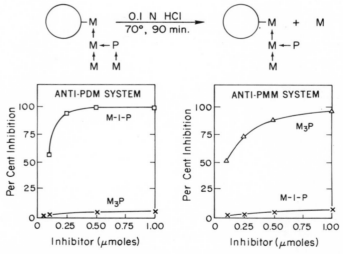

Fig. 10. Chemical "transformation" of the phosphodiester mannan (PDM) antigen to a phosphomonoester mannan (PMM) antigen. The two inhibition curves show the change in specificity for the homologous precipitin reactions from one that is inhibited by α-D-mannose 1-phosphate (M-1-P) to one that is inhibited by mannotriose phosphate (M_3P). Data from Raschke and Ballou (37).

which short 1→2- and 1→3-linked units are built into a linear chain via occasional 1→6-linkages. Such a polysaccharide would have very few end groups, and consequently certain internal linkages of the mannan, or other nonmannan components of the cell surface, may assume importance as immunogens.

IV. Genetic Control of Yeast-Mannan Structure

A. MAPPING THE FIRST GENE CONCERNED WITH MANNAN BIOSYNTHESIS

During a survey of yeast mannans by the acetolysis fingerprinting technique we noted a striking difference between *S. cerevisiae* S288C, a strain closely related to common bakers' yeast, and *S. cerevisiae* 4484–24D, which was originally derived from strain A364A. Mannan from strain S288C had a relatively low phosphate content, and the acetolysis pattern showed four components: mannose, mannobiose, mannotriose, and mannotetraose (2). On the other hand, strain 4484–24D mannan had a higher phosphate content, and the fingerprint contained mannose, mannobiose, mannotriose, and a mannotriose phosphate, thus showing a chemotype very similar to that of *Kloe. brevis* (6,7,40). In the subsequent immunochemical analysis of these two yeasts we observed that the mannotetraose unit was the major determinant on the S288C cell and that the α-D-mannosyl-1-phosphate group was the major determinant on the 4484–24D cell. Clearly this system provided an opportunity to study the genetic regulation of mannan structure.

The general approach was as follows (41): Haploid strains of opposite mating type (a or α) and with complementing auxotrophic markers (i.e., *ade1* and *ade2*) were mixed and allowed to grow on a rich medium, during which mating and diploid formation occurred. Hybrid diploids were selected by their ability to grow on minimal medium lacking adenine, since the defective adenine genes complement each other. When transferred to a special medium, the selected hybrid diploids sporulated. Four-spored asci were dissected by micromanipulation, and the individual spores were allowed to grow up. Segregation of the various markers, such as mating type and adenine requirement, were scored in the usual way (41), whereas segregation of mannan chemotype was scored by an agglutination test using anti–S288C serum (which was specific for the mannotetraose side chain) or anti–*Kloe. brevis* serum (which was specific for the mannosylphos-phate side chain).

Isogenic haploid strains derived from strain S288C, designated *S. cerevisiae* X2180–1A(a) and X2180–1B(α) and shown to possess man-nan of the B-chemotype (Fig. 4), were crossed with strain 4484–24D haploids of appropriate mating type. Surprisingly, all hybrids isolated from this mating reacted in the agglutination assay like the S288C strain and failed to react with antiserum directed against the man-nosylphosphate determinant. Thus the mannan was not a mixture of the two chemotypes, and we concluded that there was an epistatic relationship between the expression of the two types, with the S288C mannan appearing to be dominant (40). On sporulation of such hy-brids, the two mannan chemotypes segregated 2:2 in the haploid clones obtained by dissecting the tetrads. This indicates that the dif-ference between the two strains was under the control of a single gene. From tetrad analysis of a large number of different crosses, in which the mannan chemotype was followed with respect to other known markers, it was found that the mannan gene was tightly linked to the centromere on chromosome V (Table I) (40). Since this was the first gene concerned with mannan synthesis to be placed on the yeast chromosome map, we designated it *mnn1*.

Comparison of the two mannan chemotypes of these strains (Figs. 4A and B) suggests that this gene must be involved with the synthesis of an $\alpha 1 \rightarrow 3$-mannosyltransferase that catalyzes the addition of the $\alpha 1 \rightarrow 3$-linked D-mannose unit to the ends of trisaccharide units in the mannan molecule. Moreover, the presence of this gene modifies the expression of the mannosylphosphate determinant on the cell surface, apparently because it converts these units to $\alpha 1 \rightarrow 3$-mannobiosylphos-phate groups (32). The biochemical mechanism of this interaction was first suggested from our studies on mannan mutants (42), described in the next section.

Spencer, Gorin, and Rank (43) reported in an earlier study, related to the above, that a difference in mannan structure, detectable in the nuclear-magnetic-resonance spectrum, segregated as though under the control of a single gene. Although Spencer et al. did not succeed in mapping this gene, their analysis suggests that it was concerned with the synthesis of the same $\alpha 1 \rightarrow 3$-mannosyl linkage that we have in-vestigated (40).

TABLE I

Tetrad Segregations for *mnn1* with Respect to Standard Genetic Markers[a]

Chromo-some	Genetic marker	Tetrads scored	Parental ditype	Nonparental ditype	Tetratype
I	*ade1*	58	39	47	12
III	*a/α*	61	10	16	35
IV[b]	*trp1*	41	19	22	0
V[c]	*ura3*	88	78	0	10
VI	*his2*	5	1	3	1
VII	*leu1*	66	25	40	1
VIII	*arg4*	18	6	9	3
IX	*his5*	8	1	1	6
X	*ilv3*	3	2	1	0
XI	*met14*	3	2	1	0
XII	*asp5*	43	12	14	17
XIII	*lys7*	8	0	2	6
XIV	*pet8*	6	3	3	0
XV	*ade2*	89	17	9	63
XVI	*tyr7*	4	1	1	2

[a] From Antalis, Fogel, and Ballou (40).

[b] The absence of tetratype asci with respect to *trp1*, known to be linked to the centromere on chromosome IV, indicates that the *mnn1* locus is also centromere-linked.

[c] The absence of NPD asci with respect to *ura3*, and therefore first division segregants, indicates that the *mnn1* locus is linked to *ura3*.

B. ISOLATION OF YEAST-MANNAN MUTANTS

Mannan mutants can be obtained with relative ease using a selection procedure based on the failure of the mutant cells to be agglutinated by specific antiserum (42). From the nonagglutinated cells that remain in suspension individual haploid clones were selected, and the nature of the mutations were determined by chemical characterization of the mannans they synthesized. Mutants isolated from strains X2180–1A and X2180–1B were of three types, all lacking the mannotetraose side chain, two of which gave the fingerprints shown in Figure 11. One class had a chemotype very similar to that of strain 4484–24D and was also strongly agglutinated by anti–*Kloe. brevis* serum, indicative of the presence of mannosylphosphate determinants in the mannan. We assumed that this was an *mnn1* mutant, and in agreement

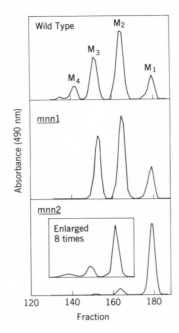

Fig. 11. Comparison of the acetolysis fingerprints of mannans from *S. cerevisiae* X2180 wild type with those of an *mnn1* mutant ($\alpha 1 \rightarrow 3$-mannosyltransferase negative) and an *mnn2* mutant ($\alpha 1 \rightarrow 2$-mannosyltransferase I negative). The apparent "leakiness" of the *mnn2* mutant is partly accounted for by the release of oligosaccharides attached to serine and threonine, the formation of which does not seem to be affected in this mutant. The acidic (phosphorylated) side chains from the *mnn1* mutant mannan are not shown. From Raschke et al. (42).

with this assumption the hybrid diploid formed in a cross with strain 4484-24D had the phenotype of the latter. Thus the X2180 *mnn1* mutation appears to be allelic with the presumed defective gene in strain 4484-24D. On the other hand, a cross between X2180 wild type and the X2180 *mnn1* mutant had the phenotype of the former parent, and the mutant phenotype segregated 2:2 with respect to the wild type, which confirmed the assumption that a single gene was involved (44).

Two other classes of mutants were obtained, one of which made a linear unsubstituted $\alpha 1 \rightarrow 6$-linked mannan and is presumed to lack the $\alpha 1 \rightarrow 2$-mannosyltransferase that adds the first mannose unit to the backbone to form the side chains. We have designated this as an *mnn2* mutant. A second mutant type, called *mnn3* and apparently defective in the α1→2-mannosyltransferase that adds the second mannose side-chain unit, made mannan that yielded mainly mannose and mannobiose on acetolysis (42).

These three classes of mutants are those expected if the synthesis of the mannotetraose side chain involves the action of three different mannosyltransferases. They appear to involve structural genes, and all three types complement each other such that the various hybrids make the parent wild-type mannan. No mutant was obtained that lacked mannan, and at present we are inclined to feel that such mutants may be unable to survive the isolation procedure. A mannanless mutant of *Candida periphelosum*, selected for its Gram-negative property, was reported to be unusually fragile (45).

A fourth type of mannan mutant (*mnn4*) was obtained by mutagenesis of strain 4484–24D (phenotypically *mnn1*) and selection for cells lacking the mannosylphosphate determinant (44). This mutant makes mannan with an extremely low phosphate content and is presumed to lack the mannosylphosphate transferase that adds this group to the mannotriose side chains in a reaction similar to that described by Kozak and Bretthauer (46). This is considered to be an *mnn1*, *mnn4* double mutant. When it is crossed with strain X2180 and the resulting hybrid is sporulated, some of the tetrads segregate 2:2 with the immunochemically determined phenotypes of strains 4484–24D and X2180 as a result of recombination of the *mnn4* locus (nonparental ditypes). Such X2180-type recombinants carry the defective *mnn4* gene, and in this manner it has been possible to construct the single mutant (47). As discussed in the next section, the *mnn4* mutation has an important effect on the binding of alcian blue dye and is readily detectable by that test. The chemotype representations of the mannans made by the various *S. cerevisiae* mutants we have isolated are illustrated in Figure 12.

C. A REVISED STRUCTURE OF *S. CEREVISIAE* X2180 WILD-TYPE MANNAN

The striking observation that the hybrid diploid formed by crossing strain X2180 with 4484–24D had the X2180 mannan chemotype was first interpreted to mean that the presence of the α1→3-mannosyltransferase in some way prevented expression of the mannosylphosphate transferase (40,44). This seemed reasonable because the hybrid cells

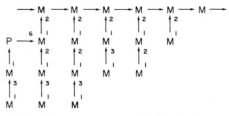

S. cerevisiae X2180 'wild type'

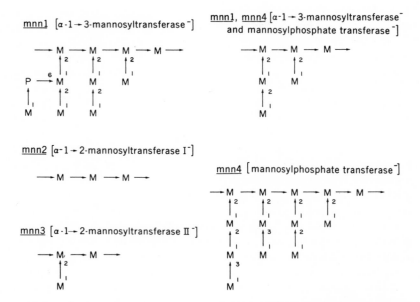

Fig. 12. Illustration of the mannan structures produced by *S. cerevisiae* X2180 wild type and the various mannan mutants. The presumed enzymatic defects in the mutants are indicated in brackets, although direct assays for the various enzymes have not been carried out. In earlier publications we have not included the mannobiosylphosphate side chain as a part of the wild-type structure (40) even though its presence in this mannan was clearly demonstrated (6). Data from Raschke et al. (42), Ballou, Kern, and Raschke (44), and Yeh and Ballou (47).

failed to agglutinate with the antiserum that was directed against the latter determinant, and the isolated mannan reacted poorly with the same serum. However, though the amount of phosphate was reduced in the hybrid, it was still significant.

We have reinvestigated the X2180 mannan, with particular emphasis on the structural role of the phosphate. In a preliminary study (17), the significance of which is only now fully appreciated, it had been shown that a minor phosphomannan fraction of S288C mannan (which is closely related to X2180) gave both mannose and mannobiose on mild acid hydrolysis of the phosphodiester linkage, rather than exclusively mannose as is the case with 4484–24D mannan. Moreover, the phosphorylated fragments from the acetolysis of the S288C phosphomannan contained a mixture of mannotriose phosphate and mannotetraose phosphate. Thus in this mannan the phosphate must be involved in a diester linkage connecting the disaccharide units with trisaccharide and tetrasaccharide side chains.

The disaccharide has now been shown to have an $\alpha1\rightarrow3$-linkage (32). This important finding suggests that the absence of the mannosylphosphate antigenic determinant in cells containing an active $\alpha1\rightarrow3$-mannosyltransferase is due not to a failure of the mannosylphosphate transferase to act but rather to the conversion of the mannosylphosphate groups to mannobiosylphosphate groups by action of the $\alpha1\rightarrow3$-mannosyltransferase. Thus the latter enzyme appears to utilize as acceptors both the mannotriose side chains and the mannosylphosphate groups attached thereto. In addition, some of the mannobiose side chains become glycosylated with $\alpha1\rightarrow3$-linked units. This alteration simultaneously creates a new antigenic determinant and eliminates an existing one, but on different parts of the molecule. The reduced phosphate content of X2180 mannan may result from the conversion of potential acceptor mannotriose units to mannotetraose ones. At present we assume that the terminal $\alpha1\rightarrow3$-mannobiose units that are attached to phosphate and those that are part of the tetrasaccharide side chain are antigenically very similar.

D. BINDING OF ALCIAN-BLUE DYE BY YEAST CELLS

Alcian blue is a positively charged phthalocyanine dye that binds to anions. In acid solution certain yeast cells bind the dye strongly while others do not, and the binding property has been shown by Friis and Ottolenghi (48) to segregate as though under the control of a single gene. Thus it may involve a single enzyme difference. It was also suggested (48) that the binding ability was correlated with the presence of phosphate in the cell-wall mannan. We have investigated this matter, making use of our various mannan mutants (44). In agreement with the earlier conclusions we found that strain 4484–24D and the X2180 *mnn1* mutants, both of which make a cell-wall phosphomannan, bound the dye very strongly. On the other hand, the *mnn1*, *mnn4* double mutant (4484–24D–1) failed completely to bind the dye. Mutants of the *mnn2* class are interesting because, presumably, they do make an active mannosylphosphate transferase. However, being unable to make the mannan side chains to which the mannosylphosphate group would be attached, they also fail to bind the dye.

The binding of alcian blue by the wild-type X2180 cells is somewhat anomalous, since some cultures react weakly whereas others bind almost to the same extent as strain 4484–24D. Moreover, the phosphate content of the X2180 mannan varies with different preparations, the mannose-to-phosphate ratios ranging from 50 to 150. We suspect that in our cultures some other factor may be segregating that regulates the amount of phosphate incorporated into the mannan much as the amounts of the different side chains seem to be regulated. At present it seems possible that the binding of alcian blue may be sensitive both to the location of the phosphate and the nature of its linkage in the mannan. There also may be a concentration factor that enhances dye binding when the phosphate groups are located together in a particular mannan fraction. Although we can say with certainty that phosphate is required for dye binding, the factors that determine the extent of binding by a phosphomannan are still undefined. Further investigation of this system may provide an effective probe with which to learn something about the organization of the acidic and neutral mannan fractions in the cell wall.

V. Enzymes that Degrade Yeast Mannan

Whole yeast cells and isolated cell walls are readily solubilized by the action of certain bacterial enzyme preparations as well as by the crude snail-gut enzyme (1). The activities most important for cell lysis are the $\beta1\rightarrow3$- and $\beta1\rightarrow6$-glucanases, but the degradation of the mannan component has not been thoroughly studied in these systems. Most α-mannosidase preparations, such as that from jack beans, have relatively little activity on yeast mannan. One of the first mannan-specific degradative enzymes to be isolated was the "phosphomannanase" of Lampen (49), an activity found in culture filtrates of *Bacillus circulans* and shown to enhance the lytic action of snail-gut enzyme on yeast cells. From a study of the enzyme Lampen concluded that it hydrolyzed mannosidic linkages in the mannan that were contiguous to phosphodiester linkages. From such a result it could be inferred that phosphodiester groups crosslink the mannan and stabilize the cell wall. However, as discussed below, it is still a debatable point whether *S. cerevisiae* cell-wall mannan is crosslinked by phosphate. Thus the action of the "phosphomannanase" enzyme remains uncertain.

In our own studies we have utilized bacterial enzymes that are secreted by soil organisms isolated by enrichment on media containing mannan as the sole carbon source. With bakers' yeast mannan as the carbon source we obtained an organism that secretes an *exo-α*-mannanase that acts preferentially to hydrolyze the $\alpha1\rightarrow2$- and $\alpha1\rightarrow3$-linked mannose units in the side chains (22–24). Under appropriate conditions one can recover the resistant core that represents the backbone of the mannan. As we have noted (23), the enzyme does act on $\alpha1\rightarrow6$-linked oligosaccharides, but since there is only one such end in a mannan molecule and many side chains, the recovery of essentially intact, stripped-down backbone is possible. The properties and acetolysis pattern of the backbone material obtained from *S. cerevisiae* mannan support this conclusion. Its molecular weight is about what one would expect, knowing the size and degree of branching of the starting mannan, and the acetolysis pattern shows essentially only mannose.

The action of the same enzyme on some other mannans gives a different kind of result. For example, the side chains of *Kluyv. lactis* mannan that are substituted with *N*-acetyl-D-glucosamine are not hydrolyzed, so the resulting product lacks only the unsubstituted side chains (8). Such a mannan gives an acetolysis pattern showing mannose and Man$_4$GNAc (Fig. 7). The phosphate-containing side chains in mannans partially resist the action of this enzyme (6), as do terminal galactose and *β*-linked D-mannose units (50). In the latter case a mild acid hydrolysis has been used to remove such units so that the *exo-α*-mannanase can continue its digestion on the internally linked α-D-mannose units.

Cultures of the organism that secrete this enzyme have been supplied to many different laboratories around the world, but to my knowledge no one has reported an extensive purification of the enzyme or successful efforts to fractionate the different enzyme activities.

We have recently isolated another kind of mannanase-secreting organism by enrichment on the mannan obtained from the *S. cerevisiae* X2180 *mnn2* mutant (51). Since this is an unsubstituted $\alpha1\rightarrow6$-linked mannan, we expected to obtain an *endo-*$\alpha1\rightarrow6$-mannanase, and this was the result. The enzyme can be recovered from the culture filtrate by ammonium sulfate precipitation, and the action of this preparation on the same mannan leads to the formation of a homologous series of $\alpha1\rightarrow6$-mannooligosaccharides. The enzyme has no apparent activity on the branched wild-type X2180 mannan.

Mannan-protein complexes are subject to degradation by Pronase (1,15,17), and this can be followed by a decrease in the size of the native molecule or by a loss of activity in the case of mannan-invertase (52) or the sexual agglutination factors of *H. wingei* (11). *Kloe. brevis* phosphomannan yields mannan subunits that have a molecular weight of about 25,000 and contain about 15 moles of amino acids per mannan chain (17). The failure to obtain more extensive release of amino acids may reflect inhibition of the enzyme action by short oligosaccharides that are attached to serine and threonine, which together make up one-third or more of the amino acids in the glycopeptide residue (15). The subunits obtained by Pronase digestion contain about 1 mole of glucosamine and 1 mole of aspartic acid, which is consistent with the idea that a large polysaccharide chain is attached via the amino sugar to asparagine. Since many glycoproteins have at least two glucosamine units at the linkage region (53), these ratios may be only approximately correct. The utilization of subtilisin digestion for releasing the *H. wingei* 5-factor from the cell wall suggests that peptide bonds are involved in holding this macromolecule on the cell surface.

VI. Some Thoughts about Mannan Biosynthesis

Progress in working out the pathway of mannan biosynthesis has been slow. Since it is a polysaccharide–protein, the inhibition of its synthesis by cycloheximide is understandable (54). Incorporation of [^{14}C]mannose from GDP-[^{14}C]mannose into endogenous mannan by particulate enzyme preparations has been observed (55), and analysis of the product by acetolysis indicated that the mannose was incorporated into all units of the side chains, rather than just onto the ends.

Evidence of the intermediate formation of a lipid-bound mannosyl-phosphate group has been obtained (56), but oligosaccharide-bound intermediates have not been detected. The transfer of mannosylphosphate groups from GDP–mannose to endogenous acceptors by particulate fractions from *Hansenula* species, which produce an exocellular phosphomannan, has been demonstrated (46). Since yeast mannans contain long polysaccharide chains attached to asparagine and short oligosaccharides attached to serine, the idea that the latter were precursors of the former was investigated, but no support for this postulate was found (57).

As the problem stands today, the main obstacle appears to be that of obtaining cell-free systems that carry out reactions clearly concerned with mannan biosynthesis and utilizing exogenous acceptors. From what we know of mannan structure one can visualize a mechanism in which the protein or short polypeptides are assembled, mannooligo-saccharides are then built up on the serine and threonine units, and the longer polysaccharide chains are formed by the addition first of *N*-acetyl-D-glucosamine to asparagine followed by the stepwise addition of mannose units. Alternative schemes by which this could occur are shown in Figure 13. Subsequently the polymer could be modified by addition to the side chains of mannosylphosphate or *N*-acetyl-D-glucosamine units. Finally, there would appear to be the need for a crosslinking step to build up the larger macromolecular forms of mannan from the presumed subunits.

A central question is whether the polysaccharide chain is formed by the stepwise addition of single mannose units or by the formation of side-chain units on a carrier followed by their polymerization to form the $1\rightarrow6$-linked backbone. A useful analogy here is to the different mechanisms for synthesis of the *Salmonella* lipopolysaccharide core and 0-antigen (58). The mannan mutants we have obtained (Fig. 12) do give some hints concerning the biosynthetic process. First, since the *mnn2* mutants make an unsubstituted backbone of approximately normal length (51), the presumed $\alpha1\rightarrow6$-polymerase must be able to act on monosaccharide precursors alone to build up a polymer of substantial size. Such mutants appear to have a normal amount of the mannan-protein complex in the cell wall, so the fine structure of the polysaccharide component must not be important for the translocation of the macromolecule or for regulation of the total amount of mannan that is made.

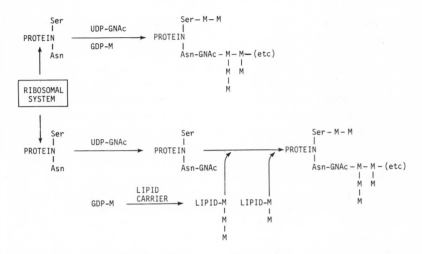

Fig. 13.   Illustration of two alternative mechanisms for mannan biosynthesis, one emphasizing the stepwise addition of single sugar units to the growing chain and the other emphasizing the intermediate formation of lipid-bound oligosaccharides that are transferred to the growing chain by an $\alpha 1 \rightarrow 6$-polymerase.

Second, we note that the *mnn4* mutants, which have essentially no phosphate in the mannan, have an otherwise normal wall structure. This could mean that crosslinking of mannan by phosphodiester groups does not occur normally or is not important for maintaining the structural integrity of the wall. Distinction must be made between exocellular phosphomannans and the phosphorylated mannan of the cell wall. Some yeasts secrete mannans that have a high phosphate content and consist of oligosaccharides with three to five mannose units built into a polymer by way of phosphodiester bridges between C1 of the reducing mannose unit and position 6 of a mannose unit in the next oligosaccharide in the polymer (1,59,60). Cell-wall mannans, on the other hand, can usually be resolved by DEAE-column chromatography into a neutral fraction and fractions of variable phosphate content (6). The phosphate is in diester linkage between C1 of mannose or mannobiose units and position 6 of a mannose unit in the side chain of the polysaccharide. Thus in the exocellular phosphomannans the phosphate is clearly involved in polymerizing small units into polymers, much as in teichoic acids (58), whereas in the cell-wall mannan it does not appear to serve this same role. More specifically, neither *S. cerevisiae* nor *Kloe. brevis* wild-type mannan appears to be crosslinked by phosphate; mild acid hydrolysis of these mannans releases mannose and/or mannobiose equivalent to the amount of phosphate in the mannan and simultaneously converts the phosphodiesters to monoester groups (6). Thus one ligand to phosphate is a small sugar unit. By analogy with the crosslinking mechanism for peptidoglycan biosynthesis in bacteria (58), it is an attractive idea that the mannosylphosphoryl diester units might be the activated equivalent of the D-alanyl-D-alanine group, which provides the energy for peptidoglycan crosslinking outside the cell. However, at this time there is no evidence that such a reaction occurs in yeast mannan synthesis.

Third, we find that the structural mannan-protein complex and the mannan enzymes have their mannans altered in a coordinate fashion in the mannan mutants (61). Thus the same biosynthetic machinery must be involved in making both kinds of mannan. Since the mutant mannan-invertase appears to have an unaltered enzymatic activity with respect to the wild-type enzyme, the mannan component cannot play an important part in modulating the enzyme activity. This conclusion agrees with the fact that invertases of widely differing mannan contents have been isolated from different yeasts (52,62).

Fourth, while the *mnn2* mutants cannot add the first $\alpha1{\rightarrow}2$-linked mannose unit to the backbone, they do form the disaccharide, trisaccharide, and tetrasaccharide units that are attached to serine and threonine (51). Thus there must be at least one other mannosyltransferase enzyme for which we have not detected a mutant, namely, one involved in attaching the first mannose to these amino acids. On this point, one wonders whether there might be a temporal separation in the reactions that are involved in the addition of the mannose to these amino acids and those reactions involved in the synthesis of the polysaccharide chains and their attachment to asparagine. For example, might the short chains be added in some early part of the biosynthetic process, serving perhaps to direct the translocation of the protein units, with the polysaccharide chains being added later at the moment of secretion through the membrane? Thus it may be operationally useful to think of "early," "middle," and "late" stage glycosylation reactions that occur as separated steps in the processing of the mannan-protein molecules.

Finally, an important question for investigation will be what regulates the amounts of the different side chains in a mannan. As already

pointed out, the ratios are fairly constant in a particular strain, regardless of conditions of growth, whether haploid or diploid, and independent of mating type (26). Interestingly, in the *mnn1* mutants of
S. cerevisiae that lack the tetrasaccharide side chain the amount of
trisaccharide side chain is increased by an amount corresponding to
the amount of mannotetraose unit that is normally present in the wild-
type mannan (42). In a mutant of *Kluyv. lactis* that lacks the *N*-acetyl-
D-glucosamine–containing side chain the mannotetraose side chain is
increased by an amount that is almost exactly equivalent to the
Man_4GNAc that was not formed (61). In the *mnn1*-type mutant of this
yeast that cannot make the mannotetraose side chain or the Man_4GNAc,
the mannotriose side chain is increased by the sum of the Man_4GNAc
and the mannotetraose side chain normally found. Thus there seems
to be a programmed amount of the longer side chains that are to be
formed, and if they cannot be made because of enzymatic defects, the
incompleted precursors accumulate as shorter side chains in the polysaccharide. Regulation of this kind could occur simply by fixing the
activities of the different transferases, which then work in an "assembly
line" manner, with the final product appearing loaded with different
amounts of "extras" depending on the specific enzymes that were
available for "late-stage" modification of the mannan-protein complex. On the other hand, a more subtle form of regulation might be
imposed if a fixed sequence of side chains occurred in the mannan,
which would lead in turn to acceptor modulation of transferase activity. In view of the results obtained with the mannan mutants this
seems unlikely because the operation of such a mechanism should
interfere with the formation of altered polysaccharides of large size
and would be expected to have more dramatic consequences than we
have observed.

Acknowledgments

I am particularly indebted in these studies to the collaborative participation of Drs. Y.-C. Lee, T. S. Stewart, G. H. Jones, J. Kocourek,
T. R. Thieme, T. N. Cawley, and W. C. Raschke; to the help and
advice of Professors S. Fogel and R. K. Mortimer; and to the assistance
of fifty or so unnamed graduate students who carried out many of the
initial studies of yeast mannan structure in a graduate laboratory
course in the department.

This work was supported by the National Science Foundation (grants GB-19199 and GB-35229X) and by the U.S. Public Health Service (grant AM884).

References

1. Phaff, H. J., in *The Yeasts*, Vol. II, A. H. Rose and J. S. Harrison, Eds., Academic Press, New York, 1971, p. 135.
2. Lee, Y.-C., and Ballou, C. E., *Biochemistry*, *4*, 257 (1965).
3. Gorin, P. A. J., and Perlin, A. S., *Can. J. Chem.*, *34*, 1796 (1956).
4. Stewart, T. S., and Ballou, C. E., *Biochemistry*, *7*, 1855 (1968).
5. Kocourek, J., and Ballou, C. E., *J. Bacteriol.*, *100*, 1175 (1969).
6. Thieme, T. R., and Ballou, C. E., *Biochemistry*, *10*, 4121 (1971).
7. Cawley, T. N., and Ballou, C. E., *J. Bacteriol.*, *111*, 690 (1972).
8. Raschke, W. C., and Ballou, C. E., *Biochemistry*, *11*, 3807 (1972).
9. Raschke, W. C., Ph.D. thesis, University of California, Berkeley, 1972.
10. Lampen, J. O., *Antonie van Leeuwenhoek*, *34*, 1 (1968).
11. Crandall, M. A., and Brock, T. D., *Bacteriol. Rev.*, *32*, 139 (1968).
12. Yen, P. H., and Ballou, C. E., *J. Biol. Chem.*, *248*, 8316 (1973).
13. Haworth, W. N., Hirst, E. L., and Isherwood, F. A., *J. Chem. Soc.*, 784 (1973).
14. Peat, S., Whelan, W. J., and Edwards, T. E., *J. Chem. Soc.*, 29 (1961).
15. Sentandreu, R., and Northcote, D. H., *Biochem. J.*, *109*, 419 (1968).
16. Neumann, N. P., and Lampen, J. O., *Biochemistry*, *6*, 468 (1967).
17. Thieme, T. R., and Ballou, C. E., *Biochemistry*, *11*, 1115 (1972).
18. Gascon, S., Neumann, N. P., and Lampen, J. O., *J. Biol. Chem.*, *243*, 1573 (1968).
19. Taylor, N. W., and Orton, W. L., *Arch. Biochem. Biophys.*, *126*, 912 (1968).
20. Nickerson, W. J., Falcone, G., and Kessler, G., *Symp. Soc. Gen. Physiol.*, *205*, (1961).
21. Peat, S., Turvey, J. R., and Doyle, D., *J. Chem. Soc.*, 3918 (1961).
22. Jones, G. H., and Ballou, C. E., *J. Biol. Chem.*, *243*, 2442 (1968).
23. Jones, G. H., and Ballou, C. E., *J. Biol. Chem.*, *244*, 1043 (1969).
24. Jones, G. H., and Ballou, C. E., *J. Biol. Chem.*, *244*, 1052 (1969).
25. Stewart, T. S., Mendershausen, P. B., and Ballou, C. E., *Biochemistry*, *7*, 1843 (1968).
26. Thieme, T. R., and Ballou, C. E., *Biochem. Biophys. Res. Commun.*, *39*, 621 (1970).
27. Hakomori, S., *J. Biochem.* (*Tokyo*), *55*, 205 (1964).
28. Björndal, H., Hellerqvist, C. G., Lindberg, B., and Svensson, S., *Angew. Chem. Int. Edit.*, *9*, 610 (1970).
29. Hall, L. D., *Adv. Carbohyd. Chem.*, *19*, 51 (1964).
30. Gorin, P. A. J., and Spencer, J. F. T., *Adv. Appl. Microbiol.*, *13*, 25 (1970).
31. Guthrie, R. D., and McCarthy, J. F., *Adv. Carbohydr. Chem.*, *22*, 11 (1967).
32. Rosenfeld, L., and Ballou, C. E., unpublished data.
33. Lindberg, B., *Acta Chem. Scand.*, *3*, 1153 (1949).

34. Tsuchiya, T., Fukazawa, Y., Sato, I., Kawakita, S., Yonezawa, M., and Yamase, Y., *Yokohama Med. Bull.*, *9*, 359 (1958).
35. Hasenclever, H. F., and Mitchell, W. O., *Sabouradia*, *3*, 288 (1964).
36. Ballou, C. E., *J. Biol. Chem.*, *245*, 1197 (1970).
37. Raschke, W. C., and Ballou, C. E., *Biochemistry*, *10*, 4130 (1971).
38. Suzuki, S., Sunayama, H., and Saito, T., *Jap. J. Microbiol.*, *12*, 19 (1968).
39. Gorin, P. A. J., and Spencer, J. F. T., *Adv. Appl. Microbiol.*, *12*, 25 (1970).
40. Antalis, C., Fogel, S., and Ballou, C. E., *J. Biol. Chem.*, *248*, 4655 (1973).
41. Fink, G. R., *Methods Enzymol.*, *17*, 59 (1970).
42. Raschke, W. C., Kern, K. A., Antalis, C., and Ballou, C. E., *J. Biol. Chem.*, *248*, 4660 (1973).
43. Spencer, J. F. T., Gorin, P. A. J., and Rank, G. H., *Can. J. Microbiol.*, *17*, 1451 (1971).
44. Ballou, C. E., Kern, K. A., and Raschke, W. C., *J. Biol. Chem.*, *248*, 4667 (1973).
45. Harada, Y., Ono, J., and Nagasawa, T., *Abstracts IVth Int. Fermentation Symp.*, *Kyoto, Japan, 1972*, p. 84 (1972).
46. Kozak, L. P., and Bretthauer, R. K., *Biochemistry*, *9*, 1115 (1970).
47. Yeh, Y.-F., and Ballou, C. E., unpublished data.
48. Friis, J., and Ottolenghi, P., *C. R. Trav. Lab. Carlsberg*, *37*, 327 (1970).
49. McLellan, W. L., Jr., and Lampen, J. O., *J. Bacteriol.*, *95*, 967 (1967).
50. Gorin, P. A. J., Spencer, J. F. T., and Eveleigh, D. E., *Carbohydr. Res.*, *11*, 387 (1969).
51. Nakajima, T., and Ballou, C. E., unpublished data.
52. Neumann, N. P., and Lampen, J. O., *Biochemistry*, *8*, 3552 (1969).
53. Pazur, J. H., and Aronson, N. N., Jr., *Adv. Carbohydr. Chem.*, *27*, 301 (1972).
54. Elorza, M. V., and Sentandreu, R., *Biochem. Biophys. Res. Commun.*, *36*, 741 (1969).
55. Behrens, N. H., and Cabib, E., *J. Biol. Chem.*, *243*, 502 (1968).
56. Tanner, W., *Biochem. Biophys. Res. Commun.*, *35*, 144 (1969).
57. Sentandreu, R., and Northcote, D. H., *Biochem. J.*, *115*, 231 (1969).
58. Nikaido, H., and Hassid, W. Z., *Adv. Carbohyd. Chem.*, *26*, 351 (1971).
59. Slodki, M. E., *Biochim. Biophys. Acta*, *69*, 96 (1963).
60. Bretthauer, R. K., Kaczorowski, G. J., and Weise, M. J., *Biochemistry*, *12*, 1251 (1973).
61. Smith, W. L., and Ballou, C. E., unpublished data.
62. Greiling, H., Vögele, P., Kisters, R., and Ohlenbusch, H. D., *Z. Physiol. Chem.*, *350*, 517 (1969).

THE NEUROPHYSINS

By ESTHER BRESLOW, *New York*

CONTENTS

271

I. Introduction

The biological activities associated with the posterior pituitary hormones oxytocin and vasopressin were originally attributed to a protein of 20,000–30,000 molecular weight that contained equal amounts of oxytocic and pressor (vasopressin) activity (1–3). The characterization of this protein by van Dyke et al. (3) led to its appellation as "van Dyke protein." Ultimately it was established that the two hormonal activities belonged to separate molecules (4–6) that were, in fact, nonapeptides (7–11). Acher and coworkers (12,13) demonstrated that van Dyke protein could be resolved by a variety of procedures into the nonapeptide hormones and a protein that had a much higher molecular weight which they named "neurophysin." The procedures by which neurophysin could be separated from the hormones included precipitation of the protein by trichloroacetic acid (which left the hormones in solution) and electrodialysis, indicating that the nature of the association between the hormones and neurophysin was noncovalent. Purified neurophysin was shown to be capable of recombining with the hormones (14,15).

Actually neurophysin is not a single protein in most species but appears most generally to represent a family of closely related proteins. Neurophysins have been isolated from the codfish (16) and from a number of mammalian species, including sheep, cow, pig, man, rat, and the guinea pig (13,14,17–29). With the exception of the guinea pig (29) and perhaps the rat (30–32), all mammalian species appear to have at least two significant electrophoretically demonstrable neurophysins, and in species such as the cow and pig, when isolation is performed under the correct conditions, several of the neurophysins have been shown to represent proteins of distinctly different primary structure rather than artifacts of preparation (20,21,33).

The principal biological function of the neurophysins appears to be their role as carrier proteins for oxytocin and vasopressin within the hypothalamic–neurohypophyseal system (34), although they may have ancillary activities as well (see Section III). While there is no evidence indicating an enzymatic function for the neurophysins, they are relevant to enzyme chemistry for a variety of reasons. First, they provide uniquely simple models for the study of specific protein–protein or protein–peptide interactions, uncomplicated by subsequent catalytic steps; what is known about the structure of neurophysin and the molecular basis of neurophysin–hormone interaction will provide a principal focus for this review. Second, the neurophysins are of evolutionary interest. As will be shown in Section V, their primary structure is marked by sequences that are duplicated within the same polypeptide chain. What is the significance of the duplication? Is there more than one binding site per chain? In terms of conformation, do the duplicated regions have the same three-dimensional structure or are they modified by interactions with nonduplicated segments? Third, evidence has been presented by Sachs et al. (35–37) that the hormones (vasopressin in particular) are synthesized on the ribosome as part of hormonally inactive, high-molecular-weight precursors that are subsequently cleaved to yield the active hormones. Is there any relationship between the neurophysins and the hormone precursors? Finally, the neurophysins are of interest as model proteins to which a variety of physicochemical techniques can be applied in the analysis of their structure and interactions. As will be seen, the neurophysins are low-molecular-weight proteins (approximately 10,000 daltons per monomer) and of an amino acid composition that allows such techniques as circular dichroism and nuclear magnetic resonance (NMR) spectroscopy to be applied and interpreted without excessive ambiguity. Thus a paucity of optically active aromatic residues and a high disulfide content renders the neurophysins unique models for the study of disulfide optical activity. In NMR studies the presence of typically only a single tyrosine, three phenylalanine residues, and no tryptophan allows changes in the aromatic region on binding hormone to be relatively easily interpreted. Although complexes of the neurophysins with the hormones have been crystallized (19), these crystals have not yet led to any structural data. It is certainly conceivable, however, that solution studies of neurophysin and its complexes will ultimately be paralleled by X-ray-diffraction analysis.

By and large, neurophysin studies are in their infancy. This chapter will attempt to present a picture of the field as a whole, but will selectively emphasize those studies that are most relevant to the topics cited above.

II. Multiplicity of the Neurophysins; Nomenclature

The best characterized neurophysins (NP) are those from the cow and pig. Three principal porcine (21,23) and three principal bovine (20,38) neurophysins have been described when isolation is performed under conditions that minimize proteolytic degradation (see Section IV). Following the nomenclature of Uttenthal and Hope (21), the three porcine neurophysins are labeled I, II, and III in order of decreasing anodic electrophoretic mobility in starch gel at pH 8.1. The three principal bovine neurophysins, following the nomenclature of Rauch, Hollenberg, and Hope (20), are I, C, and II in order of decreasing anodic mobility at pH 8. Of the porcine neurophysins, I and II are the principal components (21). Of the bovine neurophysins, II > I > C represents the relative percentage of each (20). Although the relative mobilities of the bovine neurophysins are not altered when electrophoresis is carried out at pH 8.5 or 9.5 in polyacrylamide gels (38), at least one other band, migrating between NP–II and NP–C, has been observed in unfractionated bovine neurophysins under these conditions (38,39) and may represent an additional bovine neurophysin (Fig. 1).

Because both bovine and porcine neurophysins have been isolated by a number of different laboratories, it is worthwhile to relate the different nomenclatures used. In this chapter we shall retain the nomenclature introduced by Hope and coworkers. For comparison with porcine-neurophysin nomenclature used by other investigators it should be noted that the "W-S peptide" of Wuu and Saffran (22), peptide 7D6 of Rudman et al. (40), and peptide-II of Friesen and Astwood (41) are apparently all identical with porcine NP–I. As analyzed by Cheng and Friesen (23), peptide III (41) and peptide IV (23) are analogous to porcine NP–II and porcine NP–III, respectively, although the amino acid composition reported for peptide III (23,27) differs significantly from that reported for porcine NP–II (21).

In early isolations of the bovine neurophysins (42,43) the number of protein fractions that were isolated from the posterior pituitary and found to be capable of binding oxytocin and vasopressin exceeded the number of bovine neurophysins now known to be present. This multiplicity of neurophysins was shown by Dean, Hollenberg, and Hope (44) to arise from the proteolytic degradation of native neurophysins during isolation. Nonetheless, because some early studies utilized these

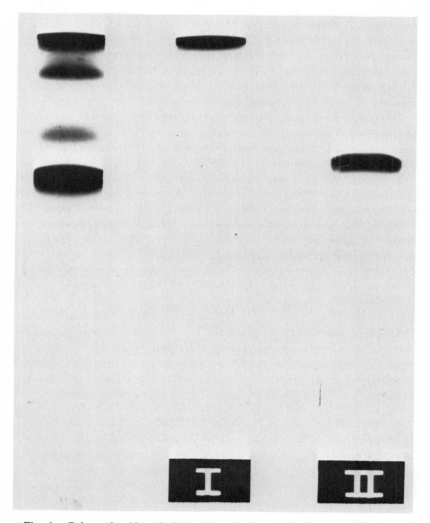

Fig. 1. Polyacrylamide gel electrophoresis of crude bovine neurophysin prepared by 0.1 N HCl extraction of acetone-desiccated bovine posterior pituitary glands (far left) and of purified bovine NP–I (I) and NP–II (II). Direction of migration is to the top of the gel; the buffer front coincides with the leading edge of bovine NP–I. The band immediately below NP–I in the crude preparation is NP–C. The band immediately above NP–II is unidentified. From Breslow et al. (38).

partially degraded neurophysins, it is worthwhile to note that fraction B of Breslow and Abrash (42) is bovine NP–II and fraction 3 of Hollenberg and Hope (43), also called neurophysin-M, is probably a derivative of bovine NP–II. Fraction C of Breslow and Abrash (42) is bovine NP–II from which the first two N-terminal residues have been removed (38) and is not to be confused with bovine NP–C defined by Rauch et al. (20). Fraction E of Breslow and Abrash is probably a mixture of partially degraded neurophysins (38).

In species other than the cow and pig the neurophysins have been insufficiently characterized to warrant the use of any particular nomenclature. There are conflicting reports in the literature as to the multiplicity of neurophysins in the rat. A single principal and perhaps one minor rat neurophysin were observed electrophoretically by Norström et al. (30,45); Coy and Wuu (31) found one major and two minor rat neurophysins. However, Burford, Jones, and Pickering (25) reported the presence of three rat neurophysins (two of which are major components) whose electrophoretic resolution is dependent on the concentration of bromophenol blue used as a tracking dye (32). Only one neurophysin has been found so far in the guinea pig (29).

III. Biology of the Neurophysins

Much of what is known about the localization of neurophysin, oxytocin, and vasopressin within the hypothalamic–neurohypophyseal system, and the biosynthesis and release of these factors, has been reviewed relatively recently (36,37). It is appropriate here only to review the principal aspects of this story and to bring the facts up to date.

A. PROBABLE COMPARTMENTALIZATION OF THE HORMONES IN SEPARATE NEURONS AND NEUROSECRETORY GRANULES

Oxytocin and vasopressin are typically isolated as complexes of neurophysin (van Dyke protein) from the posterior pituitary. Bargmann and Scharrer (46,47) postulated that the "posterior pituitary" hormones were actually synthesized in the supraoptic and paraventricular nuclei of the hypothalamus, from which they traveled within neurosecretory granules along the nerve axons to the posterior pituitary, there to be released into the blood on stimulation of the neuron. This hypothesis has been confirmed in many different investigations (cf. refs. 36, 37, 48–50). In addition, a variety of stimuli have been shown, *in vivo*, to lead to the release of vasopressin independently from oxytocin and of oxytocin independently from vasopressin (36,37,51–55). Such data give strong support to the thesis that oxytocin and vasopressin are synthesized and stored within separate neurons differing in the stimuli to which they respond. Moreover, partial fractionation of posterior pituitary neurosecretory granules by sucrose-density-gradient centrifugation indicates that granules of different densities differ in their oxytocin/vasopressin ratios (56–60), which in turn suggests that some granules may contain only oxytocin and others only vasopressin.

B. ASSOCIATION OF NEUROPHYSIN WITH THE HORMONES IN NEUROSECRETORY GRANULES

The neurophysins have been shown to be located together with the hormones in isolated neurosecretory granules (59–62); the concentration of neurophysin within the granules is very high, about 0.02 M (63,64). There is no firm evidence of any extragranular neurophysin within the neurons *in vivo*, although it has been reported (56,60) that 10–20% of the *hormone* content of the neurons does not sediment with the granule fraction. Moreover, like the hormones, the neurophysins are synthesized within the hypothalamus and travel down the axon to the posterior pituitary (45,65–68); a rate of movement of 2–3 mm/hr has been estimated in the rat (45). This is shown in Figure 2, where the distribution of neurophysin (measured with fluorescent antibody) along the pituitary stalk of a dog is shown 20 hr after the stalk was pinched by surgery. Between the hypothalamus and the crush there is an accumulation of neurophysin. Distal to the crush is an area devoid of neurophysin, while neurophysin reappears in the region of the stalk entering the posterior pituitary. The data support a one-way transport of neurophysin from the hypothalamus.

The question arises as to the ratio of neurophysin to the hormones within the neurosecretory granules and whether, like the hormones, the different neurophysins within a species are segregated within different granules. (It should be emphasized here, as discussed further subsequently, that purified neurophysins do not differ significantly among themselves in their affinities for oxytocin and vasopressin. Each neurophysin can bind oxytocin and vasopressin; the binding ability of isolated neurophysins therefore cannot be used as an index of whether they are associated with a particular hormone *in vivo*.) Dean, Hope, and Kazic (59) studied the distribution of bovine NP–I and NP–II

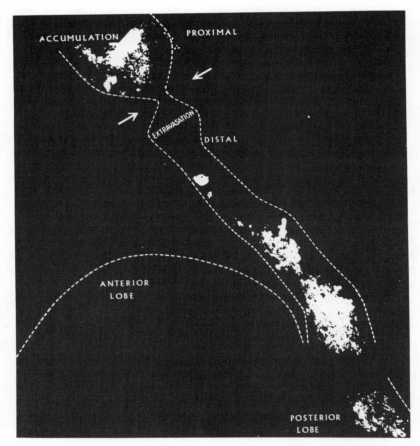

Fig. 2. Distribution of neûrophysin fluorescence in the pituitary stalk both proximal and distal to the stalk injury and in the posterior lobe of the pituitary, 20 hr after operation. From Alvarez-Buylla et al. (66).

in partially fractionated neurosecretory granules as a function of the oxytocin and vasopressin content. Their results suggest that bovine NP–I is found within oxytocin-containing granules and bovine NP–II within vasopressin-containing granules. The observed ratio of hormone to neurophysin was 2 moles of oxytocin per 20,000 g of NP–I and 2 moles of arginine vasopressin per 20,000 g of NP–II (69); using a monomeric molecular weight of about 10,000 for each bovine neurophysin (see below) this represents a 1:1 complex of each hormone with bovine neurophysin. Some question can be raised as to the quantitative validity of these studies, however, since the amount of NP–C was not quantitated, but was apparently included in the NP–I estimate.

Studies of the localization of porcine NP–II within the pig hypothalamus using fluorescent antibodies specific for porcine NP–II demonstrated that it was located in the supraoptic nucleus (70). Because vasopressin originally appeared to be synthesized predominantly in the supraoptic nucleus and oxytocin in the paraventricular nucleus, these results were originally interpreted as indicating the *in vivo* association of porcine NP–II with lysine vasopressin. This was a particularly attractive hypothesis since porcine NP–II may differ from most other neurophysins in its monomeric molecular weight (see Table I) and the pig differs from most other mammals in that it has lysine, rather than arginine, vasopressin; thus an unusual vasopressin appeared to be associated in evolution with an unusual neurophysin (70). More recently, however, neurosecretory granules from porcine posterior pituitary have been partially fractionated by centrifugation in a sucrose-density gradient (73). In this study antibodies for both porcine NP–I and porcine NP–II were prepared. The distribution of oxytocin and lysine vasopressin within the partially fractionated granules (as measured by bioassay) was compared with the distribution of the two neurophysins as measured by immunoassay. The results indicated that porcine NP–I, rather than porcine NP–II, followed the distribution of lysine vasopressin within the granules, and that porcine NP–II followed the distribution of oxytocin. The alternative original conclusions, based on the histochemical localization of porcine NP–II in the supraoptic nucleus of the pig hypothalamus, were explained on the basis of the fact that in the pig (74) the supraoptic nucleus is the source of both oxytocin and lysine vasopressin. Recent studies by Zimmerman et al. (75) in the cow also suggest that NP–I and NP–II are each synthesized in both the supraoptic and paraventricular nuclei.

Additional evidence for the association of lysine vasopressin with porcine NP–I is found in the studies of Friesen and Astwood (41), who showed that, of the different porcine neurophysins, peptide-II (porcine NP–I) undergoes the most dramatic decrease in pituitary concentration during prolonged dehydration. In addition, Johnston et al. (74) report that blood levels of porcine NP–I increase much more

TABLE I

Amino Acid Compositions of Some Mammalian Neurophysins[a]

Amino acid	Porcine				Bovine			Human	
	NP-I[b]	NP-II[e]	NP-II[d]	NP-III[e]	NP-I[e]	NP-II[e,f]	NP-C[g]	NP-I[d]	NP-II[d]
Lys	2	4	4	2	2	2	2	3	2
His					1			1	1
Arg	5	6	6	6	4	7	5	5	6
Asp	6	12	9	5	7	5	6	8	5
Thr	2	3	2	2	2	2	2	2	2
Ser	7	9	6	7	5	6	7	5	6
Glu	13	15	11	12	11	14	14	11	13
Pro	7	14	10	6	9	8	9	9	5
Gly	13	20	14	13	14	16	16	15	15
Ala	7	12	9	8	9	6	7	12	8
Half-Cys[h]	14	18	16	12	14	14	14	14	14
Val	2	4	4	2	3	4	3	4	3
Met	1			1		1	1		1
Ile	2	3	2	2	2	2	2	1	1
Leu	7	9	7	7	6	6	7	8	6
Tyr	1	2	1	1	1	1	1	1	1
Phe	3	6	4	3	3	3	3	3	4
NH₃	6	11		7	5	7	7		
Minimum molecular weight	9521	14,020	9908	9214	9560	10,041	10,347	10,333	9641

[a] Data are reported as moles of amino acid (to the nearest integer) per minimum molecular weight. For bovine NP-I, porcine NP-I, and bovine NP-II the minimum molecular weight is known to be the molecular weight of the monomer (33,71,72).
[b] Data of Wuu et al. (71).
[c] Data of Uttenthal and Hope (21).
[d] Data of Cheng and Friesen (27).
[e] Data of Breslow et al. (38).
[f] Data of Walter et al. (72).
[g] Data of Rauch et al. (20).
[h] All cysteine residues in Bovine NP-I, bovine NP-II, and porcine NP-I are known to be in disulfide form.

than do those of porcine NP–II during hemorrhage. Since lysine vasopressin is released during dehydration and hemorrhage and the mechanism of hormone release is via exocytosis (see below), both the hormone and its accompanying neurophysin should enter the bloodstream. The finding of an increased porcine NP–I concentration in blood during hemorrhage and a decreased pituitary concentration during dehydration therefore suggests that both lysine vasopressin and porcine NP–I are to be found in the same neurosecretory granules.

Evidence has been presented (25) that, of the three neurophysins found in the rat by some investigators (32), the most anionic binds arginine vasopressin, particularly since Brattleboro rats homozygous for diabetes insipidus lack this neurophysin (25). On intracisternal injection with [^{35}S]cysteine the ratio of label in "A," the most anionic protein, to "B," the second most anionic neurophysin, was the same as the lysine vasopressin/oxytocin ratio, which also supports the thesis that oxytocin is associated with protein B (25).

From the above, it seems probable that the different neurophysins are to be found in different neurons and neurosecretory granules associated with a particular hormone—although the cell bodies of the different neurons may not be segregated into discrete regions of the hypothalamus. However, further data are needed to establish more conclusively which neurophysin is associated with which hormone. This is particularly true in the case of oxytocin and its associated neurophysin because arguments based solely on differences among the granular fractions in their hormone/neurophysin ratios are not entirely convincing when the potential contributions of minor neurophysin components are also considered (see also Section III.C). Indeed, the presence of at least three neurophysins in the cow, pig, and rat introduces a note of confusion into the "one neurophysin, one hormone" theory. In the cow at least the most minor neurophysin, NP–C, has an amino acid composition (Table I) that suggests tentatively that it is not a breakdown product of NP–I and NP–II, and the same may well be true for porcine NP–III and the third rat neurophysin. Are these neurophysins associated with either oxytocin or vasopressin, or are they associated with as yet unidentified hormones, chemically related to oxytocin and vasopressin? Pickup et al. (73) noted that, in the pig, the lysine vasopressin/NP–I and oxytocin/NP–II ratios were close to 2 in isolated granule fractions. Assuming one binding site per neurophysin molecule for hormone (see Section VI.A), they suggested that the excess hormone, particularly lysine vasopressin, might be associated with porcine NP–III. Burford, Jones, and Pickering (25) have noted that 8-arginine oxytocin (vasotocin) is found in some mammalian embryos and suggested, among other possibilities, that the third neurophysin found in many species might be related to fetal vasotocin. However, not only does the existence of three neurophysins in some species give rise to questions as to the validity of a "one hormone, one neurophysin" story, but even more serious questions are raised by the guinea pig. Here only a single neurophysin has as yet been seen by a wide variety of techniques (29), despite the fact that guinea pigs have both oxytocin and vasopressin. If this observation is sustained, it should provide important insights into the significance of multiple neurophysins in other species.

C. RELATIONSHIP BETWEEN NEUROPHYSIN AND HORMONE RELEASE

Two general mechanisms have been proposed for the release of oxytocin and vasopressin from the neurons in which they are stored: hormone dissociation and exocytosis. Both mechanisms utilize the observation by Douglas and Poisner (76,77) that release is accompanied by, and dependent on, an increase in intraneuronal Ca^{2+} concentration. The hormone-dissociation hypothesis was elaborated by Thorn and Smith (78,79) and Ginsburg et al. (64,80,81) and is based on reports (78–80) that Ca^{2+} ion, at concentrations comparable to those attained within the neuron on stimulation, weakened the affinity of neurophysin for oxytocin and vasopressin. According to this hypothesis, the intraneuronal entry of Ca^{2+} that attends nerve stimulation dissociates the hormones from neurophysin so that they diffuse from within the neurosecretory granules into the extragranular space and out of the neuron. The hormone-dissociation hypothesis also allows for the presence of an extragranular neurophysin–hormone complex that could be directly dissociated by Ca^{2+} (64) and account for a "readily releasable" hormone pool (36,37,78). In addition to studies that demonstrated an apparent effect of Ca^{2+} on the affinity of neurophysin for the hormone, Burford, Ginsburg, and Thomas (82,83) reported an apparent effect of Ca^{2+} on polymerizing equilibria in porcine neurophysin.

Evidence against the hormone-dissociation hypothesis has come from several different lines of investigation. First, studies of purified bovine neurophysins have shown that Ca^{2+}, over a wide range of concentrations (up to 0.1 M) has no significant effect on hormone affinity (38, 39,84); in addition, no Ca^{2+} binding to bovine neurophysin was demonstrable by potentiometric titration (38). Although it can be argued that porcine and bovine neurophysins may differ in their response to Ca^{2+}, it should be noted that in those studies where effects of Ca^{2+} on porcine neurophysin–hormone interaction were claimed no interaction of Ca^{2+} with either neurophysin or the hormones alone could be detected (64,80); this is not a thermodynamically possible situation. More significantly, any hormone-release mechanism can be assumed to be common to both the pig and the cow. Second, positive evidence has been obtained that exocytosis is the mechanism of release. The distribution of neurosecretory granules within the neuron and the concentration of neurosecretory granules at the neuronal membrane, as observed by electron microscopy (85), provides strong associative evidence for exocytosis. Moreover, isolated pituitary glands *in vitro*, in Ca^{2+}-containing solutions, stimulated electrically or by high K^+ concentration, release hormone and neurophysin into the medium in parallel (36,68,86,87). The ratio in which neurophysin and hormone are released in this system has not been unequivocally established, but the release of neurophysin and hormone is *not* accompanied by the release of enzymes known to be in the cytoplasm external to the neurosecretory granules (88). Such data are best explained by assuming an exocytotic mechanism by which the granules empty their total contents external to the cell membrane. The precise role of Ca^{2+} is uncertain, but Ca^{2+} is associated with many exocytotic phenomena (89) and may act to stimulate an actinomyosin-like ATPase (90), contraction possibly being involved in the exocytotic process (91,92).

Evidence that exocytosis is the mechanism of hormone release has been obtained *in vivo* as well as *in vitro*. A number of studies (68,93–99) have demonstrated by radioimmunoassay that neurophysin is both present in the blood normally and is released into the blood under conditions known to lead to vasopressin release. However, *in vivo* studies (68,93) generally indicate a lesser release of neurophysin than of vasopressin; it has been variously suggested that this may be due to slower diffusion of neurophysin than of vasopressin across the capillary bed (36,37), a preferential release of vasopressin from granules with high vasopressin/neurophysin ratios (73), or a superimposition (100) of an exocytotic mechanism on a principally hormone-dissociation mechanism (81). Interestingly, most neurophysin assays (93–95, 97) show no significant rise in blood neurophysin after milking or suckling, a known inducer of oxytocin release. However, a rise in serum neurophysin levels has been reported at parturition (94,95,97) and during pregnancy (101), which are also oxytocin-related events. Robinson, Zimmerman, and Frantz (97) prepared antibodies that were specific for bovine NP–I and NP–II and gave little cross-reactivity with the nonhomologous neurophysin. Use of these antibodies to assay for release of bovine NP–I and NP–II in the cow indicated that bovine NP–II was released only by stimuli leading to vasopressin release, apparently confirming its association with vasopressin in bovine neurons. The serum level of bovine NP–I was elevated at parturition, but it was also elevated under some conditions leading to the release of vasopressin. Some qualifications about this antibody system are discussed in Section V.D.

D. FATE AND FUNCTION OF RELEASED NEUROPHYSIN

Relatively little is known about the metabolism of neurophysin released into the bloodstream or any biological role it might play. From the magnitude of the binding constants of neurophysin for oxytocin and vasopressin at physiological pH, Breslow and Walter (102) concluded that release of the bovine neurophysin–hormone complex into the blood would lead to complete dissociation of the complex since the circulating concentrations of the hormone are orders of magnitude below the dissociation constant of the complex. A similar conclusion was in fact first reached by Ginsburg and Ireland (103) and indicates that neurophysin does not compete with receptor for hormone. Ginsburg and Jayasena (104) report finding proteins that react with antibodies to neurophysin in serum and in target tissues for the hormones (kidney, uterus, and mammary gland), but not in liver, spleen, brain, or skeletal muscle. Fawcett, Powell, and Sachs (68) find a significant concentration of labeled neurophysin present as protein in the kidney. Robinson et al. (98) have found neurophysin-like protein in the anterior pituitary, muscle, and kidney, but not in other organs.

Lipolytic activity has been attributed to human neurophysin (28, 105) and to porcine NP–I (40). Porcine NP–I, at concentrations of 0.1 μg/ml or greater, produced lipolysis in slices of adipose tissue from guinea pig and chicken, but not in the rat, hamster, cat, or opossum;

lesser concentrations of an apparently related smaller peptide produced similar effects. In the intact rabbit and monkey total porcine-neurophysin doses of 0.01 and 2.5 mg, respectively, produced increases in free plasma fatty acid concentration associated with increases in blood glucose and decreases in plasma amino acid concentration (40). Rudman et al. suggest that lipolysis is a secondary effect of neurophysin; that is, neurophysin might have a hormonal activity independent of its effect on lipolysis (40). It should be noted here that the concentration of porcine NP–I necessary to produce an effect in slices of adipose tissue may be low enough to be in the physiological range, but studies elsewhere indicate that resting plasma neurophysin levels, measured by immunoassay, vary from 0.2 ng/ml in the cow (97) to 3–11 ng/ml in other species (94–96,99) and, except in severe hemorrhage (94), are generally not increased more than tenfold under conditions leading to hormone release. Thus some question can be raised as to the significance of neurophysin lipolytic activity. This is reinforced by the finding (93) that no increase in plasma neurophysin levels occurs during starvation. What appear to be needed here are studies of the physiological effects of neurophysins in their homologous species since the above data indicate a species difference in response to neurophysin. A similar criticism can be made of the studies of Robinson et al. (106), who observed a natriuretic effect of bovine NP–I, but not of bovine NP–II, in the dog; here again, neurophysin was not tested in the species from which it was isolated.

For other postulated biologic effects of neurophysin, the reader is referred to Cheng and Friesen (93).

E. BIOSYNTHESIS OF THE NEUROPHYSINS; POSSIBLE RELATIONSHIP TO HORMONE BIOSYNTHESIS

On the basis of data such as those shown in Figure 2 and elsewhere (36,37), the neurophysins are known to be synthesized in the cell bodies of the hypothalamus. Sachs et al. (65) have been able to grow neurons of the supraoptic nuclei of guinea-pig hypothalamus in organ culture for 15 days and demonstrate *de novo* biosynthesis of both vasopressin and neurophysin. Evidence is accumulating that, like the biosynthesis of vasopressin (35–37), the biosynthesis of neurophysin involves a precursor. Specifically, studies have been performed in the guinea pig (107) and in the dog (108) in which *in vivo* pulses of labeled amino acids were given into the third ventricle followed by isolation of the hypothalamic median eminence. Half the recovered tissue was immediately assayed for labeled neurophysin and vasopressin; the rest, prior to assay, was subjected to *in vitro* "chase" of label with nonlabeled amino acids for several hours in the presence of inhibitors of *de novo* protein synthesis. Both neurophysin and hormone isolated without benefit of *in vitro* "chase" exhibited a markedly lower number of total counts than that isolated after incubation with nonlabeled amino acids. Such results are most readily explained by assuming that much of the initial label was incorporated into precursors of vasopressin and neurophysin that were subsequently cleaved by proteolysis during the "chase," liberating vasopressin and neurophysin.

The above studies, coupled with other results such as the demonstration that Brattleboro rats homozygous for diabetes insipidus also do not produce one of the principal neurophysins (25,41), support the concept of a correlation between vasopressin and neurophysin biosynthesis. In addition, they suggest the possibility of a common precursor for each neurophysin and the hormone with which it is normally found in neurosecretory granules (37). Thus the noncovalent interaction between hormone and neurophysin may be analogous to the same interactions existent in a common precursor, in much the same manner as the interactions between RNase-S-peptide and RNase-S-protein are the same as those found in the intact RNase A molecule (109).

IV. Isolation and Purification of the Neurophysins

The neurophysins are most generally, and probably most conveniently, prepared on a large scale by isolation of the crude neurophysin–hormone complex from acetone-desiccated posterior pituitary glands (19–21,38). Isolation procedures generally take advantage of the fact that neurophysins are remarkably stable proteins, apparently undergoing irreversible changes only under conditions favoring disulfide interchange, such as the presence of mercaptans or very alkaline pH (84). [However, neurophysins will also slowly denature if kept in solution at neutral pH over a long period, presumably due to disulfide interchange under these conditions as well (82).] The resistance of the neurophysins to acid is extremely useful in their preparation. Dean et al. (44) showed that extraction of posterior pituitary acetone powder under conditions of relatively strong acidity (0.1 N HCl) irreversibly denatured enzymes responsible for neurophysin proteolysis and allowed isolation of the native proteins; such proteolysis occurred (see Section II) when isolation was performed under the more mild conditions (pH $\sim$ 4) used in prior investigations (3,14,42,43).

After extraction of the powder with 0.1 N HCl, the crude neurophysin–hormone complexes are separated from insoluble protein and then generally precipitated by salt at pH 4 (19–21,38); crude neurophysin is subsequently separated from both the hormones and high-molecular-weight contaminants by gel filtration in 0.1 or 1 M HCOOH on Sephadex G–75 (19–21,38), a procedure that takes advantage of the lower affinity of neurophysin for the hormones at low pH. The high-molecular-weight contaminant fractions that coprecipitate with the neurophysin–hormone complexes have not been extensively charac-terized except to find that they do not contain protein cross-reactive with antibody against neurophysin (67) and are a mixture of proteins of very different molecular weights, most of which are insoluble at neutral pH after separation from the complex (C. J. Menendez-Botet and E. Breslow, unpublished observations). The mixture of neurophysins resulting from the gel-filtration step is then resolved into its indi-vidual components either by ion-exchange chromatography (19–21, 38), isoelectric focusing (39,110), preparative gel electrophoresis (26), or a combination of these techniques.

Alternative methods of purifying the neurophysins have been used which do not proceed via initial isolation of the neurophysin–hormone complex (22,40). Although these may be advantageous in those spe-cific instances where they have already been applied, they are not directly applicable to the isolation of neurophysins from new species; this is in contrast to methods that proceed via isolation of the neurophysin–hormone complex, which appears always to be insoluble at pH 4 at high salt concentration. In this respect it is of interest that neurophysin complexes with vasopressin require a higher concentration of salt to "salt out" at pH 4 than do the corresponding complexes with oxytocin (103,111).

The general procedure of isolating neurophysins from acetone-desic-cated posterior pituitary may have one disadvantage. Mylroie and Koenig (112) have recently presented evidence suggesting that neu-rophysin exists in neurosecretory granules as lipoprotein. In particular, protein isolated from neurosecretory granules by sonication in Triton

X–100, of amino acid composition and electrophoretic mobility like that of neurophysin, was found to have an appreciable lipid content. The lipid was lost when neurophysin was prepared from fresh glands in the absence of Triton X–100 or from acetone-desiccated glands. Comparisons of the chemical properties of lipid-free and lipid-associated neurophysins are necessary to assess the significance of this interesting observation.

The reader is referred to a recent review (113) for more detailed methods of neurophysin purification.

V. Composition and Covalent Structure of the Neurophysins

A. HOMOLOGY AMONG THE NEUROPHYSINS

The low-molecular-weight and distinctive amino acid composition of the neurophysins are useful in their identification (Table I). Typical monomeric molecular weights are 9000–10,000 (22,38,71,72); the only possible exception so far is porcine NP–II, to which a monomeric weight of either 14,000 (21) or 10,000 (27) has been attributed. Neurophysins are extremely rich in disulfides, most containing six to seven disulfide groups per monomer. Of the aromatic amino acids they contain a single tyrosine per monomer (the only exception again may be porcine NP–II, which may contain two tyrosines, as shown in Table I), no tryptophan, and two to four phenylalanine residues. The high cystine and low aromatic content of the neurophysins generates a 260/280 absorbance ratio that is greater than unity (38,43). Most neurophysins contain a single methionine and no histidine, but there are exceptions, such as bovine NP–I, which contains a single histidine and no methionine (20,38), and a human neurophysin, which contains one histidine and one methionine (27).

Figure 3 shows the primary structure and disulfide pairing of bovine NP–II as determined by Walter et al. (72,114). This is the only neurophysin for which both the sequence and disulfide pairing have been determined. The extent to which disulfide pairing involves sequentially adjacent half-cystines is of particular interest. In addition to bovine NP–II, bovine NP–I has been definitively sequenced through the first 50 residues (33) and tentatively sequenced through residue 75 (R. Walter, J. D. Capra, and E. Breslow, unpublished observations). Porcine NP–I has been sequenced by Wuu, Crumm, and Saffran (71),

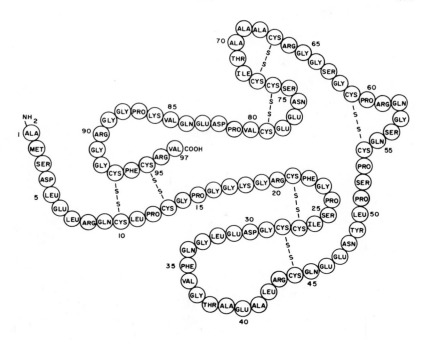

Fig. 3. Covalent structure of bovine NP–II. Numbers indicate positions of amino acid residues starting from the α-amino terminus. More recent studies (D. H. Schlesinger, J. D. Capra, and R. Walter, unpublished data) indicate that position 84 is Glu, not Gln. From Schlesinger, Frangione, and Walter (114).

and the first 17 residues of one of the human neurophysins have been sequenced by Foss, Sletten, and Trygstad (28).

Figure 4 shows and compares the primary structures of all neurophysins sequenced to date. A number of features are immediately apparent. First, as might be expected from the similarity in amino acid composition of the different neurophysins, a high degree of sequence homology is also present. The degree of homology between porcine NP–I and bovine NP–II is 87% (33). The first 75 residues of porcine NP–I and bovine NP–II show 95% homology, and the first 75 residues of porcine NP–I and bovine NP–I also show 95% homology (these numbers are based on the assumption that the apparent Gly-Cys inversion at positions 28–29 of the porcine neurophysin is an artifact). However, the first 75 residues of bovine NP–II and bovine NP–I show only 90%

5   10   15   20   25

Bovine NP-II   H₂N-ALA-MET-SER-ASP-LEU-GLU-ARG-GLN-CYS-LEU-PRO-CYS-GLY-PRO-GLY-GLY-LYS-GLY-ARG-CYS-PHE-GLY-PRO-SER-ILE-CYS-CYS

Bovine NP-I   VAL-LEU_____   ASP-VAL_____

Porcine NP-I   GLY

30   35   40   45   50   55

Human   ALA-PRO_____   ASP-VAL_____   LYS_____

Bovine NP-II   GLY-ASP-GLU-LEU-GLY-GLN-PHE-VAL-GLY-THR-ALA-GLU-ALA-LEU-ARG-CYS-GLN-GLU-GLU-ASN-TYR-LEU-PRO-SER-PRO-CYS-GLN-SER-GLY

Bovine NP-I   CYS_____

Porcine NP-I   CYS_____

60   65   70   75   80   85

Bovine NP-II   GLN-ARG-PRO-CYS-GLY-SER-GLY-GLY-ARG-CYS-ALA-ALA-ALA-THR-ILE-CYS-CYS-SER-ASN-GLU-CYS-VAL-PRO-ASP-GLU-GLU-VAL-LYS

Bovine NP-I   LYS_____   GLY_____   ASN

Porcine NP-I   LYS_____   GLY_____   ASN-ASP_____   SER_____   THR-GLU-[   ]

90   95

Bovine NP-II   PRO-GLY-GLY-ARG-[   ]-GLY-GLY-CYS-PHE-CYS-ARG-VAL-COOH

Porcine NP-I   GLU-CYS_____   GLU_____   ALA-SER_____   LEU

Fig. 4. Complete primary structure of bovine NP-II (72,114) and porcine NP-I (71) together with the partial sequence of bovine NP-I (ref. 33 and R. Walter, J. D. Capra, and E. Breslow, unpublished observation) and a human neurophysin (28). The sequence of bovine NP-II is specifically shown. Residues on the other neurophysins that differ from bovine NP-II are also explicitly given, but identical residues are indicated by the solid lines. Brackets indicate apparent sequence deletions determined by aligning the different sequences to obtain maximum homology (33).

homology. Of the residues that are particularly constant in amount among these neurophysins (Table I), the single tyrosine, the two isoleucines, and at least two of the three phenylalanine residues are constant in position.* Most surprisingly, however, of the 14 half-cystines found in each neurophysin, four show important differences in positions Porcine NP–I and bovine NP–I each have a half-cystine at residue 34 that is not present in bovine NP–II, and comparison of bovine NP–II and porcine NP–I indicates that differences in half-cystine position are also found at the carboxyl end of the chains. Therefore overall identities in amino acid composition appear in this case to mask differences in structure. A corollary to this is that the absence of methionine in bovine NP–I and the presence of histidine is not a reflection of a methionine → histidine substitution. Although the histidine in bovine NP–I has not yet been definitively located, it clearly differs in position from the single methionine of porcine NP–I and bovine NP–II (Fig. 4).

The significant differences in disulfide position between porcine NP–I and bovine NP–I on one hand and bovine NP–II on the other are particularly surprising since, as will be seen, bovine NP–I and NP–II are remarkably similar in their physical, chemical, and binding properties. Moreover, the half-cystine in position 34 of both porcine NP–I and bovine NP–I will be seen to lie very close in sequence to carboxyl groups attributed to the binding site of bovine NP–II. Differences in half-cystine position among the different neurophysins can be shown to necessitate differences in the pairing of at least two disulfide groups. Presumably the different disulfide pairings are compatible with similar conformations at the binding sites. R. Walter and D. Schlesinger (personal communication) believe that the disulfide bonds in bovine NP–I and porcine NP–I are identical. This hypothesis presupposes that all cysteine residues occupy the same positions in both proteins, as has already been found for the half-cystines within the first 75 residues. With this as background, two alternatives of disulfide bond pairing for bovine NP–I and porcine NP–I are considered by

---

*It is of interest that human neurophysin differs from bovine and porcine neurophysins in having only one isoleucine residue. As seen in Figure 5, the two isoleucine residues of bovine NP–II and porcine NP–I are found in homologous positions within the internally duplicated regions of the protein. The presence of a single isoleucine in both human neurophysins suggests that the same change in one of the internally duplicated segments has occurred in both human proteins.

Walter and Schlesinger. The first is Cys_{10}–Cys_{85}, Cys_{13}–Cys_{34}, Cys_{21}–Cys_{27}, Cys_{28}–Cys_{44}, Cys_{54}–Cys_{61}, Cys_{67}–Cys_{73}, and Cys_{74}–Cys_{79}. The alternative would involve disulfide bridges between Cys_{10} and Cys_{34}, and between Cys_{13} and Cys_{85}.

Another feature that is apparent from Figure 4 is that the first 75 residues of the protein are relatively invariant and that most of the differences among the different neurophysins lie near the carboxyl terminus; this suggests that the carboxyl terminus of the protein is not important to binding hormone. Of the differences present among the first 75 residues of the protein, it is particularly interesting from an evolutionary standpoint that porcine NP–I and bovine NP–II are identical at residues 2, 3, 6, and 7 and differ here from bovine NP–I, while porcine NP–I and bovine NP–I differ from bovine NP–II in positions 34, 59, 71, and 75, where they are identical with each other. The Cys → Gln substitution in position 34 of bovine NP–II represents a three-base mutation, and the Gly → Thr change in position 71 is a two-base mutation. On the other hand, differences in positions 2, 3, 6, and 7 between porcine NP–I and bovine NP–I all represent single-base changes. These data, coupled with the previously cited degree of homology among the different neurophysins and the differences in disulfide pairing between porcine and bovine NP–I on one hand and bovine NP–II on the other, suggest a closer relationship between porcine NP–I and bovine NP–I than between porcine NP–I and bovine NP–II. This is particularly surprising in that porcine NP–I and bovine NP–I differ in the hormones with which they are supposed to be associated *in vivo*, whereas porcine NP–I and bovine NP–II are both presumed to be associated with vasopressin (see Section III.B).

Any evolutionary scheme to account for relationships among the neurophysins is probably premature. However, as cited elsewhere (33), it is likely that an ancestor common to the pig and cow carried a gene that coded for a single neurophysin and that duplicated to give two independent neurophysin genes prior to species divergence. It might be additionally suggested that the two neurophysins of the common ancestor resemble bovine NP–I (oxytocin-carrying) and porcine NP–I (vasopressin-carrying), respectively, structural differences between the two at the amino-terminus having arisen through independent mutation. Transmission of both genes to the cow and subsequent mutation of the vasopressin-associated neurophysin at positions including 34, 59, 71, and 75 would give the proteins we recognize as bovine NP–I and bovine NP–II. In the pig, the vasopressin-associated neurophysin has not mutated at these positions although there is insufficient data to judge what changes might have occurred within the other porcine neurophysins. It should be stressed that this scheme is strictly predicated on the assumption that the hormone-carrier assignment for each neurophysin is correct and that the hormone associated with a particular neurophysin does not change from oxytocin to vasopressin during evolution. Moreover, this scheme is limited in two respects. First it does not give a teleologic explanation of the major structural differences between porcine NP–I and bovine NP–II, both of which appear to be vasopressin-carriers. Second, it postulates that these structural differences arose subsequent to divergence of the cow and the pig and therefore occurred over a relatively short time period. Clearly the sequence of other neurophysins and a more definitive assignment of the carrier role of each neurophysin are essential to a better understanding of neurophysin evolution and of any functional significance of structural changes.

B. INTERNAL DUPLICATION IN THE NEUROPHYSINS

A remarkable feature of neurophysin structure is that every neurophysin sequenced to date shows stretches of internal duplication within the same polypeptide chain (33). In particular (Fig. 5) residues 12–31 and residues 60–77 within the same chain show approximately 60% homology; these duplicated regions lie on either side of a non-duplicated, or "unique," region that appears to contain at least one of the "active-site" residues, Tyr_{49} (see Section VII.F). Evidence of additional internal duplication in bovine NP–II can be seen in residues 87–95, which show 66% homology with residues 15–22 (33). Capra et al. (33) have suggested that the principal duplication arose early in evolution from unequal crossover of a gene that coded for a neurophysin-like protein comprising approximately the first 60 residues of the neurophysins we see now. Clearly this internal duplication preceded the gene duplication, referred to above, that gave rise to multiple neurophysins in a single species. It will be of interest to ascertain whether neurophysin-like proteins that contain no internal duplication can be found in more primitive species or whether partially degraded neurophysins, in which one of the duplicated segments has been lost, retain any ability to bind oxytocin or vasopressin.

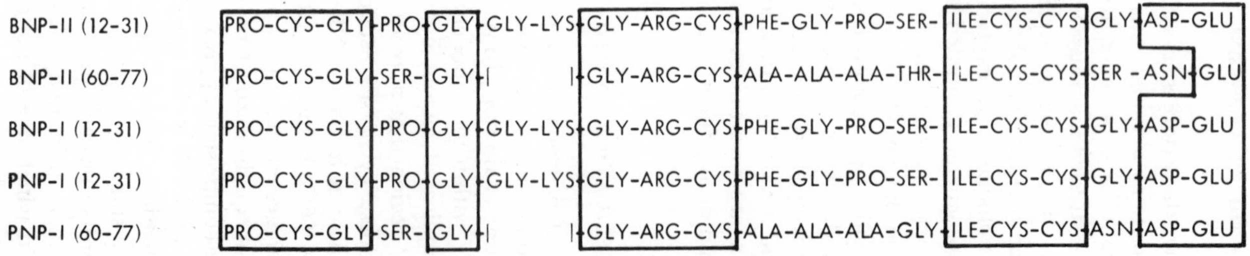

Fig. 5. Evidence of repeating sequence patterns in bovine NP-II and porcine NP-I. Residues 12–31 and 60–77 have been aligned and identities are outlined. Brackets represent apparent deletions. From Capra et al. (33), modified to allow for re-identification of residue 76 in bovine NP-II (114).

294

C. POSSIBLE HOMOLOGIES AMONG THE NEUROPHYSINS, OXYTOCIN, AND VASOPRESSIN

Capra et al. (33) have also suggested that there is a homology, albeit weak, between the neurophysins and the posterior pituitary hormones. For example, residues 6–14 and 54–62 of neurophysin show 30% homology with arginine vasopressin or oxytocin. Although this degree of homology seems to be of little statistical import, it takes on added significance with the almost inexplicable observation that residues 11–13 of bovine NP–II, bovine NP–I, and porcine NP–I (Leu-Pro-Cys) are the exact inverse of residues 6–8 of oxytocin, residues 59–61 of bovine NP–II (Arg-Pro-Cys) are the exact inverse of residues 6–8 of arginine vasopressin, and residues 59–61 of porcine NP–I and bovine NP–I (Lys-Pro-Cys) are the inverse of residues 6–8 of lysine vasopressin. If these inversions are included as former homologies, changed during evolution, the degree of homology between residues 6–14 or 54–62 of neurophysin and the hormones increases to 55%. Although we know of no simple genetic mechanism that allows an inversion of amino acids in the sequence, it seems possible from these data that the hormones and the neurophysins might share a common evolutionary precursor in addition to a common biosynthetic precursor.

D. RELATIONSHIP BETWEEN NEUROPHYSIN STRUCTURE AND ANTIGENICITY

A number of immunoassay procedures have been developed for the neurophysins (97–99,115), but the close structural homology among the neurophysins has made difficult the preparation of antibodies completely specific for any one neurophysin. In general the specificity of antibodies prepared has varied with the investigation. Livett, Uttenthal, and Hope (70) prepared antibodies against porcine NP–II that showed little cross-reactivity with porcine NP–I, suggesting that the lack of cross-reactivity was due to the larger size of porcine NP–II. However, Cheng and Friesen (115) found that rabbit antibodies prepared against porcine peptides II and III (porcine NP–I and NP–II) cross-reacted with the other antigen as well as with a serum component of higher molecular weight than the porcine neurophysins. Robinson et al. (97, 98) have claimed the preparation of antibodies to bovine NP–I that show little cross-reactivity with bovine NP–II and of antibodies to bovine NP–II that show little cross-reactivity with bovine NP–I; a note of caution with respect to these studies should be added, however, since the method used to prepare bovine NP–I and NP–II was that shown elsewhere (20) not to separate bovine NP–I from bovine NP–C. Antibodies made to unfractionated bovine neurophysin show cross-reactivity with posterior pituitary extracts from all mammals, but not with that from birds or fish, including cod (99).

VI. Conformation of the Neurophysins

A. QUATERNARY STRUCTURE

Abundant evidence exists that neurophysins are a self-associating system, a fact that explains the discrepancy in molecular weight reported by different investigators. In unfractionated porcine neurophysins frontal and zonal analyses of elution patterns from Sephadex G–75 were interpreted in terms of a rapidly polymerizing equilibrium that was shifted by lysine vasopressin in favor of the higher-molecular-weight components (82). Also seen in this study was a slow and possibly irreversible polymerization of neurophysin allowed to remain in solution for 24 hr; this may represent disulfide exchange. Burford et al. (83) observed polymerization of porcine neurophysin in the ultracentrifuge; however, reported effects of Ca^{2+} on this equilibrium are doubtful in view of the low Ca^{2+}/neurophysin ratio used. Uttenthal and Hope (21) studied the sedimentation velocity and average molecular weights of porcine NP–I and NP–II as a function of protein concentration. At pH 4.6, increasing the protein concentration from 1.5 to 4 mg/ml increased the average molecular weight of porcine NP–I from 17,000 to 20,640 and that of porcine NP–II from 24,000 to 30,370. Since the minimal molecular weights of these proteins (Table I) are 9400 and 14,000 respectively, and sequence data (Fig. 4) indicate that the monomer molecular weight of porcine NP–I is 9400, the sedimentation data indicate aggregation to dimer and possibly to higher-molecular-weight species under these conditions.

Similar evidence for polymerizing equilibria is also found in the bovine neurophysins (38,111). In particular, an increase in $s_{20,w}$ of bovine NP–II is seen on increasing the protein concentration, and a meniscus-depletion sedimentation-equilibrium study (Fig. 6) gives a clear indication of a change in molecular weight with change in protein concentration. Under the conditions used in Figure 6 only mono-

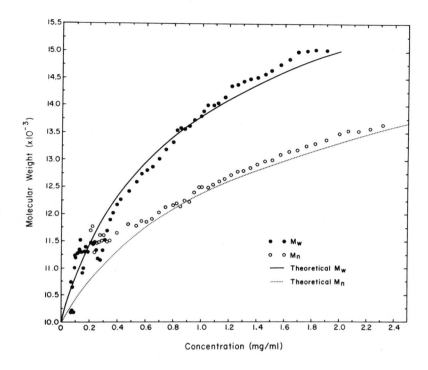

Fig. 6. Results of meniscus-depletion sedimentation-equilibrium run of bovine NP–II at 25°C, pH 8.2. Points represent observed values of the weight-average molecular weight (M_w) and number-average molecular weight (M_n) at each concentration. Theoretical lines represent values calculated for a monomer $\rightleftharpoons$ dimer equilibrium with an association constant $5.01 \times 10^3 \ M^{-1}$. From Breslow et al. (38).

mer and dimer were seen and an apparent equilibrium constant ($5 \times 10^3 \ M^{-1}$) for a monomer $\rightleftharpoons$ dimer equilibrium was calculated. However, despite the fact that concentration-dependent changes in $s_{20,w}$ indicate at least some degree of equilibrium between monomer and dimer, it is possible that irreversibly formed dimer or other aggregates are also present in solution. Few detailed studies of bovine NP–I have been carried out, but weight-average molecular weights of 15,000 and 19,000 have been reported under different conditions (19,38), again suggesting monomer $\rightleftharpoons$ oligomer equilibria.

B. SECONDARY STRUCTURE

Circular dichroism data (38,84,116) have not yet been fully ex-
ploited in terms of neurophysin structure. One of the reasons for this
is that the high disulfide content of the neurophysins (see below) leads
to near-ultraviolet ellipticity bands that dominate the near-ultraviolet
circular dichroism spectrum and can be expected to make major but
incalculable far-ultraviolet contributions as well (117). Thus the far-
ultraviolet spectrum may suffer more than the usual contributions
from nonpeptide chromophores. Nevertheless, several aspects of bovine-
neurophysin structure are clear from the far-ultraviolet ellipticity data.
First, bovine NP–I and NP–II have very similar (but not identical)
far-ultraviolet circular dichroism spectra (Fig. 7) characterized by remarkably weak ellipticities (38). Analysis of the far-ultraviolet data
by the method of Greenfield and Fasman (118) suggests that very

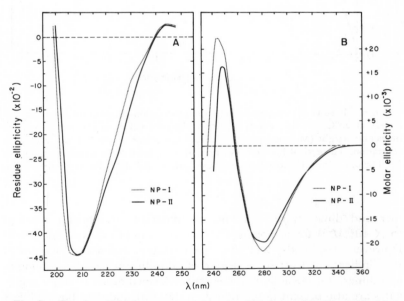

Fig. 7.   Circular dichroism spectra of bovine NP–I and NP–II at pH 6.2 in
0.16 M KCl: (A) far-ultraviolet spectra calculated as the ellipticity (deg cm²/
decimole) per amino acid residue; (B) near-ultraviolet spectra calculated as ellipticity (deg cm²/decimole) per mole of protein. From Breslow et al. (38).

little α-helix is present ($\sim 5\%$) (84,116). [Based on the primary structure of bovine NP–II, Schlesinger, Frangione, and Walter (114) predict a maximum possible α-helix content of 20% and a probable content of 15%.] The Greenfield–Fasman analysis, however, is particularly unsatisfactory for neurophysin; using their parameters, no postulated combination of α-helix, β-structure, and random-coil structure fits the far-ultraviolet circular dichroism data over a significant wavelength range (E. Breslow, unpublished data). However, the data do suggest that a significant amount of β-structure may be present (116). This conclusion is supported by analysis of the circular dichroism data using the parameters of Saxena and Wetlaufer (119); good fit of the observed circular dichroism spectrum of bovine NP–II between 230 and 200 nm is obtained assuming 5% α-helix, 40% β-structure, and 55% random-coil structure, although the fit above 230 nm is less satisfactory (E. Breslow, unpublished data).

C. INTERACTION OF BOVINE NEUROPHYSINS WITH H$^+$ IONS; CONFORMATIONAL IMPLICATIONS

The H$^+$-ion titration curves of bovine NP–II and NP–I have been studied (38). Bovine NP–II titration data, when analyzed using the Linderstrøm-Lang distributed charge titration theory, suggest that all titratable groups are freely available to protons and have intrinsic pK values within the normal range. A more cursory analysis of the titration curve of bovine NP–I also gives no indication of any buried groups and further indicates that the single histidine of bovine NP–I titrates normally. Cohen et al. (120) have presented NMR evidence that also indicates that the histidine of bovine NP–I titrates normally. However, there are some puzzling features to the titration data. First, the intrinsic pK of the side-chain carboxyl groups of bovine NP–I is 0.2 pH unit lower than that of bovine NP–II; this situation may reflect a relatively larger number of aspartic carboxyls in bovine NP–I (which have a lower intrinsic pK than glutamic carboxyls) or it may be an indication of more profound differences between the two proteins. Another feature of the titration data that suggests that not all groups are entirely "normal" is the fact that the pK of the single neurophysin tyrosine, when it is nitrated (116,121) is significantly higher than expected for a nitrotyrosine in a protein (116). These data suggest that it may be under the influence of a nearby negatively charged carboxyl group (see Section VII.F) or perhaps in a partially hydrophobic environment. The question arises as to why the phenolic groups of the nitrated tyrosine ($pK' = 7.45$) should appear to be titrating abnormally when the unmodified tyrosine appears to titrate normally when analyzed by the Linderstrøm-Lang treatment. Either a conformational change occurs in the protein between pH 7 and 10 or else the Linderstrøm-Lang model is inappropriate in this case. In either event the pK of the nitrotyrosine of the nitrated protein will be seen to be of particular significance in interpreting the binding of hormones and peptides (see Section VII.F).

Also hidden in the rather smooth titration curve of bovine NP–II is the fact that conformational changes occur in the pH region near pH 4.5 (116). This is seen most clearly by a change in the circular dichroism spectrum of the nitrotyrosine residue of nitrated bovine NP–II in this pH region, but it can also be seen by careful examination of the circular dichroism spectrum of the native protein as a function of pH. Similar changes with pH are also seen with nitrated bovine NP–I (S. L. Lundt and E. Breslow, unpublished observations). Since the change occurs in a region of low net charge on the protein (the pI of bovine NP–II is 4.95), the results suggest that titration of a specific carboxyl is responsible for the change in the circular dichroism. It is tempting to speculate that the conformational change is a local change near the tyrosine and that the carboxyl responsible is the same group responsible for elevating the pK of the nitrotyrosine. However, far-ultraviolet circular dichroism studies suggest that general changes in secondary structure accompany the change in tyrosine environment, implying that this change may be the result of a general backbone rearrangement.

D. CONFORMATION AND CHEMISTRY OF NEUROPHYSIN DISULFIDES

The near-ultraviolet circular dichroism spectra of the bovine neurophysins (Fig. 7) contain two bands, at 248 and 280 nm, which have been assigned to disulfide bonds (38,84,116). Ionization of the single bovine NP–II tyrosine is without effect on these bands; since there is no tryptophan in neurophysin, the assignment of disulfides is completely secure for the negative 280-nm band and fairly secure for the 248-nm band. The only alternative assignment is to attribute the 248-nm band to peptide-bond chromophores, but preliminary resolution of this band from its neighbors suggests that it is centered near

245 nm and therefore probably not assignable to carbonyl transitions (E. Breslow, unpublished observations). It is interesting to note that bovine NP–I, which differs in several of its disulfides from bovine NP–II (see Section V.A), shows a generally similar 280-nm band to NP–II but has a 248-nm band that is distinctly more intense and blue-shifted than that in NP–II (38). It is not clear whether these differences are a reflection of different disulfide ellipticities in the two proteins or of different contributions from peptide-bond chromophores in the 230–240-nm region (Fig. 7).

Disulfide optical activity is a reflection of the screw sense of the disulfide and the —C—S—S—C— dihedral angle (122). The sign of the longest wavelength disulfide-ellipticity band is indicative of the screw sense, and, for dihedral angles that are not larger than 90 degrees, a negative long-wavelength band signifies a left-handed screw sense (122). For a disulfide restricted to one particular screw sense an ellipticity of approximately 7000 degrees can be expected (123). Thus a possible interpretation of the near-ultraviolet disulfide-ellipticity bands of neurophysin is that there is a net *excess* of three disulfide groups in the left-handed screw sense; that is, four disulfide groups can be envisioned as equally divided between left- and right-handed screws and the other three as fixed in a left-handed screw.

The optical activity of neurophysin disulfides is markedly dependent on the conformation of the protein. The two-banded near-ultraviolet spectrum is changed to one with a single negative band at 270 nm in $5\ M$ guanidine or $8\ M$ urea (124), which indicates that those disulfides, which are maintained in a relatively fixed configuration in the native protein, are allowed different conformations in the denatured protein.

A remarkable feature of bovine-neurophysin disulfides is their great sensitivity to reduction by mercaptans (125). In the presence of only 1 mole of dithiothreitol (DTT) per mole of neurophysin *in the absence of urea* bovine neurophysin undergoes rapid conformational changes leading to almost complete obliteration of the 248-nm circular dichroism band and diminution of the 280-nm band to approximately half its original ellipticity (Fig. 8). Increasing molar aliquots DTT can be shown to give almost stoichiometric reduction of the neurophysin disulfides (C. J. Menendez-Botet and E. Breslow, unpublished observations) coupled with complete loss of the 280-nm band (84). Although circular dichroism changes attending partial or complete reduction of bovine NP–II with DTT suggest that a two-state process is involved,

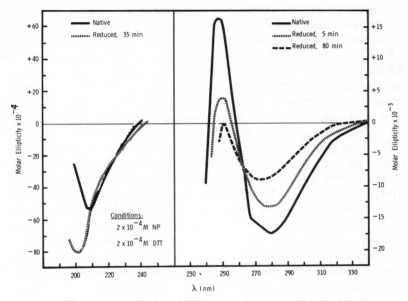

Fig. 8. Effect of dithiothreitol (DTT) on the circular dichroism spectrum of bovine NP–II at pH 8 in 0.16 M KCl. Ellipticities (deg cm²/decimole) are calculated per mole of protein. From C. J. Menendez-Botet and E. Breslow, unpublished observations.

a two-state process is contraindicated by data (C. J. Menendez-Botet and E. Breslow, unpublished observations) indicating that partially reduced bovine NP–II is a complex mixture of partially reduced species; thus reduction does *not* proceed via an all-or-none unzipping of the protein disulfides or selective reduction of particular disulfides. These facts have been demonstrated by analysis of the gel-electrophoresis patterns of partially reduced carboxymethylated bovine NP–II preparations (Fig. 9). Protein reduced with 1 mole of DTT per mole of neurophysin and then treated with iodoacetic acid shows a mixture of carboxymethylated products and an uncarboxymethylated product. Carboxymethylation after reduction with 4 moles of DDT per mole of neurophysin gives only a mixture of carboxymethylated products, all of which show greater reduction than products resulting from the addition of 1 mole of DTT. These results indicate that there are a number of equally reducible disulfides in the native state and/or that

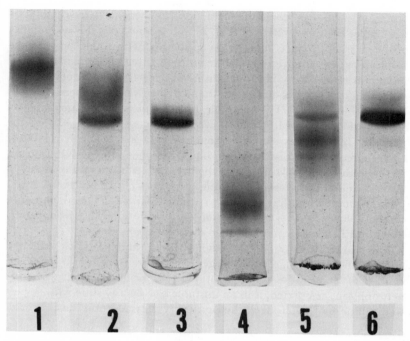

Fig. 9. Pattern of reduction of bovine NP–II as indicated by gel electrophoresis at pH 9.5. Direction of migration (toward the anode) is toward the bottom of the gel. Native bovine NP–II is shown in gels 3 and 6. Bovine NP–II reduced with 1 and 4 moles of DTT per mole and then carboxymethylated with iodoacetic acid is shown in gels 5 and 4, respectively, the greater mobility in gel 4 indicating a larger degree of alkylation. Protein reduced with 1 and 4 moles of DTT per mole and then reacted with iodoacetamide is shown in gels 2 and 1, respectively; the slower mobility indicates a higher degree of unfolding in the reduced carboxamido-methylated protein. From C. J. Menendez-Botet and E. Breslow, unpublished observation.

partial reduction of bovine NP–II is accompanied by extensive disulfide interchange. The existence of disulfide interchange has been confirmed by reduction in the presence of less than 1 mole of DTT per mole of neurophysin; loss of the 248-nm ellipticity band is not stoichiometrically related to the amount of DTT added, but, for example, is more than 70% complete when only 0.5 mole of DTT has been added per mole of NP–II (C. J. Menendez-Botet and E. Breslow, unpublished observations). Marked unfolding can also be shown to

occur on partial or complete reduction, as evidenced by major far-
ultraviolet ellipticity changes (Fig. 8) and a markedly slower mobility
of the partially reduced carboxamido-methylated protein on gel electrophoresis (Fig. 9). Presumably the unfolding that accompanies partial
reduction renders any "buried" disulfide accessible to reduction.

Preliminary studies indicated that bovine NP–II reduced with as
little as 1 mole of DTT per mole of neurophysin was incapable of
renaturation (125). More recent studies indicate that 20–30% of the
partially reduced protein can reoxidize to a species with relatively
native circular dichroism and binding properties. Bovine NP–II completely reduced in 8 M urea and then reoxidized shows no recovery of
native circular dichroism or binding properties (125), but there is preliminary indication that complete reduction with DTT in the absence
of urea allows 20–30% renaturation on reoxidation (C. J. Menendez-
Botet and E. Breslow, unpublished observations).

The marked susceptibility of bovine NP–II (and NP–I) to reduction
and disulfide exchange, and the general failure to obtain a high percentage of renaturation on reoxidation raises the question as to whether
neurophysin, as isolated, is in an unstable free-energy state with respect
to disulfide pairing. Indeed the possibility of a neurophysin precursor

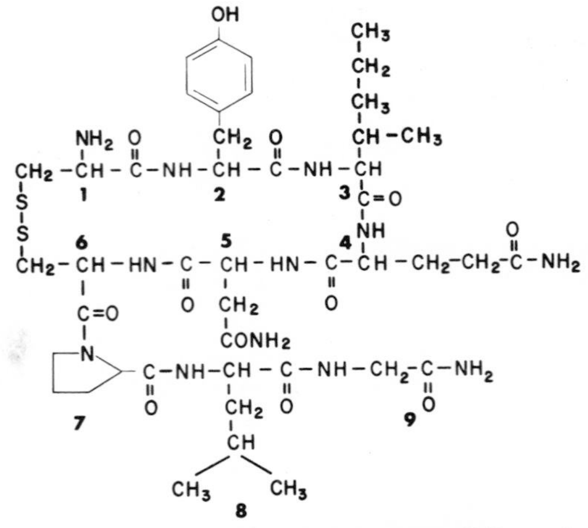

Fig. 10.   Structure of oxytocin. Lysine vasopressin differs only in that it has a
phenylalanine in position 3 and lysine in position 8. Arginine vasopressin also has
a phenylalanine in position 3, but arginine in position 8.

(see Section III.E) suggests that disulfide pairing may be achieved during biosynthesis when, like insulin, neurophysin is a part of a larger molecule. In this respect, since it has been suggested that the hormones may share a common precursor with the neurophysins, we have studied the effect of added hormone or hormone analogs on the reduction and reoxidation of neurophysin. Peptide analogs of the hormones that bind to neurophysin (see Section VII.D), but do not contain disulfides, protect neurophysin against ready reduction (125), but neither these peptides nor the hormones increase the percentage of native protein generated by the reoxidation of reduced neurophysin (C. J. Menendez-Botet and E. Breslow, unpublished observations). These results indicate that kinetic, rather than equilibrium, considerations govern the reduction–reoxidation behavior of neurophysin and therefore do not shed any light on whether neurophysin, as isolated, is in its lowest free energy state.

VII. Interaction of the Neurophysins with Oxytocin, Vasopressin, and Peptide Analogs of the Hormones

A. NUMBER OF BINDING SITES; DO THE TWO HORMONES BIND TO THE SAME SITES?

Oxytocin and vasopressin are very similar in structure, differing only in positions 3 and 8 (Fig. 10). Nonetheless, the number of binding sites attributed to neurophysin for oxytocin relative to vasopressin, as well as the absolute number of sites available for either hormone, has varied with the source of the neurophysin and the particular investigation. The composition of the original van Dyke protein from the ox (3) suggested that approximately equal quantities of oxytocin and arginine vasopressin were precipitated as a complex with neurophysins. Ginsburg and Ireland (24) studied the binding of oxytocin and arginine vasopressin to unfractionated bovine neurophysin by dialysis and concluded that a maximum of 7 moles of oxytocin and 4 moles of arginine vasopressin were bound per 25,000 g of bovine neurophysin at pH 5.8. (Using an average molecular weight of 10,000 per monomer for bovine neurophysin, this reduces to 2.8 moles of oxytocin and 1.6 moles of arginine vasopressin per monomer.) Similar studies on binding to a porcine neurophysin (64,80) suggested that 14 moles of oxytocin and 4 moles of lysine vasopressin were bound per 25,000 g. Although, as will be seen, these figures are undoubtedly too high, they suggest that neurophysin has a greater capacity for binding oxytocin than it does for binding vasopressin. Pickering (16) has interpreted his binding studies with cod neurophysins also to indicate a greater capacity for the binding of oxytocin than of vasopressin. In these studies cod neurophysin was found to be capable of binding approximately 2.2 moles of either oxytocin or 8-arginine oxytocin per 14,000 g, but only 1.1 moles of arginine vasopressin per 14,000 g. All these studies used biological activity as the basis of estimating hormone concentration. Studies using radioactive hormones have allowed binding capacities and binding affinities of the neurophysins to be more accurately determined, although the matter of binding capacity is by no means completely resolved. Breslow and Abrash (42) studied the binding of labeled oxytocin to bovine NP fraction E (see Section II) by thin-film equilibrium dialysis and found that approximately 2 moles of oxytocin were bound per 25,000 g of neurophysin at pH 5.8. Recalculating these data on the basis of a monomeric molecular weight of 10,000, the results give 0.9 binding site for oxytocin per monomer with a binding constant at pH 5.8 of 1.4×10^5 (102). In this study the binding of oxytocin was found to be competitive with that of lysine vasopressin and it was assumed that both hormones were bound to the same site(s). Competition of nonlabeled lysine vasopressin with labeled oxytocin, using the assumption of an equal number of oxytocin- and lysine-vasopressin-binding sites gave a binding constant for lysine vasopressin that was just slightly lower (by approximately 20%) than that found for oxytocin. Breslow and Walter (102) used the same technique to study the binding of labeled oxytocin and labeled lysine vasopressin to purified bovine NP–II near pH 7.4. The data were interpreted to indicate that a single site for either hormone was present per monomer, with binding constants at pH 7.4 of 1.2×10^4 for oxytocin and 8×10^3 for lysine vasopressin. Competition studies confirmed that unlabeled oxytocin displaced labeled lysine vasopressin; binding constants derived on the assumption that competition between oxytocin and lysine vasopressin was complete gave relative binding constants for the two hormones that agreed with those calculated directly, again supporting the theory that the two hormones competed for a single binding site. The affinities of bovine NP–I for the two hormones were found to be similar to those of bovine NP–II.

More recently Camier et al. (39) have studied the binding of labeled oxytocin and vasopressin to bovine NP–I and NP–II over a wider range of hormone concentrations than those used by Breslow and Walter (102). Their studies confirmed the similarity between the two neurophysins in their binding constants and the presence of a single oxytocin-binding site on both proteins, but gave a somewhat lower binding constant near pH 5.8 (4×10^4) than that found by Breslow and Abrash (42), a discrepancy that may be attributable to the different buffer conditions used. However, Camier et al. report the presence of two thermodynamically equivalent sites for lysine vasopressin on both neurophysins, each of which has the same affinity for lysine vasopressin as the single oxytocin site has for oxytocin (Fig. 11). A possible relationship between the internal duplication in the neurophysins and the two lysine vasopressin sites was invoked. Competition studies between lysine vasopressin and oxytocin (39) were interpreted to indicate that inhibition of the binding of one hormone by the other was "noncompetitive" and that binding of the two hormones may occur to two different sites that only partially overlap. However, these competition studies can be shown to have been incorrectly analyzed since the concentration of unbound competitor was not held constant over the course of the binding isotherms; the data merit reinterpretation.

In an effort to resolve the discrepancy between the number of lysine-vasopressin-binding sites found in the two sets of dialysis studies, Breslow, Weis, and Menendez-Botet (111) studied changes in proton equilibrium and circular dichroism that accompanied addition of increasing molar ratios of hormones to bovine NP–II. Briefly, circular dichroism studies at pH 6.2 of both native and mononitrated bovine

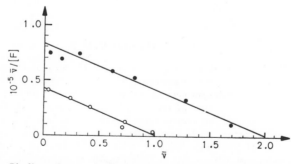

Fig. 11.   Binding of oxytocin (open circles) and lysine vasopressin (solid circles) at 25°C and pH 5.68 to bovine NP–I plotted according to the method of Scatchard. From Camier et al. (39).

NP–II indicated that ellipticity changes accompanying the addition of either hormone were almost complete at 1 mole/mole, although slightly more so for oxytocin than for lysine vasopressin (Fig. 12). Changes in proton equilibria presented the same picture. These results are incompatible with two thermodynamically equivalent sites for lysine vasopressin; they allow only a slightly weaker binding of lysine vasopressin than of oxytocin or the presence of a second site for lysine vasopressin that is considerably weaker (less than one-fifteenth the affinity) than the first. Interestingly, equilibrium dialysis studies (126) have recently confirmed the presence of only a single site for lysine vasopressin on *nitrated* bovine neurophysin.

Estimates of the hormone-binding capacity of the bovine neurophysins have also been derived from the hormonal activity of crystalline

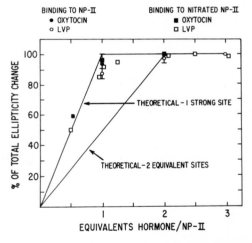

Fig. 12. Circular dichroic titration of bovine NP–II and mononitrated bovine NP–II with oxytocin and lysine vasopressin. Data points represent the ellipticity change seen at each hormone/protein ratio relative to that seen on saturation with hormone. Conditions: 2×10^{-4} M protein, pH 6.2 in 0.16 M KCl. Data shown were obtained at 350 nm for mononitrated bovine NP–II and at 245, 280, and 295 nm for native bovine NP–II from difference circular dichroism spectra. For native bovine NP–II individual points represent the average of the data at the three wavelengths, with the average deviation shown as an error bar. Theoretical lines represent those expected for infinitely strong binding to one site or to two equivalent sites and are independent of any assumptions as to the nature of the sites. From Breslow, Weis, and Menendez-Botet (111).

and amorphous complexes of the neurophysins with the hormones (19,20), but these data are difficult to relate to hormone-binding properties in solution because it cannot be ascertained that the solid complexes contain neither occluded free hormone nor free protein. Clearly additional studies are therefore necessary to define completely the number and relative affinity of vasopressin binding sites per native bovine-neurophysin monomer. However, there is an accumulating body of evidence that the single oxytocin-binding site of each neurophysin monomer is also the principal vasopressin-binding site and that the binding of each hormone to this site, though probably not identical in all aspects, is very similar. First, circular dichroism (111) and absorbance (127) changes accompanying the binding of both hormones to native bovine neurophysin (see Section VII.B) are strikingly similar; the small differences between the hormones that are found can be attributed to minor differences in the orientation of interacting residues without necessarily invoking major differences in the binding sites (111). Second, the binding of both hormones to neurophysin derivatives in which the single protein tyrosine is mononitrated leads to identical changes in the pK and in the circular dichroism spectrum of the nitrotyrosine (111,116). (It is relevant to note here that small quantitative differences in nitrotyrosine circular dichroism changes brought about by oxytocin and lysine vasopressin in one study cited have since been shown to be lost when samples of both hormones are more highly purified.) Third, changes in NP–II sedimentation velocity brought about by the binding of both hormones (see Section VII.B) are identical (111). Fourth, the pH dependence of the binding of both hormones (39,103) approximates a bell-shaped curve with a maximum near pH 5.5 (Fig. 13). Although some differences in the width of the curves for the two hormones may be present (see Section VII.C), the data suggest that ionizable groups with the same pK_a are involved in the binding of both hormones. Finally, peptides containing the first two to three residues of the hormones can be shown to lead to almost the same changes in circular dichroism, absorbance, proton equilibria, and sedimentation velocity on binding to either native or mononitrated bovine neurophysin as do the hormones, taking into account differences in binding constants and α-NH$_2$ pK_a (38,84,111,127). This effect is independent of whether residue 3 is phenylalanine (as in vasopressin), isoleucine (as in oxytocin), or not present; moreover, there appears to be general agreement that neurophysin has only a single binding site

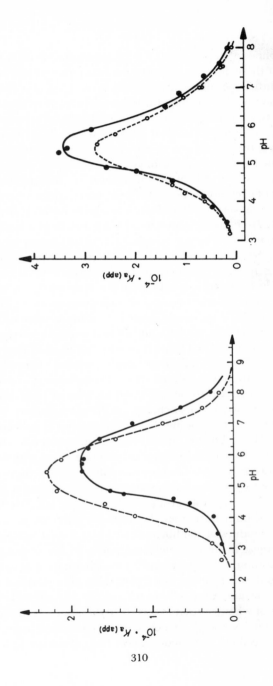

Fig. 13. The pH dependence of the binding to bovine neurophysins of oxytocin (left) and lysine vasopressin (right). Solid circles, bovine NP–I; open circles, bovine NP–II. From Camier et al. (39).

310

for such peptides (38,111,126). Such data, some of which will be presented in greater detail below, suggest that the binding of both hormones is due principally to the interaction of hormone residues 1–3 with a particular set of residues on the protein and that the observed differences between the binding of both hormones are due to slight differences in the orientation of residues 1–3 with respect to the protein arising from differences in hormone conformation or from secondary binding interactions brought about by differences between the hormones in position 8.

B. EFFECTS OF BINDING ON CONFORMATION

The binding of oxytocin, lysine vasopressin, or tripeptide analogs of the first three residues of the hormones leads to an increase in the sedimentation velocity of bovine NP–II that is significantly greater than can be accounted for by the weight of the bound peptide. Under conditions of saturation with peptide, values of $s_{20,w}$ increase approximately 25%, and the observed $s_{20,w}$ value increases with increasing protein concentration, as does the $s_{20,w}$ value of the uncomplexed protein (111). These results indicate that either the conformation of neurophysin is altered by binding and/or that the purported monomer $\rightleftharpoons$ dimer equilibrium is shifted in favor of dimer by binding. However, the dependence on protein concentration of the $s_{20,w}$ value of the saturated protein indicates that binding is not the exclusive domain of either monomer or dimer. Gel-filtration studies of porcine neurophysin have also suggested that lysine vasopressin increases the monomer $\rightleftharpoons$ oligomer equilibrium in favor of more highly polymerized species (82).

Large changes in circular dichroism accompany hormone and peptide binding to native neurophysin (84,116), both in the near- and the far-ultraviolet (Fig. 14). In the near-ultraviolet, the principal changes seen are (a) an increased negative ellipticity above 295 nm, which has been assigned partly to the disulfides of the protein (84) and partly to the disulfides of the hormone (111); (b) an increased positive ellipticity centered near 280 nm, which has been assigned chiefly to the tyrosine of the hormone (116) with a small contribution from the tyrosine of the protein (116); (c) an increased positive ellipticity near 240 nm, which has been assigned principally to disulfides (116). Changes in the far-ultraviolet spectrum have not been definitively assigned, but it has been suggested (116) that side-chain, rather than peptide-bond, chromophores are the principal contributors since the

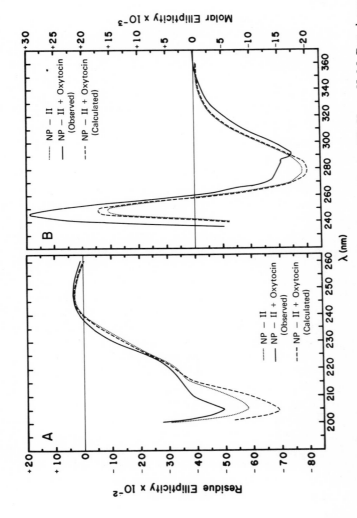

Fig. 14. Circular dichroism changes attending the binding of oxytocin to bovine NP-II at pH 6.2. Dotted curves, bovine NP-II alone; solid curves, observed spectra of bovine NP-II in the presence of 1.4 equivalents of oxytocin; dashed curves, calculated spectrum for bovine NP-II and oxytocin in the absence of interaction. From Breslow and Weis (116). More recent spectra, obtained with purer oxytocin, show a greater degree of fine structure in the hormone-complex in the 280-nm region (E. Breslow, unpublished observation).

peptide-bond $n \rightarrow \pi^*$ transition-wavelength region (see 225-nm region in Fig. 14) is relatively unaffected by binding. These data, however, by no means preclude conformational transitions involving changes in secondary structure; changes in secondary structure may occur which give internal cancellation in ellipticities at 225 nm or involve no *net* change in the percentage of any one kind of backbone structure.

Changes in circular dichroism spectra generated by the binding of lysine vasopressin and oxytocin to native neurophysin are very similar, but nonidentical. The principal differences between changes generated by the binding of lysine vasopressin relative to those observed on oxytocin binding are a 30% decrease in the magnitude of changes near 280 nm (accompanied by a lesser degree of fine structure in the complex at 287 nm) and a 30–40% increase in the magnitude of changes near 240 nm (111).

The environment of the single neurophysin tyrosine is clearly changed by binding. This has been demonstrated by studies of the effect of binding on a neurophysin derivative in which the tyrosine is mononitrated (116,121) and by NMR studies, and is discussed in detail in Section VII.F.

C. EVIDENCE SUPPORTING THE FORMATION OF AN AMINO-CARBOXYLATE ELECTROSTATIC BOND ON BINDING

Stouffer, Hope, and du Vigneaud (128) were the first to note that substitution of the α-amino group of oxytocin by hydrogen (as in deaminooxytocin) led to loss of binding of the hormone to neurophysin under their conditions, although deaminooxytocin is more potent than oxytocin in hormonal activity. A role of the α-amino group of the hormones in binding was also suggested by Ginsburg and Ireland (103), who noted that the pH dependence of binding was consistent with binding of the protonated α-amino group of the hormones to an unprotonated carboxyl of the protein. Marked changes in H^+-ion equilibria also accompany binding (as would be expected for a pH-dependent reaction); protons are liberated on binding of oxytocin to bovine NP–II below pH 5.2 and protons are consumed on binding above this pH (38). Although the precise interpretation of some of the changes in proton equilibria in binding is confused by the insolubility of the oxytocin–NP–II complex in the pH region 4–6, the data when considered together with the fact that the pK_a of the hormone α-amino group is 6.3 (129) are compatible with the sharing of a proton between the hormone α-amino group and a protein carboxyl group.

The protein carboxyl involved is almost certainly a side-chain carboxyl. This follows from the extent of proton displacement accompanying the binding of oxytocin or peptides to bovine NP–II below pH 5 (38), but is particularly evident from changes accompanying the binding of lysine vasopressin since the lysine vasopressin–NP–II complex is soluble under titration conditions (111). In particular, proton-displacement data suggest that an intrinsic pK value of 4.3 can be assigned to the binding carboxyl (E. Breslow, unpublished observations). Similar conclusions are apparent from the midpoint of the acid limb of the pH-dependence curve of lysine vasopressin binding (Fig. 13) and apparent pK values of 4.05 and 4.65 have been assigned to the ligand carboxyls for oxytocin of bovine NP–II and NP–I, respectively (39). Such pK_a values are clearly more compatible with carboxylate side chains than with the free terminal α-carboxyl group of the peptide chain (p$K = 3$–3.5). Moreover, it has been pointed out (116) that the high degree of variability among the neurophysins in length and in sequence near the carboxyl terminus (33) (see Fig. 4) makes it unlikely that the binding carboxyl is the α-carboxyl. Recently, in fact, Walter and Hoffman (130) have reported that they have been able to selectively couple the α-amino group of arginine vasopressin with a specific carboxyl of bovine NP–II in the presence of a water-soluble carbodiimide. This carboxyl, which may well represent the carboxyl that noncovalently binds to the α-amino group in the absence of a coupling agent, has been identified as either Asp_{30} or Glu_{31}. Interestingly, both these residues lie in one of the internally duplicated regions of the peptide chain (see Fig. 5).

The identification of the participating carboxyl of neurophysin as a side-chain carboxyl rules out the simplest potential biosynthetic relationship between neurophysin and the hormones. Such a relationship would envision a precursor in which the α-carboxyl group of neurophysin and the α-amino group of the hormones were linked in a peptide bond that was subsequently hydrolyzed, with noncovalent interactions between the α-amino and α-carboxyl groups in the complex replacing the covalent bond of the precursor. However, the data do not eliminate the possibility of a common precursor for neurophysin and the hormones, and it might be noted that in α-chymotrypsin (131) there is an ion pair between the α-amino group of the terminal Ile_{16} and the

β-COO$^-$ of Asp$_{194}$, the α-amino group of Ile$_{16}$ apparently being derived from a covalent peptide bond between this α-amino group and the α-carboxyl of Arg$_{15}$ in the precursor, chymotrypsinogen.

A question that arises when considering the electrostatic interaction between the α-NH$_3^+$ of the hormones and COO$^-$ of the protein is whether it is mandatory that an ion pair between the NH$_3^+$ and unprotonated COO$^-$ be present in order for binding to occur; this question is particularly relevant in trying to give precise interpretation to pH–binding profiles, for example. Hope and Walti (132) showed that, under the conditions of their study, no binding occurred between neurophysin and a derivative of oxytocin in which the α-amino group was replaced by a hydroxyl group. Although this indicates that binding between the carboxyl of neurophysin and the α-amino group of the hormone is primarily electrostatic (a fact also shown by the pH dependence of the reaction), insufficient studies were performed to determine whether any binding at all could occur to the hydroxyl analog. Recent NMR studies (133,134) and circular dichroism studies (E. Breslow, unpublished observations) in fact indicate that the binding of oxytocin and certain peptides to bovine NP–II may occur at very low pH (pH 1–2.5) with slightly different *qualitative* features attributable to the complex than those seen at neutral pH. A possible interpretation of these data is that some binding may occur to give a complex in which both the α-amino and carboxyl groups are protonated. It is interesting to note in this respect that Glasel et al. (135) have recently demonstrated the binding of deaminooxytocin to bovine NP–II at pH 2.5 (0.45 M KCl) by deuteron NMR spectroscopy, although no binding constant was determined. Such results suggest that, in the presence of sufficient secondary binding interactions involving residues other than the α-amino and carboxyl, binding can occur in the absence of an amino–carboxyl ion pair, at least if the carboxyl is protonated.

Camier et al. (39) have presented data indicating that the pH dependence of oxytocin and lysine vasopressin binding to bovine NP–II is not identical below pH 5 and that the pH dependence of oxytocin binding below pH 5 differs somewhat for the two bovine neurophysins (Fig. 13). Although these results may reflect a difference in the NP–II carboxyl to which oxytocin and lysine vasopressin bind or a difference in the pK of the carboxyl that binds to oxytocin in bovine NP–I relative to NP–II, an alternative interpretation lies in potential differences among the various complexes in their ability to exist in the presence of a proton on the neurophysin carboxyl that participates in the ion pair. (Another potential explanation of differences in the effects of pH on binding lies in the known solubility differences among the various complexes in the pH region 3–6. For example, the lesser solubility of the oxytocin-NP-II complex than that of the complex with lysine vasopressin would give stronger binding of oxytocin than of lysine vasopressin under conditions of pH and protein concentration where a precipitate of the oxytocin complex was present).

D. IDENTIFICATION OF OTHER HORMONE RESIDUES INVOLVED IN BINDING

The magnitude of the neurophysin–hormone binding constant clearly indicates that more is involved in the interaction than an amino–carboxylate electrostatic bond; such a bond alone would have little stability in aqueous solution. Ginsburg and Ireland (103) suggested that the amino–carboxylate interaction might be stabilized by secondary interactions, such as disulfide exchange or "lipophilic" (hydrophobic) interactions. There is no evidence to support disulfide interchange between the hormone and the protein; in addition, the ready reversibility of the binding reaction at acid pH (as on chromatography in acid) and the apparently instant completion of the binding reaction as judged by changes in proton equilibria or circular dichroism (E. Breslow, unpublished observations) suggest that no typical disulfide-interchange mechanism can be operative.

On the other hand, there is considerable evidence supporting the presence of hydrophobic interactions between the hormones and neurophysin. Breslow and Abrash (42) studied the binding of a series of oxytocin analogs to bovine fraction BC and E neurophysins. Substitution of the tyrosine in position 2 by D-tyrosine, glycine, or isoleucine reduced binding by a factor of at least 100, whereas substitution by phenylalanine led to almost no loss in binding. Since the 2-isoleucine derivative is hormonally active (and therefore not of vastly different conformation than the native hormone), the results were attributed to the direct participation of Tyr_2 of the hormones in binding. In position 3, substitution of the oxytocin isoleucine by glycine led to an approximately thirtyfold decrease in binding, whereas substitution by phenylalanine (as in lysine vasopressin) had almost no effect on binding. Together, these results were interpreted to indicate that the neurophysin–hormone interaction involved an electrostatic amino–carboxylate bond in the midst of a hydrophobic region provided in part by the side chains in positions 2 and 3. The apparently specific requirement for an aromatic residue in position 2 suggested that this residue might participate via π-π interactions with an aromatic residue on neurophysin.

The role of Tyr_2 in binding has been confirmed in a number of studies. Furth and Hope (121) interpreted ultraviolet difference spectra arising from the interaction of arginine vasopressin with native and modified neurophysin as indicative of insertion of the bound Tyr_2 into a hydrophobic binding region. Similar data have more recently been obtained by Griffin, Alazard, and Cohen (127). Circular dichroism studies of the binding of small peptides to neurophysin indicate that the principal source of the 270- to 280-nm circular dichroism changes accompanying hormone binding is Tyr_2 of the hormone (116). In Figure 15 the effects of binding Met-Tyr-Phe-NH_2 and S-methyl-Cys-Phe-Ile-NH_2 are compared. As evidenced by the figure, the binding of Met-Tyr-Phe-NH_2 qualitatively leads to the same circular dichroism changes in the near-ultraviolet as does that of oxytocin (cf. Fig. 14). Substitution of the tyrosine in position 2 by phenylalanine (which leads to no change in binding) results in almost a total loss of the 270- to 280-nm circular dichroism changes; since phenylalanine does not have any 280-nm electronic transitions, these data indicate that Tyr_2 of the hormones and of the peptides is significantly constrained when bound to neurophysin.

Nuclear-magnetic-resonance studies also support the direct participation of the Tyr_2 ring protons in binding. Balaram et al. (133,134, 136) showed that ring protons both *ortho* and *meta* to the Tyr_2 hydroxyl were broadened on binding; in the case of the binding peptide Ala-Tyr-Phe-NH_2 (Fig. 16) the *ortho* and *meta* ring protons of Tyr_2 were differentially broadened with respect to each other, a fact that in this instance can only be explained by invoking dipolar interactions between the peptide tyrosine ring protons and protons on the protein. The extent to which the *ortho* and *meta* protons of Tyr_2 are differentially broadened with respect to one another varies with the nature of the bound peptide. For example, the Tyr_2 *ortho* and *meta* protons of Met-Tyr-Phe-NH_2 are broadened to the same extent on binding to bovine NP–II (133,134). These results indicate that the nature of the side chain in position 1 influences the relative orientation of position 2.

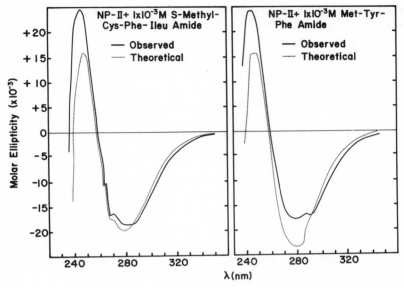

Fig. 15. Near-ultraviolet circular dichroism spectra of bovine NP–II in the presence of 1×10^{-3} M S-methyl-cysteinyl-phenylalanyl-isoleucine amide (left) and methionyl-tyrosyl-phenylalanine amide (right). "Theoretical" curves represent calculated spectra for a solution of bovine NP–II that is 1×10^{-3} M in the appropriate peptide in the absence of interaction. Both peptides bind to bovine NP–II with equal affinity. From Breslow and Weis (116).

Cohen and coworkers (120,137) observed broadening of the *ortho* and *meta* Tyr$_2$ ring protons of oxytocin and lysine vasopressin on binding to bovine NP–II and suggested that this was associated with considerable restriction of motion of the Tyr$_2$ ring in the bound state—a probability in keeping with conclusions from circular dichroism studies (116) as well as with the fact (39) that iodination of the hormone tyrosine prevents binding. Studies of the line broadening observed for both the *ortho* and *meta* protons of Tyr$_2$ of either oxytocin or lysine vasopressin on addition of increasing quantities of bovine NP–II (pH 6.7 in D$_2$O, 0.1 M NaCl) indicated a linear dependence of $1/T_2$ versus ligand/ neurophysin ratio with similar slopes for both hormones (137). The effect of temperature on $1/T_2$ in this system was interpreted as indicating a fast exchange between free and bound forms of both hormones (137). However, plots of $1/T_2$ for lysine vasopressin Tyr$_2$ protons

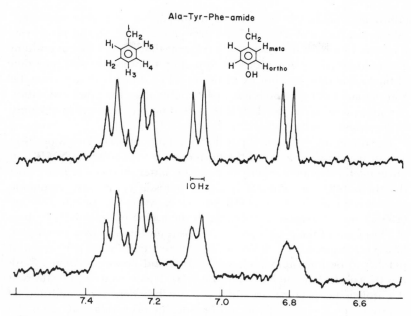

Fig. 16. The 250-MHz NMR spectrum of the aromatic protons of alanyl-tyrosyl-phenylalanine amide. Upper trace, 2.3×10^{-3} M peptide in D_2O; lower trace, 0.09 equivalent of bovine NP–II added, pH 6.5. Abscissa is in ppm downfield from the reference DSS. Tyrosine *ortho* ring protons are at 6.8 ppm and the *meta* ring protons are at 7.08 ppm. Phenylalanine ring protons are located between 7.2 and 7.4 ppm. From Balaram, Bothner-By, and Breslow (136).

decreased more abruptly with temperature than those of oxytocin and were interpreted (137) as suggesting that the Tyr_2 ring of oxytocin is more rigidly bound to neurophysin than that of lysine vasopressin, in accord with the lesser circular dichroism changes near 280 nm (Section VII.B) produced by the Tyr_2 of lysine vasopressin than that of oxytocin. It is noteworthy here that in other NMR studies (133,134) a slow exchange between free and bound lysine vasopressin was noted at neutral pH; the reason underlying the difference in these observations is uncertain. Also of interest is that *deuteron* NMR studies have confirmed the involvement of the Tyr_2 of oxytocin in binding (135).

Cohen et al. (120) indicate that NMR studies show no significant upfield shift of the oxytocin tyrosine ring protons on binding, as is also apparent from Figure 16 for the Tyr_2 protons of Ala-Tyr-Phe-NH_2

(136), and point out that this is incompatible with the "stacking" (π-π interaction) of Tyr_2 with an aromatic residue on the protein in the bound state. In this respect it is relevant that the tyrosine in position 2 of binding peptides can be substituted by tryptophan with retention of binding ability, but not by histidine (111). Since both are aromatic and planar, but histidine is less hydrophobic than the other aromatics, one interpretation of the data is that binding requires a planar highly hydrophobic residue in position 2.

The direct involvement of position 3 in binding, in lysine vasopressin and in tripeptide analogs of the hormones, is supported by NMR and ultraviolet-absorption spectroscopy. The latter studies (127) support the thesis (42) that Phe_3 of lysine vasopressin enters a hydrophobic environment on binding. NMR studies of lysine vasopressin binding (133) suggest a greater broadening in the region of the β-CH_2 protons of Phe_3 than of the β-CH_2 protons of Asn_5 on binding to neurophysin, however, it should be cautioned that these studies were performed at pH 1.5 to permit rapid exchange of free and bound hormone. With the tripeptide S-methyl-Cys-Phe-Ile-NH_2, NMR studies indicate that the isoleucine protons are differentially broadened on addition of neurophysin, the most significant broadening occurring with the β-CH and β-CH_3 groups and lesser broadening occurring further down the side chain (133,134). It is not rigorously possible to distinguish from these data whether the observed greater broadening of the β-CH and β-CH_3 protons in position 3 is due to dipolar relaxation by nuclei on neurophysin in the bound state or simply a restricted rotation in the bound state around bonds associated with the broadened protons. However, the data offer a potential explanation of why the binding of both oxytocin and lysine vasopressin to neurophysin is similar despite the nonidentities of these hormones in position 3. They suggest that the main determinant of binding in position 3 might be the β-carbon atom with segments of the side chain becoming less important to binding as the distance from the β-carbon is increased.

Other residues on the hormones than the α-amino group and side chains in position 2 and 3 participate in binding, the most important of which so far appears to be the half-cystine side chain of residue 1. The first study in which the side chain at position 1 was modified was that of Breslow and Abrash (42), who looked at the effect on binding of adding an additional CH_2-group between the α-amino group and the sulfur (as in 1-hemihomocysteineoxytocin). This substitution did not diminish binding (binding may actually have slightly increased), tending to minimize the role of the side chain in position 1. From studies of tripeptide analogs of residues 1–3, however (38,111), it became clear that the order of binding to native neurophysin is Gly-Tyr-Phe-NH$_2$ < Ala-Tyr-Phe-NH$_2$ < S-methyl-Cys-Tyr-Phe-NH$_2$ < Met-Tyr-Phe-NH$_2$. These data indicated an important role of the side chain in position 1, at least in tripeptide binding, and were also compatible with the lack of effect of the half-cystine → hemihomocystine conversion in position 1 of oxytocin. Proton NMR studies have since confirmed the role of the side chain of position 1 in tripeptide-binding (133,134). More recently preliminary evidence for a role of the side chain in position 1 of the hormones themselves has been obtained. Glasel et al. (135) demonstrated in deuteron NMR studies of selectively deuterated oxytocin that deuterons in the side chain of position 1 are broadened on binding much more so than are deuterons on the oxytocin tail. Of the near-ultraviolet circular dichroism changes generated in binding oxytocin (Fig. 14) or lysine vasopressin to bovine NP–II, the increased negative ellipticity above 295 nm can only be attributed to disulfide since no other residues in the protein or hormones absorb in this wavelength region (84). This ellipticity change was originally (84) attributed solely to neurophysin disulfides (because qualitatively it is also generated by binding dipeptides and tripeptides that contain no disulfides), but it has more recently been shown (111) that a difference ellipticity band near 300 nm is three times as great for the hormones as for non-disulfide-containing peptides, most probably indicating that it partly represents perturbation of the hormone disulfide on binding.

Several approaches have been applied to ascertain whether hormone residues other than 1–3 participate in binding. First, the effect on binding of modifying residues 4, 5, 8, and 9 has been studied. Studies of the binding to unfractionated bovine neurophysins of oxytocin in which position 4 was substituted by glycine suggested a slight decrease in binding with this substitution (42), whereas substitution of position 4 by ornithine increased binding to bovine NP–II, but was without major effect on binding to bovine NP–I (102). Substitution of position 5 of oxytocin by valine was without effect on binding to bovine NP–I. Substitution of position 9 by glycine diminished binding to bovine NP–II by 50% (102). Thus examination of the binding affinities of different hormone analogs has so far given evidence of only weak potential interactions at positions 4 and 9, and no evidence of the participation of position 5. Alternatively, as already indicated, circular dichroism studies suggest small differences between oxytocin and lysine vasopressin on binding to neurophysin; these differences may be due to differences in either residue 3 or 8. Position 8 is a particularly likely candidate since tripeptide-binding studies (111) show no major circular dichroism effect of an Ile→Phe substitution in position 3 on binding; however, it cannot be ascertained whether any of the small substitution effects noted with residues 4, 8, and 9 reflect a direct participation of these residues in binding or an effect on hormone conformation.

A different approach to assessing the potential contribution of residues other than 1–3 has been NMR studies. So far the most firm data here have been those of Glasel et al. (135), who showed, by deuteron NMR spectroscopy, that deuterons in position 9 of oxytocin were considerably less broadened by binding than deuterons in positions 1 and 2, which suggests that position 9 makes no important contribution to binding under the conditions employed (pH 2.5, 0.45 M KCl). The use of NMR to study the binding of selectively deuterated hormones seems a particularly promising approach to assessing the contribution of individual residues.

The other question that can be asked with respect to the possible participation of residues other than 1–3 is the following: how much of the total free energy of hormone binding do residues 1–3 contribute? This question has been answered by measuring the binding of oxytocin, lysine vasopressin, and tripeptide analogs of residues 1–3 to mononitrated bovine NP–II (and to an appreciably lesser extent to native NP–II) (111). The nitrotyrosine residue of mononitrated neurophysin is perturbed by binding hormone, and the ellipticity changes of the nitrotyrosine residue that accompany binding are dependent only on the degree of saturation of the protein (111, 116). Consequently, circular dichroism changes in the nitrotyrosine absorption region have been used to obtain binding constants for peptides and hormones. The data indicate that such peptides as S-methyl-Cys-Tyr-Phe-NH$_2$ (an analog of the first three residues of lysine vasopressin) contribute two-thirds of the binding free energy of the hormones; a similar estimate can be obtained from the relative binding of peptides and hormones to native bovine NP–I, but these results are preliminary (138). The remaining one-third of the binding free energy is therefore either derived from the direct participation in binding of residues other than

1-3, or from conformational differences between residues 1-3 in the hormone and in the tripeptide such that they are preconstrained in the hormone into a conformation favorable for binding and therefore suffer less of an entropy decrease on binding than the corresponding residues in the tripeptide.

E. QUANTITATIVE EVALUATION OF THE CONTRIBUTION OF DIFFERENT SUBSEGMENTS OF RESIDUES 1-3 OF THE HORMONE TO BINDING

The major role played by residues 1-3 in binding raises the question as to the free-energy contribution of the individual segments of residues 1-3. To begin to evaluate this, binding to nitrated bovine NP-II of a series of peptide analogs has been investigated (111); the results are summarized in Table II. In tripeptides containing Tyr in position 2 and Phe-NH$_2$ in position 3, or in dipeptides containing Tyr-NH$_2$ in position 2, change of position 1 from S-methylcysteine or methionine to glycine leads to a decrease in binding of approximately 30-fold.

TABLE II
Association Constants of Some Peptide Analogs of Residues 1-3 of
the Homones to Nitrated Bovine NP-II at pH 6.2[a]

Dipeptide	$K'(M^{-1})$	Tripeptide or tetrapeptide	$K'(M^{-1})$
Gly-Tyr-NH$_2$	5.3×10^1	Gly-Tyr-Phe-NH$_2$	2.5×10^2
Ala-Tyr-NH$_2$	2.4×10^2	Ala-Tyr-Phe-NH$_2$	1.2×10^3
α-NH$_2$-Butyryl-Tyr-NH$_2$	1.3×10^3		
N-Leu-Tyr-NH$_2$	2.4×10^3		
Met-Tyr-NH$_2$	1.7×10^3	Met-Tyr-Phe-NH$_2$	7.0×10^3
		S-Methyl-Cys-Tyr-Phe-NH$_2$	5.5×10^3
		S-Methyl-Cys-Phe-Ile-NH$_2$	4.1×10^3
Phe-Tyr-NH$_2$	1.4×10^4	Phe-His-Gly-Lys	<10
Pro-Tyr-NH$_2$	<10	N-Acetyl-Tyr-Phe-NH$_2$	<10
Met-Tyr	<10	Met-Phe-Gly	$\sim 10^2$
Leu-Gly-NH$_2$	<10		
Leu-Ala-NH$_2$	<10		
Leu-Tyr-NH$_2$	1.2×10^3	Leu-Tyr-Ile-NH$_2$	3.6×10^3
Leu-Trp-NH$_2$	$\sim 10^2$		

[a]From Breslow et al. (111).

Comparison of the binding affinities of peptides containing alanine, α-aminobutyric acid, and norleucine in position 1 indicates that the decrease in binding on changing from methionine to glycine in position 1 is due principally to loss of the β- and γ-CH_2-groups, which each contributes about 1 kcal to the binding free energy; this value is compatible with the transfer of these CH_2-groups from water to a hydrophobic environment on binding. The importance of the β-CH_2-group in the binding of peptides containing methionine in position 1 has also been confirmed by proton NMR studies (133,134). No unique role of sulfur in these methionyl peptides is indicated, since peptides with norleucine in position 1 bind as strongly as methionyl peptides. These data, together with an unexpectedly strong binding of peptides containing phenylalanine in position 1, suggest that the side chain in position 1, like those in positions 2 and 3, binds via hydrophobic interactions with residues on the protein.

Comparison of the binding of a series of dipeptides, all with Tyr-NH_2 in position 2, with binding of the corresponding tripeptides containing Phe-NH_2 or Ile-NH_2 in position 3, demonstrates that in tripeptides position 3 contributes a factor of 3–5 to the binding constant (111). This is a much lower factor than the factor of 30 attributed to position 3 in oxytocin (42) and is most readily attributed to conformational differences between hormone and tripeptide. Either position 3 is preconstrained in oxytocin into a favorable conformation for binding or it favorably affects the conformation of other residues.

Comparison of the binding free energy of such dipeptides as Met-Tyr-NH_2 with that of the hormones themselves indicates that residues 1 and 2 of the hormones alone contribute half the binding free energy of the hormones (111). It can also be demonstrated that the binding of residues 1 and 2 by neurophysin is cooperative: if residue 1 is modified in such a way that the α-amino group is lost or blocked, loss of binding at residue 2 occurs, and if residue 2 is so modified that it no longer contains tyrosine, phenylalanine, or tryptophan, simultaneous loss of binding at residue 1 occurs. For example, as shown in Table II, the single substitution of Tyr_2 by Gly_2 (compare Leu-Tyr-NH_2 with Leu-Gly-NH_2) leads to a decrease in binding that is greater than the binding constant of Gly-Tyr-NH_2 (which contains both the α-amino and Tyr_2 residues together). The data suggest that the interactions with neurophysin of the hormone α-amino group and of the hormone tyrosine are linked functions, which together act as an anchor (possibly via the induction of conformational changes in the protein) to support apolar interactions involving the side chains of hormone residues 1 and 3. From the binding data for tripeptides, interactions at the side chain in position 1 appear to be more important than those in position 3, although interactions at position 3 may assume greater significance in the hormones. An inspection of the binding data in Table II indicates that all constants are consistent with the proposed model. Moreover, the hydrophobic nature of the binding site is also supported by the weak binding of peptides in which the α-carboxyl is not amidinated.

F. IDENTIFICATION OF RESIDUES ON NEUROPHYSIN INVOLVED IN THE BINDING REACTION

Evidence of the participation of a neurophysin carboxyl in binding and its possible identity as either Asp_{30} or Glu_{31} in bovine NP–II has been already cited. Apart from the carboxyl, the group most clearly identifiable as important to the binding process is the single tyrosine (residue 49 in bovine NP–I, bovine NP–II, and porcine NP–I) found in all neurophysins. (Porcine NP–II may contain two tyrosines, but presumably one of them is analogous in position to Tyr_{49} of the other neurophysins.) The first evidence of a relationship between Tyr_{49} and binding was obtained by Furth and Hope (121), who mononitrated Tyr_{49} and demonstrated that the nitrotyrosine absorption spectrum was changed on binding arginine vasopressin in a manner suggesting that its pK_a was lowered. The change in pK was attributed to the increased proximity of a positively charged group in the bound state— possibly Arg_8 of arginine vasopressin itself or a positively charged group in the protein brought near by a conformational change. These authors also suggested that the nitrotyrosine environment became more hydrophobic in binding. Breslow and Weis (116) studied nitrated bovine NP–II by circular dichroism and spectrophotometric titration. The data indicated that arginine vasopressin lowered the apparent pK_a of the nitrotyrosine from 7.45 to 6.85 and led to marked changes in nitrotyrosine ellipticity. Identical effects were produced on binding oxytocin and tripeptide analogs of residues 1–3 of the hormones, indicating that the change in nitrotyrosine pK_a and environment was not due to Arg_8 of arginine vasopressin; it was also argued that the nitrotyrosine did *not* enter a more hydrophobic environment when hormone was bound, an argument given additional support by the diminished pK_a of the nitrotyrosine in the bound state.

NMR evidence (133,134,136) regarding the behavior of Tyr_{49} has been highly illuminating with respect to its role in binding and in interpretation of the environmental changes seen by circular dichroism and spectrophotometric titration. The *ortho* ring protons of Tyr_{49} in the uncomplexed state are slightly upfield (0.1 ppm) from their expected position, which suggests that they may be stacked with another aromatic residue or shielded in some other way. On binding of peptide analogs of residues 1–3 of the hormones (as is particularly evident with S-methyl-Cys-Phe-Ile-NH_2, which contains no tyrosine of its own) both the *ortho* and *meta* protons of Tyr_{49} move downfield, again confirming that the environment of Tyr_{49} is changed by binding. The application of the nuclear Overhauser effect to this system (136) established the fact that the change in Tyr_{49} environment was associated with the proximity of Tyr_{49} to residue 2 of binding peptides in the complexed state. (Nuclear Overhauser effects are the change in intensity of a particular proton resonance when a resonance immediately adjacent to it in space is irradiated at its own resonance frequency.) In studies of S-methyl-Cys-Phe-Ile NH_2 it was demonstrated that the phenylalanine-ring-proton resonances (which occur at approximately 7.25 ppm from the reference DSS) decreased in intensity in the presence of bovine NP–II when the system was irradiated at 1.9, 3.1, or 6.8 ppm; this effect did not occur in the absence of NP–II. Protons responsible for the 1.9- and 3.1-ppm effects were not specifically identified, but the 6.8-ppm effect was specifically attributed to the *ortho* ring protons of Tyr_{49}, these being the only protons that absorb in this region. [Nuclear Overhauser effects on the aromatic ring protons of position 2 of the bound peptide are also observed with peptides containing tyrosine in position 2, but can be demonstrated only at 1.9 and 3.1 ppm because of the overlap of the protein and peptide tyrosine-ring-proton resonances at 6.8 ppm (133,134,136).]

It remains then to account for the downfield NMR shift of the neurophysin-tyrosine protons on binding and the change in pK and ellipticity accompanying binding when the tyrosine is nitrated. First, the downfield shift is probably not due to parallel stacking of Tyr_{49} of the protein with Tyr_2 of the peptide. Such stacking effects would be expected to give upfield shifts (120), and, as shown previously, the Tyr_2 ring protons of the bound peptide also do not show an upfield shift on binding. Other arguments against the stacking of Tyr_{49} in the complexed state have been given by Balaram (133). In addition, ultraviolet difference spectra (127) suggest that while Tyr_2 of the hormone moves to a more hydrophobic region on binding, Tyr_{49} of the protein moves to a more hydrophilic environment. One possibility is that Tyr_{49} is stacked (or otherwise shielded) in the uncomplexed protein and displaced from this position directly by Tyr_2 or by a more general conformational change attending binding, the change in tyrosine position leading to the physical changes seen. Alternatively one could simply invoke the proximity of the bound peptide to account for both the NMR and circular dichroism changes observed. In this instance the major perturbant of Tyr_{49} properties would have to be either the peptide α-NH_3^+ or the phenyl ring in position 2 since modification of the side chains in positions 1 and 3 of binding peptides does not affect the nature of the binding-induced nitrotyrosine ellipticity changes (116).

The apparent proximity of Tyr_{49} to position 2 of the peptide in the complex is consistent with the fact that the pK of the nitrotyrosine is higher than expected from model compounds (see above) and is lowered by binding. The simplest interpretation of these data is that, in the uncomplexed state, the pK is raised by the proximity of the tyrosine to the same carboxyl that binds to the α-NH_3^+ and that the influence of the neutralizing α-NH_3^+ in the bound state reduces the pK to a more normal value (116). However, despite the satisfactory explanations that can be generated by assuming the proximity of Tyr_{49} of neurophysin to Tyr_2 of the hormones, confirmative distance studies are warranted. Nuclear Overhauser effects have never previously been applied to the assessment of interproton distances in proteins, and an exact estimate of the distance between Tyr_{49} and Tyr_2 is not available from these data except that they must be extremely close. Second, the remarkable constancy of the nitrotyrosine circular dichroism band in the bound state is puzzling. If Tyr_{49} and position 2 of the peptide are so close, some effect of the nature of the bound peptide on the circular dichroism spectrum of the nitrotyrosine might be expected; to date this has not been seen.

Other protein residues in addition to Tyr_{49} and the "active-site" carboxyl have been implicated either directly or indirectly in binding. Circular dichroism changes in the disulfide absorption region that attend hormone binding (see Section VII.B) are partly attributable to neurophysin disulfides because they also occur on binding small peptides that contain no disulfides (84), although with lesser amplitude (111). It is not clear, however, whether these changes result from a change in the geometry or environment of a particular protein disulfide due to its proximity to the bound peptide or to general conformational changes that may accompany binding.

Of still other protein residues that may be at the binding site or affected secondarily by binding, preliminary NMR evidence has been presented for changes in the phenylalanine ring protons of bovine NP–II on binding the peptide Met-Tyr-NH$_2$ (134). Residues at the binding site responsible for the 1.9- and 3.1-ppm nuclear Overhauser effects on the aromatic residues at position 2 of the bound peptide have not been specifically identified since they lie in the alkyl-proton region of the NMR spectrum. A direct role for the methionine in position 2 of bovine NP–II has been precluded (38) since binding occurs in bovine NP–II derivatives from which this residue has been lost. Similarly NMR studies (120) of the effect of binding on the histidine in bovine NP–I indicate only a slight shift in histidine pK, and it is suggested that this histidine plays no major role in the binding reaction. Of course, a binding role for the methionine and the histidine in bovine NP–II and NP–I, respectively, would not be anticipated since the binding properties of the two proteins are largely similar, and these residues are not common to both proteins.

G. COMPARISONS AMONG DIFFERENT NEUROPHYSINS

Almost all of the physical and binding studies of neurophysins have been carried out with bovine neurophysins, in particular with NP–II. As already cited, bovine NP–I and NP–II differ in their disulfide pairing, and some differences in their chemical and physical properties have been observed. On the whole, however, these proteins appear to be more alike than dissimilar, and the significance of the small differences in properties observed between them is not clear. In addition it can be shown (S. L. Lundt and E. Breslow, unpublished observations) that the mononitrated tyrosine of nitrated bovine NP–I behaves almost identically to the mononitrated tyrosine of bovine NP–II with respect to its circular dichroism properties and with respect to the effects of hormone binding on its circular dichroism spectrum and pK_a. Circular dichroism changes attending hormone binding to native bovine NP–I also differ in only minor detail from those attending binding to bovine NP–II. It is difficult to reconcile these data with major differences between the neurophysins in their binding sites. Future studies of the sequence and binding properties of other neurophysins will be useful in establishing which residues are most essential to the binding reaction.

VIII. Conclusion

Two concluding general comments are in order about the data so far accumulated on hormone–neurophysin interaction. First, there is a remarkable dissimilarity between the structural features of the hormones responsible for binding to neurophysin and those required for hormonal activity. This is most strikingly seen in the case of the α-amino group of the hormones, whose replacement by hydrogen leads to an increase in hormonal activity in contrast to major loss of ability to bind to neurophysin. Other contrasts between hormonal activity and binding ability have been pointed out elsewhere (102). Thus any suggestion that the hormone–neurophysin interaction may be a prototype of hormone–receptor interaction (139) is invalid. Second, the simplest explanation of the internal duplication found in the neurophysins would reside in either the presence of two equivalent binding sites per monomer or in the confinement of the binding region of neurophysin to that central region of the sequence which contains the tyrosine and is nonduplicated. In the latter instance the duplicated regions, like the constant domains of the immunoglobulins, could be envisioned as performing some auxiliary function (as, for example, binding of lipid). In general the data do not support the presence of two equivalent binding sites. Moreover, although a model has been proposed for binding to the nonduplicated region only (116), the sum of the experimental data presented here suggests that the binding site in bovine NP–II includes a residue from the nonduplicated region (Tyr_{49}) and a residue from *one* of the duplicated regions (Asp_{30} or Glu_{31}). The significance of the internal duplication remains unclear.

References

1. MacArthur, C. G., *Science, 73*, 448 (1931).
2. Rosenfeld, M., *Bull. Johns Hopkins Hosp., 66*, 398 (1940).
3. Van Dyke, H. B., Chow, B. F., Greep, R. O., and Rothen, A., *J. Pharmacol. Exp. Ther., 74*, 190 (1942).
4. Kam, O., Aldrich, T. B., Grote, I. W., Rowe, L. W., and Bugbee, E. P., *J. Am. Chem. Soc., 50*, 573 (1928).

5. Irving, G. W., Jr., Dyer, H. M., and du Vigneaud, V., *J. Am. Chem. Soc.*, *63*, 503 (1941).
6. Potts, A. M., and Gallagher, T. F., *J. Biol. Chem.*, *154*, 349 (1944).
7. Du Vigneaud, V., Ressler, C., and Tripett, S., *J. Biol. Chem.*, *205*, 949 (1953).
8. Du Vigneaud, V., Lawler, H. C., and Popenoe, E. A., *J. Am. Chem. Soc.*, *75*, 4880 (1953).
9. Acher, R., and Chauvet, J., *Biochim. Biophys. Acta*, *12*, 487 (1953).
10. Du Vigneaud, V., Pierce, J. G., Ressler, C., Swan, J. M., Roberts, C. W., Katsoyannis, P. G., and Gordon, S., *J. Am. Chem. Soc.*, *75*, 4879 (1953).
11. Du Vigneaud, V., Gish, D. T., and Katsoyannis, P. G., *J. Am. Chem. Soc.*, *76*, 4751 (1954).
12. Archer, R., Manoussos, G., and Olivry, G., *Biochim. Biophys. Acta*, *16*, 155 (1955).
13. Acher, R., Chauvet, J., and Olivry, G., *Biochim. Biophys. Acta*, *22*, 421 (1956).
14. Chauvet, J., Lenci, M., and Acher, R., *Biochim. Biophys. Acta*, *38*, 266 (1960).
15. Acher, R., Light, A., and du Vigneaud, V., *J. Biol. Chem.*, *233*, 116 (1958).
16. Pickering, B. T., *J. Endocrinol.*, *42*, 143 (1968).
17. Watkins, W. B., *J. Endocrinol.*, *53*, 331 (1972).
18. Watkins, W. B., *J. Endocrinol.*, *51*, 595 (1971).
19. Hollenberg, M. D., and Hope, D. B., *Biochem. J.*, *106*, 557 (1968).
20. Rauch, R., Hollenberg, M. D., and Hope, D. B., *Biochem. J.*, *115*, 473 (1969).
21. Uttenthal, L. O., and Hope, D. B., *Biochem. J.*, *116*, 899 (1970).
22. Wuu, T. C., and Saffran, M., *J. Biol. Chem.*, *244*, 482 (1969).
23. Cheng, K. W., and Friesen, H. G., *J. Biol. Chem.*, *246*, 7656 (1971).
24. Ginsburg, M., and Ireland, M., *J. Endocrinol.*, *32*, 187 (1965).
25. Burford, G. D., Jones, C. W., and Pickering, B. T., *Biochem. J.*, *124*, 809 (1971).
26. Coy, D. H., and Wuu, T. C., *Anal. Biochem.*, *44*, 174 (1971).
27. Cheng, K. W., and Friesen, H. G., *J. Clin. Endocrinol. Met.*, *34*, 165 (1972).
28. Foss, I., Sletten, K., and Trygstad, O., *FEBS Letters*, *30*, 151 (1973).
29. Sachs, H., Goodman, R., Shin, S., Shainberg, A., and Pearson, D., *Excerpta Medica* (in press).
30. Norström, A., Sjöstrand, J. Livett, B. G., Uttenthal, L. O., and Hope, D. B., *Biochem. J.*, *122*, 671 (1971).
31. Coy, D. H., and Wuu, T. C., *Biochim. Biophys. Acta*, *263*, 125 (1972).
32. Burford, G. D., and Pickering, B. T., *Biochem. J.*, *128*, 941 (1972).
33. Capra, D., Kehoe, J. M., Kotelchuck, D., Walter, R., and Breslow, E., *Proc. Natl. Acad. Sci., U.S.*, *69*, 431 (1972).
34. Sawyer, W. H., *Pharm. Rev.*, *13*, 225 (1961).
35. Sachs, H., and Takabatake, Y., *Endocrinology*, *75*, 943 (1964).
36. Sachs, H., *Adv. Enzymol.*, *32*, 327 (1969).
37. Sachs, H., Fawcett, P., Takabatake, Y., and Portonova, R., *Recent Prog. Hormone Res.*, *25*, 447 (1969).
38. Breslow, E., Aanning, H. L., Abrash, L., and Schmir, M., *J. Biol. Chem.*, *246*, 5179 (1971).
39. Camier, M., Alazard, R., Cohen, P., Pradelles, P., Morgat, J., and Fromageot, P., *Eur. J. Biochem.*, *32*, 207 (1973).

40. Rudman, D., Del Rio, A. E., Garcia, L. A., Barnett, J. Howard, C. H., Walker, W., and Moore, G., *Biochemistry*, *9*, 99 (1970).
41. Friesen, H. G., and Astwood, E. B., *Endocrinology*, *80*, 278 (1967).
42. Breslow, E., and Abrash, L., *Proc. Natl. Acad. Sci. U. S.*, *56*, 640 (1966).
43. Hollenberg, M. D., and Hope, D. B., *Biochem. J. 104*, 122 (1967).
44. Dean, C. R., Hollenberg, M. D., and Hope, D. B., *Biochem. J.*, *104*, 8C (1967).
45. Norström, A. and Sjöstrand, J., *J. Neurochem.*, *18*, 29 (1971).
46. Bargmann, W., and Scharrer, E., *Am. Scientist*, *39*, 255 (1951).
47. Scharrer, E., and Scharrer, B., *Recent Prog. Hormone Res.*, *10*, 183 (1954).
48. Hild, W., *Z. Zellforsch.*, *40*, 257 (1954).
49. Hild, W., and Zetler, G., *Pflügers, Arch. Ges. Physiol.*, *257*, 169 (1953).
50. Scharrer, A., and Wittenstein, G. J., *Anat. Rec.*, *112*, 387 (1952).
51. Bisset, G. W., Clark, B. J., and Errington, M. L., *J. Physiol.*, *217*, 111 (1971).
52. Bisset, G. W., Clark, B. J., and Haldar, J., *J. Physiol.*, *206*, 711 (1970).
53. Beleslin, D., Bisset, G. W., Haldar, J., and Polak, R. L., *Proc. Roy. Soc. (London)*, *B 166*, 443 (1967).
54. Schrier, R. W., Verroust, P. J., Jones, J. J. Fabian, M., Lee, J., and de Wardener, H. E., *Clin. Sci.*, *35*, 433 (1968).
55. Clark, B. J., and Rocha, E. Silva, M., *J. Physiol.*, *191*, 529 (1967).
56. Barer, R., Heller, H., and Lederis, K., *Proc. Roy. Soc. (London)*, *158*, B388 (1963).
57. La Bella, F. S., Beaulieu, G., and Reiffenstein, R., *Nature*, *193*, 173 (1962).
58. La Bella, F. S., Reiffenstein, R. J., and Beaulieu, G., *Arch. Biochem. Biophys.*, *100*, 399 (1963).
59. Dean, C. R., Hope, D. B., and Kazic, T., *Br. J. Pharmacol. Chemother.*, *34*, 192P (1968).
60. Dean, C. R., and Hope, D. B., *Biochem. J.*, *106*, 565 (1968).
61. Dean, C. R., and Hope, D. B., *Biochem. J.*, *104*, 1082 (1967).
62. Ginsburg, M., and Ireland, M., *J. Physiol.*, *169*, 114P (1963).
63. Ginsburg, M., and Ireland, M., *J. Physiol.*, *169*, 15 (1963).
64. Ginsburg, M., and Ireland, M., *J. Endocrinol.*, *35*, 289 (1966).
65. Sachs, H., Goodman, R., Osinchak, J., and McKelvy, J., *Proc. Natl. Acad. Sci., U.S.*, *68*, 2782 (1971).
66. Alvarez-Buylla, R., Livett, B. G., Uttenthal, L. O., and Hope, D. B., *Z. Zellforsch.*, *137*, 435 (1973).
67. Jones, C. W., and Pickering, B. T., *J. Physiol.*, *208*, 73P (1970).
68. Fawcett, C. P., Powell, A. E., and Sachs, H., *Endocrinology*, *83*, 1299 (1968).
69. Dean, C. R., Hope, D. B., and Kazic, T., *Br. J. Pharmacol.*, *34*, 193P (1968).
70. Livett, B. G., Uttenthal, L. O., and Hope, D. B., *Phil. Trans. Roy. Soc. (London)*. *B261*, 371 (1971).
71. Wuu, T. C., Crumm, S., and Saffran, M., *J. Biol. Chem.*, *246*, 6043 (1971).
72. Walter, R., Schlesinger, D., Schwartz, I. L., and Capra, J. D., *Biochem. Biophys. Res. Commun.*, *44*, 293 (1971).
73. Pickup, J. C., Johnston, C. I., Nakamura, S., Uttenthal, L. O., and Hope, D. B., *Biochem. J.*, *132*, 361 (1973).
74. Johnston, C. I., Pickup, J. C., Uttenthal, L. O., and Hope, D. B., *Proc. Austr. Endocrinol. Soc.*, *5th Meeting*, Abstract 41 (1972).

75. Zimmerman, E. A., Hsu, K. C., Robinson, A. G., Carmel, P. W., Frantz, A. G., and Tannenbaum, M., *Endocrinology*, *92*, 931 (1973).
76. Douglas, W. W., and Poisner, A. M., *J. Physiol.*, *172*, 1 (1964).
77. Douglas, W. W., and Poisner, A. M., *J. Physiol.*, *172*, 19 (1964).
78. Thorn, N. A., *Acta Endocrinol.*, *50*, 357 (1965).
79. Smith, M. W., and Thorn, N. A., *J. Endocrinol.*, *32*, 141 (1965).
80. Ginsburg, M., Jayasena, K., and Thomas, P. J., *J. Physiol.*, *184*, 387 (1966).
81. Ginsburg, M., *Proc. Roy. Soc. (London)*, *B170*, 27 (1968).
82. Burford, G. D., Ginsburg, M., and Thomas, P. J., *J. Physiol.*, *205*, 635 (1969).
83. Burford, G. D., Ginsburg, M., and Thomas, P. J., *Biochim. Biophys. Acta*, *229*, 730 (1971).
84. Brelsow, E., *Proc. Natl. Acad. Sci. U.S.*, *67*, 493 (1970).
85. Nagasawa, J., Douglas, W. W., and Schulz, R. A., *Nature*, *227*, 407 (1970).
86. Nordmann, J. J., Dreifuss, J. J., Legros, J. J., *Experientia*, 27, 1344 (1971).
87. Uttenthal, L. O., Livett, B. G., and Hope, D. B., *Phil. Trans. Roy. Soc. (London)*, *B261*, 379 (1971).
88. Matthews, E. K., Legros, J. J., Grau, J. D., Nordmann, J. J., and Dreifuss, J. J., *Nature New Biol.*, *241*, 86 (1973).
89. Douglas, W. W., *Br. J. Pharmacol.*, *34*, 451 (1968).
90. Puszkin, S., Nicklaus, W. J., and Berl, S., *J. Neurochem.*, *19*, 1319 (1972).
91. Poisner, A. M., and Trifara, S. M., *Mol. Pharmacol.*, *3*, 561 (1967).
92. Poisner, A. M., and Douglas, W. W., *Mol. Pharmacol.*, *4*, 531 (1968).
93. Cheng, K. W., and Friesen, H. G., *Metabolism*, *19*, 876 (1970).
94. McNeilly, A. S., Legros, J. J., and Forsling, M. J., *J. Endocrinol.*, *52*, 209 (1972).
95. McNeilly, A. S., Martin, M. J., Chard, T., and Hart, I. C., *J. Endocrinol.*, *52*, 213 (1972).
96. Burton, A. M., Forsling, M. L., and Martin, M. J., *J. Physiol.*, *217*, 23P (1971).
97. Robinson, A. G., Zimmerman, E. A., and Frantz, A. G., *Metabolism*, *20*, 1148, (1970).
98. Robinson, A. G., Zimmerman, E. A., Engleman, E. G., and Frantz, A. G., *Metabolism*, *20*, 1138 (1971).
99. Martin, M. J., Chard, T., and Landon, J., *J. Endocrinol.*, *52*, 481 (1972).
100. Thorn, N. A., in *Aspects of Neuroendocrinology*, W. Bargmann and B. Scharrer, Eds., Springer, Berlin, 1970, p. 140.
101. Legros, J. J., Franchimont, P., and Hendrick, J. C., *C. R. Soc. Biol.*, *163*, 2773 (1969).
102. Breslow, E., and Walter, R., *Mol. Pharmacol.*, *8*, 75 (1972).
103. Ginsburg, M., and Ireland, M., *J. Endocrinol.*, *30*, 131 (1964).
104. Ginsburg, M., and Jayasena, K., *J. Physiol.*, *197*, 53 (1968).
105. Trygstad, O., *Acta Endocrinol. (Kbh)*, *57*, 81 (1968).
106. Robinson, A. G., Michelis, M. F., Warms, P. C., and Davis, B. B., *Clin. Res.*, *21*, 501 (1973).
107. Sachs, H., Saito, S., and Sunde, D., *Mem. Soc. Endocrinol.*, *19*, 325 (1971).
108. Sachs, H., Shin, S., Saito, S., and Walter, R., unpublished observations.
109. Richards, F. M., and Vithayathil, P. J., *J. Biol. Chem.*, *234*, 1459 (1959).
110. Pliska, V., McKelvy, J. F., and Sachs, H., *Eur. J. Biochem.*, *28*, 110 (1972).

111. Breslow, E., Weis, J., and Menendez-Botet, C. J., *Biochemistry*, *12*, 4644 (1973).
112. Mylroie, R., and Koenig, R., *J. Histochem. Cytochem.*, *19*, 738 (1971).
113. Walter, R., and Breslow, E., in *Research Methods in Neurochemistry*, Vol. III, N. Marks and R. Rodnight, Eds., Plenum Press, New York, (in Press).
114. Schlesinger, D. H., Frangione, B., and Walter, R., *Proc. Natl. Acad. Sci. U.S.*, *69*, 3350 (1972).
115. Cheng, K. W., and Friesen, H. G., *Endocrinology*, *88*, 608 (1971).
116. Breslow, E., and Weis, J., *Biochemistry*, *11*, 3474 (1972).
117. Coleman, D. L., and Blout, E. R., *J. Am. Chem. Soc.*, *90*, 2405 (1968).
118. Greenfield, N., and Fasman, G. D., *Biochemistry*, *8*, 4108 (1969).
119. Saxena, V. P., and Wetlaufer, D. B., *Proc. Natl. Acad. Sci. U.S.*, *68*, 969 (1971).
120. Cohen, P., Griffin, J. H., Camier, M., Caizergues, M., Fromageot, P., and Cohen, J. S., *FEBS Letters*, *25*, 282 (1972).
121. Furth, A. J., and Hope, D. B., *Biochem. J.*, *116*, 545 (1970).
122. Sears, D., and Beychok, S., in *Physical Principles and Techniques of Protein Chemistry*, S. Leach, Ed., Part C, Academic Press, New York, 1973.
123. Kahn, P. C., Ph.D. thesis, Columbia University, 1972.
124. Breslow, E., and Menendez-Botet, C. J., *Abstracts of Papers, National Meeting of the American Chemical Society, New York, 1972.*
125. Menendez-Botet, C. J., and Breslow, E., *Fed. Proc.*, *31*, 485Abs. (1972).
126. Wolff, J., Camier, M., Alazard, R., and Cohen, P., manuscript in preparation.
127. Griffin, J. H., Alazard, R., and Cohen, P., *J. Biol. Chem.*, *248*, 7975 (1973).
128. Stouffer, J. E., Hope, D. B., and du Vigneaud, V., in *Perspectives in Biology*, C. F. Cori, G. Foglia, L. F. Leloir, and S. Ochoa, Eds., Elsevier, Amsterdam, 1963.
129. Breslow, E., *Biochim. Biophys. Acta*, *53*, 606 (1961).
130. Walter, R., and Hoffman, P. L., *Fed. Proc.*, *32*, 567 (1973).
131. Sigler, P. B., Blow, D. M., Matthews, B. W., and Henderson, R., *J. Biol. Mol.*, *35*, 143 (1968).
132. Hope, D. B., and Walti, M., *Biochem. J.*, *125*, 909 (1971).
133. Balaram, P., Ph.D. thesis, Carnegie-Mellon University, 1972.
134. Balaram P., Bothner-By, A. A., and Breslow, E., *Biochemistry*, *12*, 4695 (1973).
135. Glasel, J. A., Hruby, V. J., McKelvy, J. F., and Spatola, F., *J. Mol. Biol.*, *79*, 555 (1973).
136. Balaram, P., Bothner-By, A. A., and Breslow, E., *J. Am. Chem. Soc.*, *94*, 4017 (1972).
137. Alazard, R., Cohen, P., Griffin, J. H., and Cohen, J. S., manuscript in preparation.
138. Breslow, E., Aanning, H. L., Abrash, L., and Tarantino, L., *Abstracts of Papers, National Meeting of the American Chemical Society, New York, 1969*, Abstract 313.
139. Niedrich, H., and Berseck, C., *Z. Aerztl. Fortbild.* (Jena), *63*, 1161 (1969).

AUTHOR INDEX

Numbers in parenthesis are reference numbers and show that an author's work is referred to although his name is not mentioned in the text.
Numbers in *italics* indicate the pages on which the full references appear.

Aanning, H. L., 274(38), 275(38), 276 (38), 280(38), 283(38), 286(38), 287 (38), 288(38), 296(38), 297(38), 298 (38), 299(38), 300(38), 301(38), 309 (38), 311(38), 313(38), 314(38), 321 (38), 322(38), 328(38), *330*
Ab, G., 212(331), *236*
Abelson, J., 191(249), *234*
Abelson, P. H., 65(6), *88*
Abrash, L., 274(38,42), 275(38), 276, 280(38), 283(38), 286(38), 287(38, 42), 288(38), 296(38), 297(38), 298 (38), 299(38), 300(38), 301(38), 306, 307, 309(38), 311(38), 313(38), 314 (38), 316, 320, 321(38,42), 322(138), 324(24), 328(38), *330*
Acher, R., 272(9,14,15), 287(14), *330*
Ada, G. L., 54, *64*
Adams, A., 195(263), *234*
Adams, J. M., 72(45), *89*
Adelberg, E. D., 74(53), 199(299), 203 (312), 204(312), 208(312), 209(299, 312), 210(312), *89, 235, 236*
Agalarova, M. B., 144(30), *226*
Ahmad, M., 198(289), 200(289), 201 (289), 203(289), 214(289), 217(289), *235*
Alazard, R., 274(39), 283(39), 287(39), 306(39), 307(39), 308(126), 309(39, 127), 310(39), 311(126), 315(39), 317, 318(39,137), 319(137), 320 (127), 327(127), *330, 333*
Albrecht, A. M., 65(15), 70(15), 71(39), *88, 89*
Aldrich, T. B., 272(4), *329*

Alexander, R. R., 199(294), 204(294), 208(294), 214(294), *235*
Alexandrova, M. M., 189(230), 191(230), *233*
Alfsen, A., 158(107), *230*
Alkjaersig, N., 52(81), *63*
Allaudeen, H. G., 216, *237*
Allen, T. A., 12(39), 22(39), *27*
Allende, C. C., 156(124), 172(124), *230*
Allende, J. E., 156(124), 172(124), *230*
Allfrey, V. G., 32(31,32), 38, 39(32), 53(31), *61*
Allison, A. C., 2, *26*
Alvarez-Buylla, R., 277(66), 278, *331*
Ames, B. N., 75(67), 185(351,352), 199(295,297), 203(297), 204(297), 207(319), 214(295,297,319), 215(351, 352), 216(319,352), 218(295,357), *89, 235, 236, 237*
Ames, G. F., 203(313), *236*
Anai, M., 39, *62*
Anagnostopoulos, C., 200(303), 209 (303), 212(303), *235*
Anderson, D., 198(278), 212(278), *235*
Anderson, D. O., 52(78), *63*
Anderson, J. J., 200(301), 204(301), 218(301,361), *235, 237*
Anderson, J. S., 118(50), 121, *139*
Anderson, P. M., 66(23), 75(23), *88*
Anderson, W., 32(26), 37, 46(67), 48, *61*
Anderson, W. F., 213, *236*
Andersson, E., 151(75), *228*
Anatalis, C., 255(40), 256(40,42), 257 (40,42), 258(42), 259(40,42),

335

Pauling, L., 1(3), *26*
Pawelkiewicz, J., 154(92), 155(98b),
 158(102), 159(92), *229*
Pazur, J. H., 264(53), *270*
Pearson, D., 272(29), 276(29), 282(29),
 330
Peat, S., 242(14), 243(21), 244(21), *269*
Pena, A. van, 171(137), *230*
Penneys, N. S., 148(59b), *228*
Penzer, G. R., 153(85), 158(85), 172
 (85), *229*
Peri, C. di, 188(219), 193(219), *233*
Perlin, A. S., 240(3), 243(3), 247(3),
 269
Perlman, G. E., 38, *62*
Persson, B., 52(90), *63*
Perutz, M. F., 1(5), 2(5,7,9), 16, 19, *26,*
 27
Pessac, B., 51, *64*
Pestka, S., 81(101), 102(25), *90, 138*
Peterofsky, A., 143, 173(144), 179(144),
 205(23), *226, 231*
Peters, M. A., 185(204), 186(204), *233*
Peters, T. Jr., 96(12), *138*
Peterson, C. M., 22(40), *27*
Petit, J.-F., 120(58), 124(66), 125, 127
 (58), 131(57,58), 136(57,58), *139*
Petrissant, G., 158(107), *230*
Pette, D., 46(63,64,65), 47, *62*
Phaff, H. J., 240(1), 262(1), 264(1), 266
 (1), *269*
Phang, J. M., 46(68), 48(68), *62*
Phang, Y. M., 215(370), *237*
Pickering,B.T., 272(16,25,32), 276(32),
 277(67), 281(25,32), 282, 286(25),
 287(67), 306, *330, 331*
Pickup, J. C., 279(73,74), 281, 283(73),
 331
Piérard, A., 76(63), *89*
Pierce, J. G., 272(10), *330*
Pigoud, A., 148(100e), 164(100e), 169
 (100e), 183(100e), *229*
Pinchot, G. B., 32(28,29,30), 38(30),
 61
Piper, D. W., 56, *64*
Pizer, L. I., 199(293), 212(293), *235*
Plapp, R., 120(59), 127(59,69,70), 128

(59), 129(59,70), 132(69,70), 134,
 135, 136(70), 137(59), *139*
Pliska, V., 287(110), *332*
Poisner, A. M., 282, 283(91,92), *332*
Polak, R. L., 277(53), *331*
Popenoe, E. A., 272(8), *330*
Portonova, R., 273(37), 276(37), 277
 (37), 282(37), 283(37), 285(37), 286
 (37), *330*
Potts, A. M., 272(6), *330*
Powell, A. E., 277(68), 283(68), 284,
 331
Pradelles, P., 274(39), 283(39), 287(39),
 305(39), 307(39), 309(39), 310(39),
 315(39), 318(39), *330*
Praslov, V. S., 143(26), 146(26), 147
 (26), 148(26), 150(33e), *226, 227*
Preddie, E. C., 149, *228*
Printz, D. B., 200(306), *236*
Puszkin, S., 283(90), *332*

Quincey, R. V., 222(374), *238*

Radhakrishnamurthy, B., 56, *64*
Rank, G. H., 256(43), *270*
Rapenbusch, R., 146(63), 150(63), *228*
Rapenbusch, R. van, 148(56), 149(56),
 227
Rapenbusche, V. R., 143(25), 145(25),
 146(25), 147(25), *226*
Raschke, W. C., 240(8,9), 245(8), 247(8),
 251(8,9,37), 252(8), 253(8), 254(9,
 37), 256(42), 257(42), 258(42,44),
 259(42,44), 260(42,44), 262(44), 263
 (8), 268(42), *269, 270*
Raszelbach, R., 188(219), 193(210),*233*
Ratner, S., 72(47), 73(47–49), *89*
Rauch, R., 272(20), 274, 276, 280, 286
 (20), 287(20), 288(20), 296(20), 309
 (20), *330*
Raue, H. A., 212, *236*
Ravel, J. M., 178(182), *232*
Ravin, L. J., 53(96), *63*
Ray, W., 185(207), *233*
Reber, K., 32(13), 33, *61*
Ressler, C., 272(7,10), *330*
Reid, B., 149(62), *228*

SUBJECT INDEX

Acetolysis, of yeast mannans, 243, 245, 247, 249, 251, 264
N-Acetylarginine, 77
N-Acetylglucosamine, linkage to asparagine in yeast mannans, 243, 265
variability factor in yeast mannan, 246
N-Acetylglutamate, 69
Acetylglutamate kinase, 59
N-Acetylglutamate γ-semialdehyde, 70, 71, 72
N-Acetyl-γ-glutamyl phosphate, 69, 70
Acetylglutamylphosphate reductase, 70
N-Acetylornithine, 70, 71
and *E. coli* growth rates, 73, 74
growth medium, 87
Acetylornithine aminotransferase, 70, 71
effects of streptomycin and tetracycline, 81
residual repression, 82
Acetylornithine deacetylase, 71, 72
arginine antagonism, 74
transcriptional repression, 79
translation of message accumulation, 84
Acetylornithine permeation system, repression by arginine, 75
Acid phosphatase, 242
effects of polycations, 42
Affinity chromatography, tRNA synthetases, 153, 154, 155
Affinity-labeling, of aminoacyl-tRNA synthetases, 143, 144
Agarose-heparin columns, 34
Alanyl-tRNA~phosphatidylglycerol transferase, requirements, 114
substrate, donor specificity, 115, 116
alanyl-tRNA, 113, 114
arginyl-tRNA, 113, 114
lysyl-tRNA, 114
Alanyl-tRNA~UDP-MurNAc-pentapep-

tide transferase, 127, 128, 130
substrate specificity, acceptor, 134, 135
donor, 132
Albumin, arginylated, 100, 101
in vivo acceptor in arginine-transfer reaction, 108
site of arginylation, 96
of leucylation, 99
of phenylalanylation, 99
Alcian-blue, binding by yeast cells, 261
Aldolase, effects of polyanions, 41, 42
Amino acid, regulation of aminoacyl-tRNA synthetase synthesis, 219, 220, 221, 222
Amino acid biosynthesis, regulation by aminoacyl-tRNA synthetases, 212, 213, 214, 215, 216, 217
Aminoacyladenylates, intermediates of aminoacyl-tRNA synthetases, 173, 174, 175, 177, 178, 179, 180
Aminoacyl synthetases, binding to t-RNA, 50
Aminoacyl-tRNA, control of cognate synthetase levels, 220
corepressor of amino acid biosynthesis, 214, 215, 216, 217
deacylation, 186, 187
donor substrate for aminoacyl-tRNA~ protein transferase, 101, 102
for alanyl-tRNA~phosphatidylglycerol transferase, 115, 116
for lysyl-tRNA~phosphatidylglycerol transferase, 115, 116
in the arginine-transfer reaction, 102
in the biosynthesis of interpeptide bridges, 121, 130, 131
in the leucine, phenylalanine-transfer reaction, 102
Aminoacyl-tRNA biosynthesis, regulatory

359

CUMULATIVE INDEXES, VOLUMES 1–40

A. Author Index

VOL. PAGE

VOL. PAGE

B. Subject Index

VOL. PAGE